GIS Tools for Water, Wastewater, and Stormwater Systems

U. M. Shamsi

American Society of Civil Engineers
1801 Alexander Bell Drive
Reston, Virginia 20191–4400

Abstract: Data and software are the two most important tools for developing GIS applications. This book focuses on GIS data and software for water, wastewater, and stormwater systems. With appropriate GIS data and software, the possible means to manage water, wastewater, and stormwater systems are almost endless. This book is intended as a reference book on GIS data and software for water, wastewater, and stormwater system applications. It is suitable for any professional involved in the management and operation of water, wastewater, and stormwater systems.

Library of Congress Cataloging-in-Publication Data

Shamsi, Uzair M.
GIS tools for water, wastewater, and stormwater systems / Uzair M. Shamsi.
p. cm.
Includes bibliographical references and index.
ISBN 0-7844-0573-5
1. Waterworks--Data processing. 2. Sewerage--Data processing. 3. Geographic information systems. I. Title.

TD485 .S497 2002
628.1'0285--dc21 2002074497

Library of Congress Catalog Card No: 2002074497
ISBN: 0-7844-0573-5
Manufactured in the United States of America.

Dedication

To my beloved parents

Mr. Manzoor A. Shamsi and Mrs. Roha M. Shamsi

Contents

Preface

About This Book

It's a Monday morning in March 2002 in the Cleanwater City, population approximately 10,000 people. Bill, the wastewater treatment plant operator enters his office when the phone rings. The call is from a sewer customer who is complaining about his basement flooding. Bill immediately starts the City GIS and enters the customer address. GIS zooms to the resident property and shows all the sewers and manholes in the area. Bill highlights the sewer segment adjacent to the customer property, launches the work order module, and completes a work order for TV inspection. The export button saves the work order form and a map of the property and adjacent sewers in a Microsoft Word file. Bill immediately e-mails the Word file to the City's sewer cleaning contractor. The entire process, from the time the customer called, took about 15 minutes. This book presents the tools required to accomplish applications like this.

> Experts believe that in the near future, most water, wastewater, and stormwater system professionals will be using GIS in the same way they have used a word processor or spreadsheet.

More than 80% of all the information used by water, wastewater, and stormwater utility companies is geographically referenced, that is, a key element of the information is its location relative to other geographic features and objects. An information system is a framework that provides answers to questions from a data resource. A Geographic Information System (GIS) is a database of spatially distributed features and procedures used to collect, store, retrieve, analyze, and display geographic data.

The typical local government office contains hundreds of maps displaying municipal boundaries, property lines, roads, sewers, water lines, voting district boundaries, zoning areas, flood plains, school bus routes, land use, streams, topography, soil types, and so on. Paper maps, after all, have been the traditional method of storing and retrieving geographically referenced information. The sheer number, range of types, and diversity of maps used by municipalities are evidence of the importance geographically referenced information plays in our day-to-day operations. Unfortunately, the wide variety of maps and the diversity of their scales and designs at our disposal make it extremely difficult to access, use, and maximize the value of the information they contain. GIS is an integrating technology; it integrates all kinds of information and applications with a geographic component into one manageable system. GIS offers integrated solutions in the areas of planning and engineering, operation and maintenance, and finance and administration. For example, GIS can integrate multiple utilities, such as water, wastewater, and stormwater systems, in one information system.

GIS is one of the most promising technologies of the millennium. The widespread use of GIS can be exemplified by the fact that about 25% of the annual American Water Works Association (AWWA) Information Technology Conference is dominated by discussions concerning GIS applications. The time has come for all the professionals involved in the planning, design, construction, and operation of water, wastewater, and stormwater systems to enter one of the most exciting new technologies in their profession— GIS.

The Author was inspired to write this book by teaching ASCE's continuing education seminar, "GIS Applications in Water, Wastewater, and Stormwater Systems." The seminar has been attended by hundreds of water, wastewater, and stormwater professionals in major cities throughout the United States.

This book focuses on GIS data and software for water, wastewater, and stormwater systems. Data and software are the two most important requirements for developing GIS applications. The first step in developing a GIS is to acquire software to create GIS databases and layers. This requires a careful review of available software packages. People spend a lot of time in selecting the best software for their GIS applications. Data is the most important component of a GIS. Without data, you simply have a computer program, not a GIS. You should be aware of the intended use and accuracy of your GIS data. Data quality and accuracy should be evaluated in the context of the GIS application in which the data will be used.

This book will show you that with appropriate GIS data and software, the possibilities to manage your water, wastewater, and stormwater systems are almost endless. This book is intended to be a reference book on GIS data and software for water, wastewater, and stormwater system applications. It is suitable for any professional involved in the management and operation of water, wastewater, and stormwater systems. It is ideally suited for civil and environmental engineering project engineers and project managers. Other professionals, such as consultants, system analysts, GIS technicians, planners, university researchers and professors, city managers, water and sewer utility personnel, and government employees involved in GIS- and water-related work will also find this book useful.

Style Of The Book

This book has been written using the recommendations of the Accreditation Board for Engineering and Technology (ABET) of the United States and ASCE's Excellence in Civil Engineering Education (ExCEEd) program. Both of these programs recommend performance- or outcome-based learning in which learning objectives of each lecture (or chapter) are clearly stated up-front and the learning is measured in terms of achieving these learning objectives. Each chapter of this book accordingly starts with learning objectives for that chapter and ends with a chapter summary and practice problems referred to as "self evaluation." In academic settings, self evaluation problems can be used as homework assignments.

The objective of this book is to document GIS data and software tools that are appropriate for developing GIS applications for water, wastewater, and stormwater systems.

Most technical text books are written using the natural human "teaching" style called the "deductive" style, in which principles are presented before the applications. In this book, an attempt has also been made to organize the material in the natural human "learning" style called the "inductive" style, in which examples are presented before the principles. For example, the data and software examples are presented before their applications and case studies are presented before the procedures are explained. The book has plenty of pictures, diagrams, graphs, and illustrations and should cater well to the learning styles of "visual" learners—GIS, after all is defined as a visual language.

Organization Of The Book

There are 12 chapters in this book, organized as follows:

- Chapter 1. GIS Basics: GIS definitions and terminology
- Chapter 2. GIS Development Software: review of software packages to create a GIS
- Chapter 3. GIS Applications Software: review of software packages to develop GIS applications
- Chapter 4. GIS Data: types of data that are required to develop GIS applications for water, wastewater, and stormwater systems
- Chapter 5. Internet Data: water-, wastewater-, and stormwater-related GIS data on the Internet
- Chapter 6. GIS Database Design: designing a database for your GIS applications
- Chapter 7. Modeling Integration: integration of hydrology and hydraulic computer models with GIS
- Chapter 8. Water System Applications: GIS applications for water distribution systems with special emphasis on master planning
- Chapter 9. Wastewater System Applications: GIS applications for wastewater systems with special emphasis on needs analysis and utility mapping
- Chapter 10. Stormwater System Applications: GIS applications for stormwater systems with special emphasis on watershed stormwater management
- Chapter 11. Case Studies: recent case studies submitted for publication in this book by various GIS professionals showing GIS applications for water, wastewater, and stormwater systems
- Chapter 12. GIS Resources: useful resources (Web sites, books, periodicals, and so on) for finding more information on GIS data, software, and applications.

Acknowledgments

I thank the following organizations and companies for providing information for this book: American Society of Civil Engineers, American Water Works Association, Azteca Systems, CE Magazine, CH2M Hill, Chester Engineers, Computational Hydraulics International, Danish Hydraulic Institute, Environmental Systems Research Institute, Geo Spatial Solutions Magazine, Geo World Magazine, GIS Frequently Asked Questions (Census Bureau), Haestad Methods, Hansen Information Technology, Journal of American Water Resources Association, Journal of American Water Works Association, Professional Surveyor Magazine, RJN Group, Inc., USFilter, Water Environment Federation, and Water Environment & Technology Magazine. Some information presented in this book is based on my collection of papers and articles published in peer reviewed journals, trade magazines, conference proceedings, and the Internet. The authors and organizations of these publications are too numerous to be thanked individually, so I thank them all collectively without mentioning their names. Their names are, of-course, included in the Reference section.

The proposal and manuscript of this book were peer reviewed by many individuals, whose valuable suggestions were instrumental in reshaping the content and style of the book. I would like to thank Donald F. Waye of Northern Virginia Planning District Commission, Dr. Rafael G. Quimpo of the University of Pittsburgh, Dr. Shakir Husain of the Youngstown State University, Robert E. Laskey of USFilter, Inc., and several anonymous reviewers for their valuable suggestions and comments.

The case studies were submitted by 29 GIS and water professionals from five countries (United States, Canada, Germany, Turkey, Fiji Islands) in response to my "Call for Case Studies" distributed to various Internet discussion groups. My thanks to the following people for preparing them:

- Brad Roeth and Brian Roth, Stanley Consultants, Inc., Muscatine, Iowa, United States
- Chris Johnston, Andrew Boyland, and Jason Vine, Kerr Wood Leidal Associates, Ltd., North Vancouver, B.C., Canada
- Don Waye, Northern Virginia Regional Commission, Annandale, Virginia, United States
- James T. Smullen and Jennifer Angell, Camp Dresser & McKee, Inc., Edison, New Jersey, United States
- Klaus Jacobi, AHT International GmbH, Essen, Germany
- Mark Bennett, Division of Soil and Water Conservation, Virginia Department of Conservation and Recreation
- Maeve McBride and Jessica E. Leblanc, Center for Urban Water Resources Management, University of Washington, Seattle, Washington, United States

- Mustafa Aktas, Middle East Technical University, Ankara, Turkey
- Newland Agbenowski, G.V. Loganathan, and R.G. Greene, Virginia Polytechnic Institute and State University, Blacksburg, Virginia, United States
- Paula Dawe and Harald Schölzel, South Pacific Applied Geoscience Commission, Fiji Islands
- Seung Ah Byun and Brian Marengo, Camp Dresser & McKee, Inc., Philadelphia, Pennsylvania, United States
- Stu Townsley and David Ford, David Ford Consulting Engineers, Sacramento, California, United States
- Suresh Muthukrishnan, Martin Doyle, Shilpam Pandey, Nick Jokay, and Jon Harbor, Geomorphology Group, Department of Earth and Atmospheric Sciences, Purdue University, West Lafayette, Indiana, United States
- Temilayo Okeowo and W. Cully Hession, Department of Civil and Environmental Engineering, University of Vermont, Burlington, Vermont, United States

Warnings And Disclaimer

This book is intended to provide general information about Geographic Information Systems (GIS). We have made every effort to make the book as accurate as possible; however, no warranty is implied. The information is provided on an "as-is" basis. The author and ASCE Press shall have no liability or responsibility to any person or party for any loss or damage resulting from the information presented in this book.

Listing of any hardware, software, or data vendors in this book should not be considered an endorsement of any particular company or their products, and potential users should thoroughly research all available sources of hardware, software, and data and choose that which best suits their particular needs.

There are many companies that develop GIS hardware, software, data, and publications—not all of which have been mentioned in this book. Some representative companies and their main products are discussed in the book. Note that the product listings in this book are not intended to be complete; they show only representative examples.

The product costs mentioned in the book are approximate and represent average prices at the time of writing. These prices will change with time. All costs are in U.S. dollars, unless otherwise specified. Some product information is based on the product manufacturers' marketing and sales literature, brochures, and Web sites; this information was not verified for accuracy by the author or ASCE Press. The products and services mentioned in this book are trademarks of their respective companies. Author and ASCE Press do not make any claim to these products and services.

Web site URLs are current at the time of publication and may change in the future. If a URL suggested in the book does not work, the readers may want to go to the Web site's root address (home page) to search the revised link. For example, if http://edc.usgs.gov/webglis does not work, you can go to www.usgs.gov and search a revised link for webglis. Alternatively, you can try the FirstGov site (www.firstgov.gov), U.S. government's search engine.

The latest information (e.g., version, cost, Web site URL, and so on) about various data and software products discussed in this book may be obtained from the Author's Web site (www.GISApplications.com).

Chapter

1 GIS Basics

Over the last decade, GIS use has grown dramatically in government, utilities, business, and academia where it is being used for many diverse applications. Consequently, a variety of GIS terms and definitions has developed.

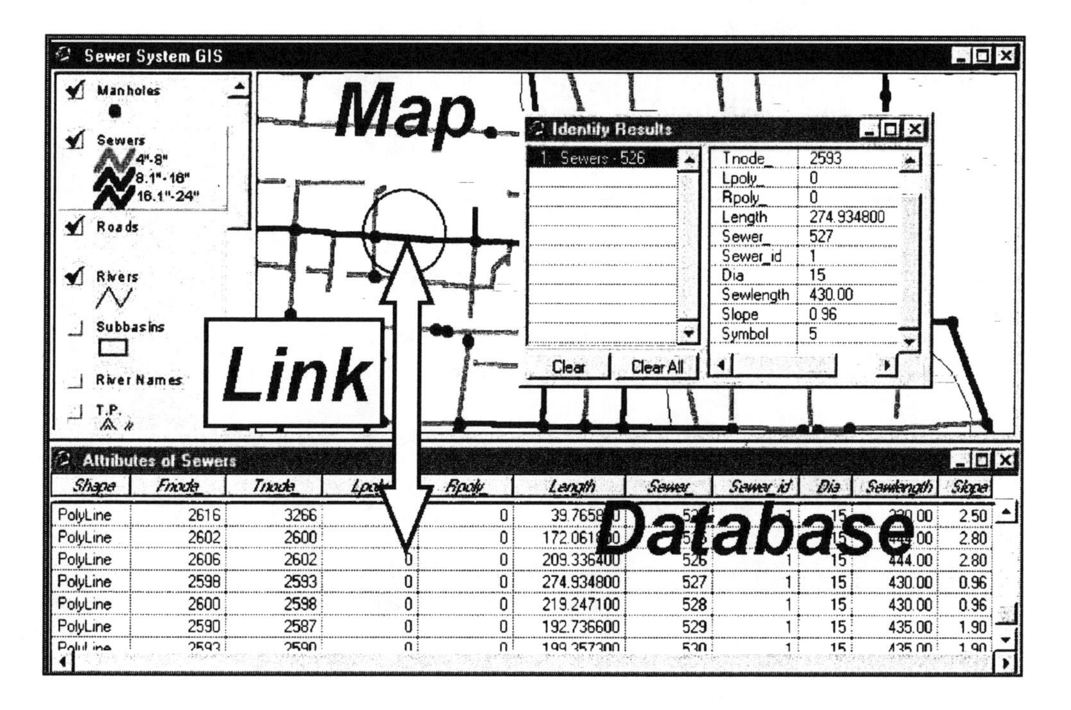

GIS Definition as a Link Between a Map and a Database

LEARNING OBJECTIVES

The learning objectives of this chapter are to understand GIS terminology and to build an adequate technical foundation for understanding subsequent chapters. Readers of this book are expected to be familiar with GIS buzz words. However, relevant GIS basics are presented here to refresh forgotten concepts and definitions[1]. Major topics discussed in this chapter include

- GIS definitions and functions
- Vector and raster GIS
- Features, attributes, and geodatabases
- Layers, coverages, and themes
- Topology
- Map projections and coordinate systems
- Differences among CADD, CAM, AM/FM, and GIS
- Enterprise-wide GIS
- GIS hardware
- GIS development steps

GIS DEFINITIONS

Geography is a discipline that looks at things within their spatial context. An information system is a sequence of operations that begins with planning and collection of data, continues with storage and analysis of the data, and ends with the use of the derived information in some decision-making process. GIS can then be defined by the equation:

GIS = Geography + Information Systems

There are three general components of geographic information:

1. *Geometry.* This component represents the real-world locations with geographic features. For example, hydrants and water mains are represented as points and lines, respectively. Data representing the geometry of features are also referred to as the geographic, spatial, or geospatial data.

2. *Attributes.* This component provides descriptive data about the geographic features, such as water main diameter.

[1] An acronym list is provided in Appendix A. Additional definitions are provided in Appendix B (Glossary). For an in-depth study of GIS essentials and more advance topics, please refer to the GIS books listed in Chapter 12 (GIS Resources).

3. *Behavior*. This component defines the rules that geographic features must follow. For example, a hydrant should not exist without a water main.

Over the last decade, GIS use has grown dramatically in government, utilities, business, and academia where it is being used for many diverse applications. Consequently, many GIS definitions have developed as various, often colorful, ways of describing the same concept. The literature presents a large number of GIS definitions. A single, reasonably inclusive definition is that a GIS consists of computer-based tools that are used to capture, store, update, manipulate, retrieve, analyze, display, print, and otherwise manage large amounts of geographic and attribute data (ESRI, 1992; Singh, 1995). Originally, these tools were developed to ease cartography, but they are currently being used for diverse applications such as, planning, facilities management, and hydrologic and hydraulic (H&H) modeling. Additional GIS definitions are given below.

- Environmental Systems Research Institute (ESRI), world's leading GIS software company offers the following GIS definitions (ESRI, 1992).
    - A computer system capable of holding and using data describing places on the earth's surface.
    - A computer-based tool for mapping and analyzing things that exist and events that happen on earth.
    - An organized collection of computer hardware, software, spatial data, and personnel designed to effectively capture, store, update, manipulate, analyze, and display all forms of geographically referenced information.
    - A technology that integrates common database operations, such as query and statistical analysis with the unique visualization and geographic analysis benefits offered by maps.
- A GIS is a special type of information system in which the database consists of observations on spatially distributed features and procedures to collect, store, retrieve, analyze, and display the spatial data (Dueker, 1987).
- A GIS is a combination of computer hardware and software that allows for the management, analysis, and mapping of infrastructure and geographic information and descriptive data with cartographic accuracy.
- A GIS is a system for organization, analysis, and rendering of spatial data used in a problem solving process.

- A GIS is an intelligent marriage between graphics and data (or maps and tables) that allows the information to be sliced, diced, and organized with the click of a mouse.

- A GIS consists of "smart maps" that can help us to find out "the why of where."

GIS can also be defined in terms of data layers because it combines layers of information about an area to aid in analyzing and studying that area. An analogy can be made between GIS and the conventional map production process because GIS data layers are analogous to mylar overlays that can be overlayed in various combinations to create various thematic maps. The layers of information that should be combined depend on one's needs: planning, H&H modeling, work-order management, or inspection and maintenance. Figure 1-1 shows how manhole, sewer, parcel, and aerial photo layers can be combined to create a sewer system GIS map for field inspections.

ESRI President Jack Dangermond says, "Knowing where things are and why is essential to rational decision making." In his welcome speech in the 1998 ESRI User Conference in San Diego (California, USA), Jack defined geography as the "science of GIS." He asserted that:

- GIS is an instrument for implementing geographic thinking,
- GIS is a visual language,
- GIS is the power of seeing,
- GIS is the framework for studying complex systems,
- GIS shows context and content,
- GIS integrates our knowledge about places, and
- GIS helps us better organize our institutions.

These are the new definitions of GIS. They are undoubtedly different from the conventional textbook definitions, yet they have beautifully captured the essence of GIS applications.

Finally, every GIS is developed for people including both the general public and the GIS professionals. The main reason for creating any GIS is to make the life of public easier. Once created, even the most sophisticated GIS with the best quality of data cannot survive if it lacks a team of trained and skilled professionals to manage and maintain it. People are, therefore, an important part of a GIS and should be included in every GIS definition.

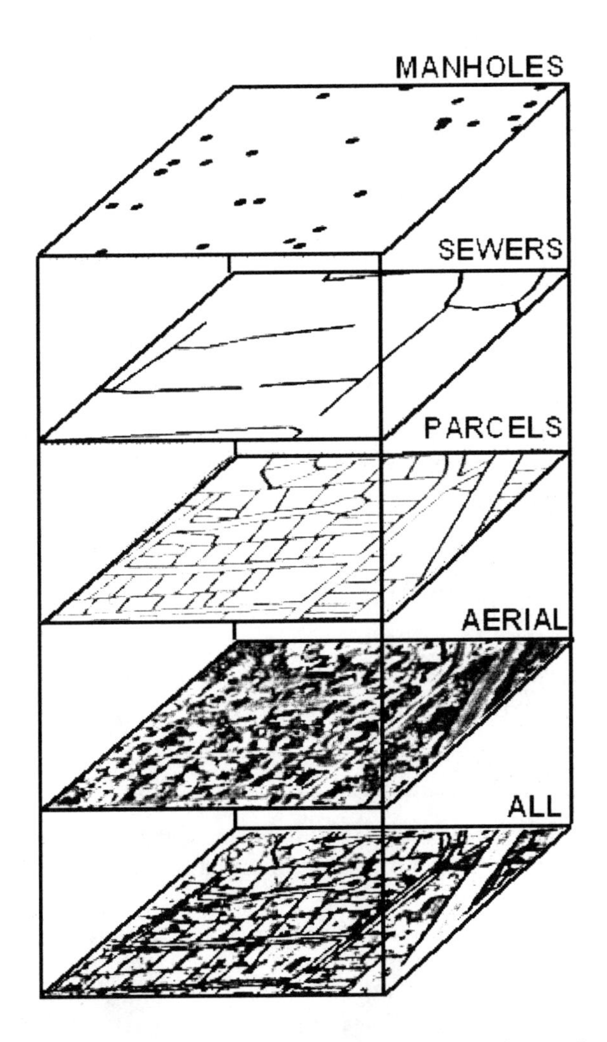

Figure 1-1. GIS Definition as a Combination of Data Layers

GIS FUNCTIONS

According to above definitions, a GIS performs seven major functions (Hay and Knapp, 1996):

1. *Data Capture*. This function enables input of both spatial (geometric) and non-spatial (attribute) data. It includes both importing data from digitizers, scanners, or image processing systems and creating new data through internal operations.

2. *Storage*. This function includes storage and management of both spatial and attribute data. Vector and raster formats described below are two basic data models for spatial data storage. In order to retain

their coordinate registration, conventional systems store the spatial data in a spatial database. Attribute data are stored in a relational database that links back to spatial features. Modern object-oriented systems, store both the spatial and attribute data in one relational database.

3. *Manipulation.* This function allows transformation of map projection, orientation, or scale to facilitate analysis across different data layers.
4. *Query.* This function provides utilities for finding specific geographic features based on their location or attribute data.
5. *Analysis.* This function relates spatial and non-spatial data, within a specified neighborhood or across multiple layers, to determine spatial conditions, spatial relations, or spatial trends.
6. *Display.* This function provides tools for computer display of geographic features and the analysis results using a variety of symbols and colors.
7. *Output.* This function allows the production of printed maps of geographic features and reports and graphs of the analysis results.

VECTOR AND RASTER GIS

As shown in Figure 1-2, there are two major formats for digital storage and manipulation of spatial information: *vector* and *raster*.

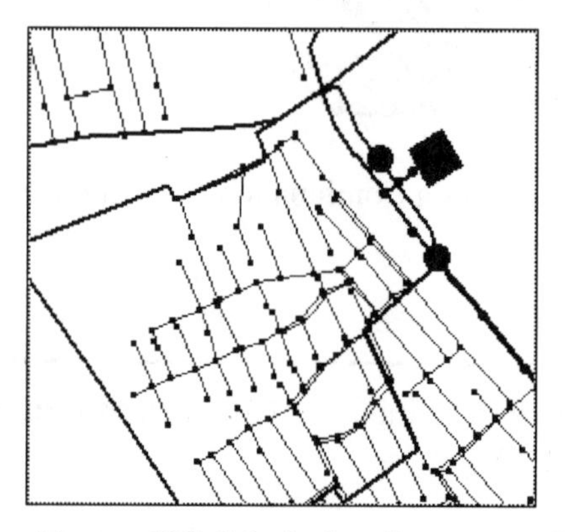

a. Vector GIS (Manholes, Sewers, and Drainage Areas)

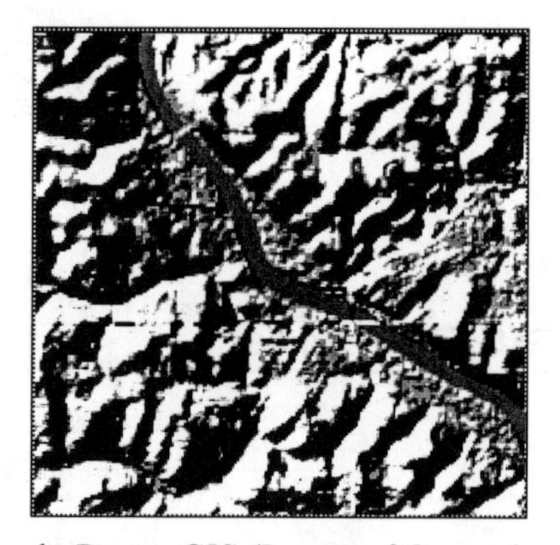

b. Raster GIS (Image of Ground Surface Elevation with Hill Shading)

Figure 1-2. Vector and Raster GIS

1. Examples of the *vector* format are maps of water mains, hydrants, and valves. In vector data, positions are stored as (x,y) coordinate pairs. Objects in vector format are represented as graphic features (points, lines, and polygons). The vector format is a topologically constructed

set of points, nodes, lines, and polygons that defines locations, boundaries, and areas. Vector data consist of a series of nodes that define line segments, which in turn are joined to form more complex features, such as networks and polygons.

2. Scanned maps, images, or aerial photographs are examples of *raster* format. Raster data are also referred to as grid, cell, or grid-cell data. In raster format, objects are represented as an image consisting of a regular grid of uniform size cells called *pixels,* each with an associated data value. Raster data sets store spatial information within a two-dimensional matrix that consists of uniformly spaced rows and columns. Each grid cell contains a value representative of the feature being depicted. Grid-cell values may be continuous, as in elevation data, or discrete, as in land use data where each grid cell value is associated with a specific land use. The resolution of raster data is dependent on the size of the grid cell (Hutchinson and Daniel, 1995). Raster images are storage hungry–they require large amounts of disk space and memory. Their file size increases as the image precision or resolution increases. The number of pixels in a 1-meter resolution image is 900 times the number of pixels in a 30-meter resolution image!

   Raster data are often less detailed and may be visually less appealing than vector data, which look like traditional, hand-drafted cartographic maps. Nevertheless, many complex spatial analyses, such as automatic land use change detection, demand a raster system. Raster data make excellent basemaps. Other advantages of raster data include three-dimensional hill shading (Figure 1-2b) and the ability to analyze nontraditional spatial features, like albedo or infrared radiation. Recent trends indicate that there is an increasing amount of interest in raster data and software systems that process them. Some GIS or image-processing software packages can perform raster-to-vector and vector-to-raster data conversions. For example, the raster data of a satellite image can be converted to vector land use/land cover data by generating polygons around all pixels with the same classification and determining spatial relationships between cells, such as adjacency or inclusion.

FEATURES, ATTRIBUTES, AND OBJECTS

As described above, a GIS contains two types of information: geographic and descriptive. The geographic data describe the location of real-world objects and entities on the earth's surface, such as water mains, sewer lines, and parcels. The descriptive data describe the characteristics of those same objects, such as installation date of a water main, sewer size, or parcel owner.

FEATURES

The graphical representations of entities in a vector type GIS are called *features*. There are four types of GIS features:

1. *Point*: For representing vaults, cleanouts, flow meters, etc.
2. *Line (arc):* For representing water mains, sewers, streams, etc.
3. *Node*: For representing valves, manholes, inlets, etc.
4. *Polygon*: For representing buildings, soils, lakes, etc.

Some GIS software, such as ArcInfo[2], has two more feature types: annotation and tic. *Annotations* are used for display text (labels) and *tics* are used for registration or geographic control points.

Point features represent objects that have discrete locations and are too small to be depicted as areas, such as manholes, wells, fire hydrants, and outfalls. Point features are also used to represent polygon interiors and label points. *Line (arc) features* represent objects that have length but are too narrow to be depicted as areas, such as roads, rivers, and pipelines. Arcs can join only at their endpoints, called *nodes*. Each arc has two nodes: a from-node and a to-node which can be used to determine the direction of the arc. *Node features* are generally used to represent point features that are connected to line features, such as sewer manholes. *Polygon features* represent objects that have area, such as lakes, watersheds, and subbasins. An illustration of the use of the above features for mapping wastewater systems is shown in Figure 1-3. This figure is an ArcView GIS screen-shot representing manholes, sewers, and subbasins as points, lines, and polygons, respectively.

ATTRIBUTES

Feature characteristics data stored in tabular format and linked to the feature by a user assigned identifier are called feature *attributes*. Standard geometric attributes, such as length, perimeter, and area, are generally supplied by the GIS software. Other descriptive attributes, such as manhole depth or sewer size, are assigned by the user.

GEODATABASE OBJECTS

ESRI's new GIS software suite called ArcGIS is built on the foundation of the geodatabase (short for "geographic database"). A *geodatabase* is a collection of spatial data stored in a relational database format. A geodatabase consists of the following:

---

[2] ArcInfo®, ArcView®, and ArcGIS® are the leading GIS software packages from ESRI. More information about these products is provided in Chapter 2 (GIS Development Software).

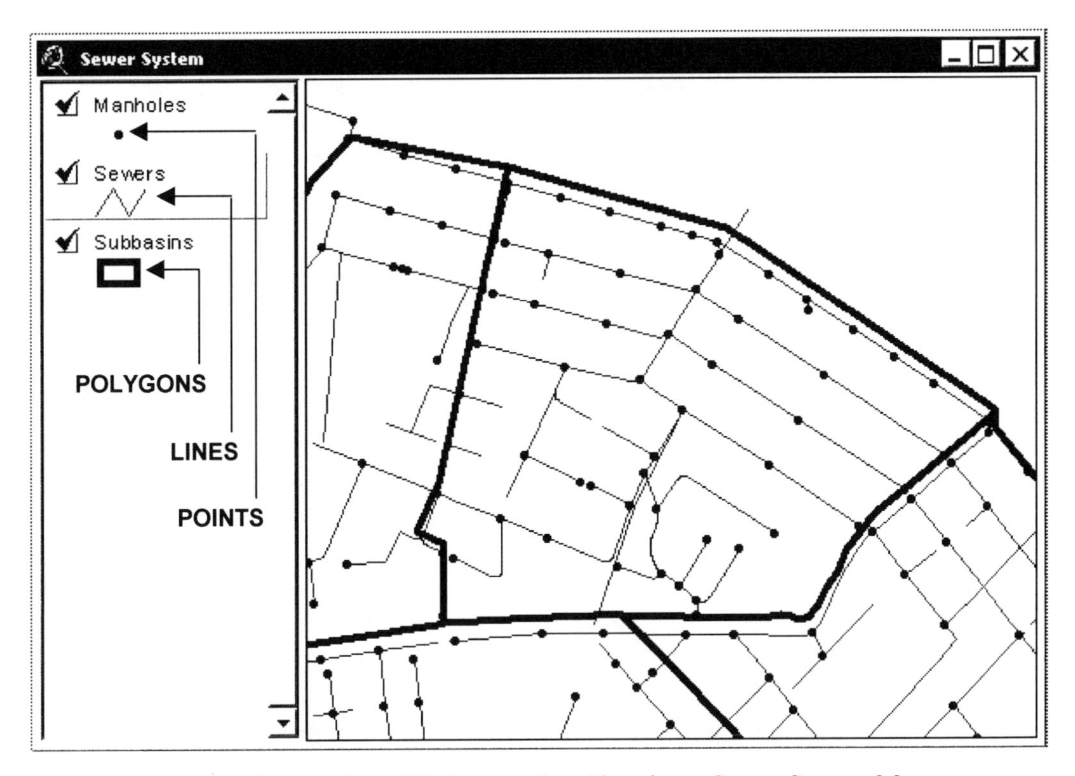

Figure 1-3. ArcView GIS Screen-shot Showing a Sewer System Map

- A storage mechanism for spatial data of all types,
- A GIS data model, and
- A series of data access components.

A geodatabase allows the user to create custom objects (manholes, sewers, subbasins, etc.) rather than simple geometric features (points, lines, and polygons). Additional geodatabase information is provided in Chapter 2 (GIS Development Software).

Layers, Coverages, and Themes

A logical set of thematic data stored in a GIS is called a *layer*. The features and attributes of thematically associated data, such as streams, watersheds, and land use, forming the basic unit of storage are called a *coverage*. Coverage is an ESRI terminology for a layer. In ArcInfo a coverage consists of a set of files containing spatial and attribute data organized within a common directory.

Map components can be logically organized into sets of layers called *themes*. A theme is a group of similar geographic features, such as hydrants (points), water mains (lines), or pressure zones (polygons). Figure 1-3 shows three sewer system themes: a point theme for manholes, a line theme for sewers, and a polygon theme for subbasins.

An ArcInfo coverage serves as a spatial data source for a theme in ArcView. In ArcView many themes can be created from the same coverage by classification of attribute data. For example, a subbasin coverage can be used to create themes of subbasin types (e.g., developed or undeveloped), subbasin slope (e.g., flat, mild, or steep), or subbasin runoff rate. In ESRI's new ArcGIS software suite, a theme and a coverage are the same thing for most practical purposes—they are both referred to as a layer.

An *overlay* operation combines two or more themes to create a new theme that contains both the spatial features and the attributes of the input themes. For example, land use and hydrologic soil group themes can be overlayed to create a theme of runoff curve numbers. The curve number and watershed themes can then be overlayed to estimate the runoff curve number for each watershed, which can then be used to estimate stormwater runoff from the watersheds.

Topology is what makes GIS so spatial. Topology is the primary difference between a GIS and a desktop mapping program.

TOPOLOGY

Topology is defined as the spatial relationship between features. Spatial relationships between connecting or adjacent features, such as the sewers upstream of a manhole or the pipes connected to a hydrant, which are so obvious to the human eye, must be explicitly defined to make the maps "intelligent." For example, our eyes can instantly identify the streams located within a watershed just by looking at a paper map. A GIS, on the other hand, must mathematically compare the stream and watershed layers to complete this task. Topology is, therefore, defined as a mathematical procedure for explicitly defining spatial relationships between features. The subject of topology is so sophisticated that it is considered a branch of mathematics.

Point, line, polygon, and node features have the following topological associations:

- Nodes make the endpoints of arcs.
- Arcs connect to each other at nodes.
- Arcs have both direction and left and right sides.

- Points mark the interiors of polygons.
- Arcs make the perimeters of polygons.

A topological GIS can determine conditions of *adjacency* (what is next to what), containment (what is enclosed by what), and *proximity* (how near something is to something else). Topological relationships allow spatial analysis functions, such as modeling flow through the connecting pipes of a water system. For example, ArcInfo 7.x GIS uses three main topological concepts (ESRI, 1990):

- *Connectivity*, which is the topological identification of the set of arcs that are connected at nodes by recording the upstream and downstream node for each arc.

- *Contiguity*, which is the topological identification of adjacent polygons by recording left and right polygons of arcs.

- *Area definition*, which consists of arcs that are connected to each other to define an area (polygon).

In ArcInfo, the topology of an arc includes its upstream and downstream nodes (fnode, tnode) and its left and right polygons (lpoly, rpoly). Similarly, a node is topologically linked to all arcs that meet at that node. Some GIS software like ArcInfo 7.x performs computationally intensive topological relationship calculations up front and stores that information as attributes (fnode, tnode, lpoly, rpoly, etc.). New data structures do not store topological information. Some new GIS software packages compute topological relationships "on-the-fly" on an as-needed basis to answer various spatial queries. Such packages include ESRI's ArcView, Bentley System's GeoGraphics®, and modern spatial relational database management systems (RDBMS) described in Chapter 3 (GIS Application Software). The benefits of this approach are that maps display faster and that data are never out of date. There is, however, one drawback to this approach: the topology errors are not encountered until a query is launched. Thus, using these new systems requires "clean" data or an add-on software that can validate, clean, and edit the data. Most major GIS software vendors, such as ESRI, Intergraph, and MapInfo, provide GIS client products for this purpose (Limp, 2000a).

MAP PROJECTIONS AND COORDINATE SYSTEMS

Because the earth is round and maps are flat, transferring locations from a curved surface to a flat surface requires some coordinate conversion. A *map projection* is a mathematical model that transforms (or projects) locations from the curved surface of the earth onto a flat sheet or two-dimensional surface in accordance with certain rules. Mercator, Robinson, and Azimuthal are some commonly used projection systems. Small-scale (1:24,000 to 1:250,000) GIS data intended for use at the state or national level are projected using a projection system appropriate for large areas, such as the Universal Transverse Mercator (UTM) projection. The UTM system divides the globe into 60 zones, each spanning six degrees of longitude. The origin of each zone is the equator and its central meridian. X and Y coordinates are stored in meters. Large-scale, local GIS data are usually projected using a State Plane Coordinate (SPC) projection.

A *datum* is a set of parameters defining a coordinate system and a set of control points with geometric properties known either through measurement or calculation. Every datum is based on a spheroid, which approximates the shape of earth. The North American Datum of 1927 (NAD27) uses the Clarke spheroid of 1866 to represent the shape of earth. Many technological advances, such as GPS, revealed problems in NAD27, and the North American Datum of 1983 (NAD83) was created to correct those deficiencies. NAD83 is based on GRS80 spheroid whose origin is located at the earth's center of mass. The NAD27 and NAD83 datum control points can be up to 500 feet apart.

Coordinates are used to represent locations on the earth's surface relative to other locations. A *coordinate system* is a reference system used to measure horizontal and vertical distances on a planimetric map. A coordinate system is usually defined by a map projection. The GIS and mapping industries use either latitude/longitude- or geodetic-based coordinate grid projections.

Since much of the information in a GIS comes from existing maps, a GIS must transform the information gathered from sources with different projections to produce a common projection. Some GIS software does not have the integrated capability to perform map projection and coordinate transformations. In that case, the user must rely on external programs. For example, Blue Marble's (*www.bluemarblegeo.com*) Geographic Calculator program is a stand-alone product than can convert coordinates from one system to another. It offers more than 12,000 predefined coordinate systems plus additional user-defined systems, datums, and units. Geographic calculator supports many popular Computer Aided Design and Drafting (CADD) and GIS formats such as, AutoCAD® DXF and DWG files, ESRI SHP file, and MapInfo® MIF file.

GIS HARDWARE

A 1992 GIS article published in *American City and County* predicted that workstations with the capability of 200–400 million instructions per second (MIPS) would be available for $10,000–$20,000 during the second half of the 1990s (Kindleberger, 1992). This prediction was correct. The 1995–2000 period did produce 200–400 MHz personal computers, but at the much lower prices of $1,000–$5,000. This article also envisioned the updating of maps in the field using a portable "slate" device with handwriting technology. This prediction was also correct. We can now run mobile GIS software, like ESRI's ArcPAD®, on palm-top and pen-based computers and personal digital assistants (PDAs). We can also use a PDA like PalmPilot™ for displaying maps and collecting GIS attribute data in the field.

The selection of appropriate GIS hardware depends on the scope of the GIS application project and the available resources. The GIS hardware can be set up as a stand-alone workstation or in a network configuration. Typical hardware requirements are listed below:

- Computer with fast processor, large memory, extensive disk space, and CD-ROM
- Large screen, high-resolution, color monitor
- Backup and storage device, such as a tape drive or writable CD-ROM
- Color printer
- Large-format color plotter
- Digitizer
- Large-format scanner
- High-speed internet connection (T1 or DSL)
- GPS receiver
- Computer server (for networked configurations)

Client-server architecture is a common computer system configuration where one main server stores and distributes all data. Workstation users, called "clients," access and manipulate the server data using client software installed on the workstation computer. For example, Orange County Geomatics/Land Information Office started to provide enterprise-wide access to GIS data in 1997 using hardware and software from Intergraph. Their Oracle database containing seven gigabytes of attribute information resided on a 200 MHz Quad Pentium Pro with 256 MB of RAM. The data could be viewed using the Intergraph's MGE product suite (Goldstein, 1997).

GIS Development Steps

There are six steps to create and implement a GIS:

1. *Data Conversion.* In this step, hard-copy maps are converted into digital files using digitization or scanning. Both the coordinates and

the attributes are captured. Table digitization is a laborious process. "Heads-up" or "on-screen" digitization is a less cumbersome alternative in which the visible features from scanned maps, digital aerial photos, or satellite imagery can be traced with a mouse on the computer monitor. The use of this process for certain specialized mapping projects, such as land use/land cover mapping, requires familiarity with photo interpretation techniques.

2. *Data Collection.* If source maps are not available for a feature, its location and attribute data must be collected using field surveys or GPS.

3. *Data Preparation.* This step makes the raw data from source maps or field/GPS surveys available for GIS use or "GIS ready." It may involve changing data formats (e.g., from DXF to Shapefile), applying map projections, georeferencing the image data, or clipping or mosaicking the aerial photos.

4. *Data Management.* This step organizes attribute data in a database management system (DBMS) format. It may require linking various relational database tables using a common field.

5. *Data Sharing.* This step takes care of data sharing needs between different departments of an organization. It may require linking interdepartmental databases to reduce redundancy and allow faster access.

6. *Data Distribution.* This step makes the GIS data available to the staff of an organization and optionally to the general public using the Internet, intranet, or wireless technology.

These steps are undertaken by data creators, such as government agencies and commercial data vendors. Some GIS users, such as most civil engineers, are users rather than creators of GIS data. These casual GIS users will not have to perform all six of the development steps.

A GIS is only a GIS if it permits spatial operations on the data (ESRI, 1992).

DIFFERENCES AMONG CADD, CAM, AM/FM, AND GIS

Non-GIS software, such as spreadsheets (e.g., Microsoft Excel), RDBMS (e.g., Microsoft Access), and drafting packages (e.g., AutoCAD) can handle simple spatial data. However, such programs cannot perform such spatial queries as how many residents are located within 100 meters of a proposed water main.

The database concept is central to a GIS and is the main difference between a GIS and a simple drafting or computer mapping system which can only produce high quality graphic output (ESRI, 1992). By the same token, a GIS is not complete without map creation and editing functions. In fact, some GIS vendors are implementing "CAD-like" editing functions in their products. Therefore, essential elements of a GIS are (1) the ability to perform spatial operations on the database and (2) map creation and editing functions. All GIS packages use some type of database for storing and manipulating data. Thus, a DBMS is an integral part of a typical GIS software. The unique method of storing and manipulating data differentiates GIS from other drafting or cartographic software (EPA, 2000). GIS might best be understood as the intersection of CADD and DBMS technologies.

CADD is used for designing and drafting new objects. A CADD layer can contain more than one feature type (points, lines, polygons, and annotations). When a multi-feature CADD file is imported into a GIS like ArcView 3.x, its feature types are segregated into separate themes. In certain cases, an automatic CADD-to-GIS conversion may require some manual editing. For example, CADD watershed boundaries may become lines in some GIS packages. These lines should be converted to polygons if the GIS application requires the estimation of watershed parameters, such as area, slope, and imperviousness.

Computer Aided Mapping (CAM) utilizes CADD technology to produce digital maps. CAM is a computerized alternative to traditional, manual, cartographic map making. CAMs are display oriented; they produce plots of selected layers of point and line features. GISs, on the other hand, are analytically oriented; they analyze relationships among point, line, and polygon features (Dueker, 1987). CAMs cannot selectively combine features between layers. For example, the user cannot select hydrants in one layer that are within 100 meters of a parcel in another layer.

Automated Mapping/Facilities Management (AM/FM) is CADD technology applied to manage utility system data. Compared to GIS, AM/FM (1) does not have topology and (2) places less emphasis on the detail and precision of its graphics and more emphasis on data storage, analysis, and reporting. In AM/FM, relationships among utility system components are defined as networks. Simply stated, GIS emphasizes location and topology, whereas AM/FM emphasizes database and connectivity.

Table 1-1 presents a comparison of CAM, AM/FM, and GIS systems. Figure 1-4 shows the difference between AM, FM, AM/FM, and GIS and how they are combined to create an integrated AM/FM/GIS system.

How to differentiate various systems? If a system can identify all the fire hydrants within 100 meters of an office building, all the catchbasins flowing into a stormwater outfall, or the location of the worst sewer pipe, then it is a GIS; otherwise, it is probably a CADD or CAM system. A system has FM capability if it can prioritize the work required to bring the worst pipe up to a minimum operating standard.

Table 1-1. Comparison of CAM, AM/FM, and GIS

Feature	*CAM*	*AM/FM*	*GIS*
Layers	Y	Y	Y
Topology	N	N	Y
Network analysis	N	Y	Y
Lines	Y	Y	Y
Nodes	N	Y	Y
Polygons (areas)	N	N	Y
Attributes	N	Y	Y
Actual locations	Y/N	Y	Y
Map intelligence	N	N	Y

Y = Yes, N = No, Y/N = Both Yes and No

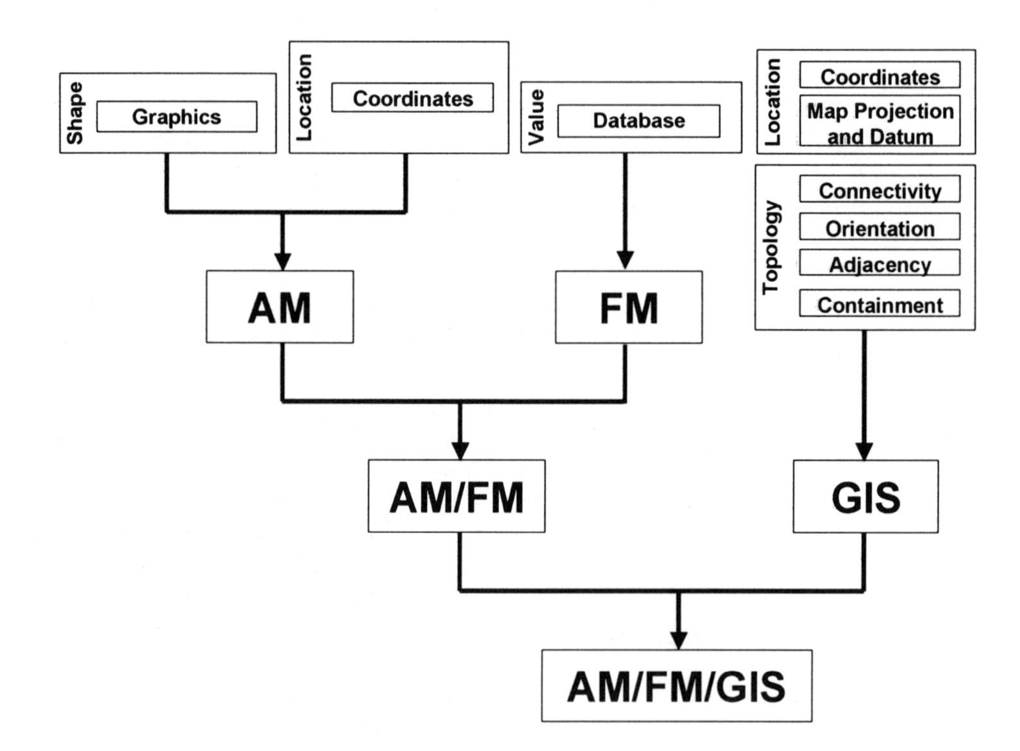

Figure 1-4. Differences Among AM, FM, and GIS

ENTERPRISE-WIDE GIS

While many local government departments have successfully automated their information management, the systems generally remain "stand alone." The integration of various department information systems remains an elusive goal because the departments usually make their own automation decisions using mutually incompatible software. GIS departments can serve as community leaders in the collection and standardization of spatial data by numerous city departments. For instance, they can create a central repository of municipal data, such as parcels; zoning; and water, wastewater, and stormwater systems. The integration of CADD, AM/FM, and GIS technologies can create a comprehensive graphic and information management system with unprecedented power and flexibility. Such "integrated cities" can be created by implementing enterprise-wide GIS projects that can link and share the information from various departments, such as water and sewer utilities, planning, public works, tax assessment, and emergency services (Robinson, 1993).

To realize the full potential of a GIS, it must be implemented on an enterprise-wide basis. The benefit/cost ratio of GIS increases with its functionality and applications. GIS applications of automated mapping return a 1:1 benefit/cost ratio. The ratio increases as the GIS use expands to more departments of the organization. Benefit/cost ratios of 4:1 can be attained when the entire organization shares the GIS information and GIS applications are maximized (Alston and Donelan, 1993).

SUMMARY

The information presented in this chapter illustrates that a GIS map is more than graphics. Every GIS map feature (point, line, or area) has both graphic properties and database attributes. As shown in the figure on the first page of this chapter, the spatial data of features and the attribute data that describe them are integrally linked. Spatial and attribute data can therefore be analyzed relative to each other in much the same way as one would analyze standard tabular data from a relational database. Thus, a GIS map can be defined as "a view of a spatial database." Alternatively, one can simply define a GIS as "a link between a map and a database" or "map + database" or "CADD + database." A GIS is unique in that it is the only computer mapping and analysis tool that provides a graphic interface for the creation and maintenance of spatial data, a database interface for the creation and maintenance of attribute data, and the capability to relate spatial and attribute data to each other through either interface. The map and database link defined in this manner is the basis of all GIS capabilities and applications, such as query and display, spatial analysis, visualization, and map composition. Chapters 2 and 3 will describe the recent developments that allow storage of both spatial and attribute data in a single RDBMS.

SELF EVALUATION

1. What is your favorite definition of GIS? Why?
2. What is the most important part of a GIS: software, data, or people? Why?
3. What type of GIS data are better: raster or vector? Why?
4. Does your GIS software have topology? Provide the reason for your answer.
5. In what ways are CADD, CAM, AM/FM, and GIS similar and different?
6. Does your GIS have the AM/FM capability? Provide the reason for your answer.

Chapter 2

GIS Development Software

GIS Software represents less than 10% of the total GIS cost in most cases, yet people spend a lot of time in selecting the best software for their GIS applications.

GIS Data Development Using a Digitizer

LEARNING OBJECTIVE

The learning objective of this chapter is to become familiar with major GIS development software products. Major topics discussed in this chapter include

- Types of software
- Vector and raster software
- GIS development and application software
- Software vendors and their GIS products
- Software selection
- Needs Analysis

TYPES OF SOFTWARE

Potential GIS users are faced with a host of questions concerning which hardware and software is best suited for their needs. Experience has shown that hardware and software costs represent only a small percentage of the total cost of GIS implementation. Data input and system management represent the largest investment of time and money. However, careful selection of software and the effective design of data capture procedures can significantly reduce the overall GIS implementation costs.

Though the art of GIS has been in existence since the 1970s, the science had been restricted to skilled GIS professionals (Jenkins, 2002). There were only a few dozen GIS software vendors before 1988. Two years later, this number had grown to at least 200 (Kindleberger, 1992). The mid 1990s witnessed the inception of a new generation of user-friendly desktop GIS packages that significantly contributed to an explosive growth of GIS applications throughout the utility industry. The Geospatial Solution Buyers Guide 2001 listed more than 500 GIS software-related companies on its Web site (URL given below). Because of the large number, all GIS software vendors cannot be listed here. Rather, representative examples of GIS software vendors and their main products are described in this chapter.

Common tasks of a GIS implementation include acquiring software to create a GIS database and layers, querying the database and performing spatial analyses, and displaying and printing maps. Once users have accomplished these core GIS capabilities, they might want to acquire application software that will allow them to do more with their GIS investment—such as develop applications that make the tasks easier, increase worker productivity, and reduce operating cost. Thus, there are two main types of software that are needed to develop water, wastewater, and stormwater applications:

1. *GIS development software.* This software is used to create the core GIS database and layers.

2. *GIS applications software.* This software is used to develop and run GIS applications, such as hydrologic and hydraulic (H&H) modeling or work-order management.

In this chapter we will focus on the GIS development software products. GIS application software products will be discussed in Chapter 3 (GIS Applications Software).

GIS Development Software

GIS development software is a geographic database management system (DBMS) (Zeiler, 1999; EPA, 2000). The database stores descriptive information about map features as attributes (ESRI, 1992). All GIS development packages use some type of database for storing and manipulating data. The GIS development software also includes tools that provide an interface between users and their geographic data. These tools are used for entering, manipulating, analyzing, and displaying the data.

Based on the level of software complexity and power, GIS software can be classified in three broad categories: professional, desktop, and viewers. Professional GIS products are the flagship products of software vendors. They are generally more complex than desktop GIS products and are appropriate for large systems (e.g., populations greater than 50,000 people) and enterprise-wide solutions. They are at the high end of the price spectrum because they are sophisticated software programs that perform complex tasks, such as image analysis, network routing, and overlay analysis. On the low end of the spectrum are free GIS viewer (data browser) tools like ESRI's ArcExplorer® and Intergraph's GeoMedia Viewer®. These programs are very easy to use but do not provide data creation or editing capability. Desktop systems are an intermediate category. They provide basic data creation and editing capabilities and are generally suitable for small systems.

Based on the format of the GIS data involved, each of the three categories of GIS software can be further classified into two categories: vector and raster. As described in Chapter 1 (GIS Basics), vector software represents geographic features as points, lines, polygons, and objects. Raster software represents geographic features in either an image or a grid format. Some packages, such as Intergraph's Modular GIS Environment (MGE), have both vector and raster GIS capability.

Careful selection of software can significantly reduce the overall GIS implementation cost.

SOFTWARE PRICE

This book lists the price, the latest version, and Web site URL for representative software products. The software price is based on the approximate average price in U.S. dollars at the time of writing. The actual software price depends on the cost of doing business (taxes, shipping, import duties, and other factors) in a certain country. The actual price will, therefore, change with time and country of purchase[1]. The prices listed in this book represent a single commercial use license. Education, government, multiple copy, and site license discounts are generally available. Academic use prices of most software packages are much lower. For example, ArcInfo 7.x, which retails for $17,000-19,000, has an academic price of approximately $5,000.

Useful Web Sites For GIS Software Listings

Geospatial Solution Buyers Guide	*www.geospatial-online.com/geospatialsolutions/*
GIS Café e-Catalog	*ecat.dacafe.com/GIS_Design_Tools/*

VECTOR GIS SOFTWARE

Table 2-1 lists selected professional (large system) vector GIS packages. Other packages in this category include GenaMap® from GenaWarehouse, FRAMME® from Intergraph, and GFIS® from IBM. Table 2-2 lists selected desktop (small system) vector GIS packages. Some programs listed in these tables can read GIS data created in other programs. For example, MGE, Geo/SQL, and MapInfo packages are compatible with the ArcView shapefile format.

[1] Always contact your local software dealer or reseller for the latest price. The updated information (version, cost, Web site URL, etc.) about various data and software products discussed in this book may be obtained from the author's Web site *www.GISApplications.com.*

Table 2-1. Vector GIS Software: Professional / Large Systems

Software	*Version / Year*	*Price Range (US$)*	*Vendor*	*Web site*
ArcGIS®	8.1/2001	1,500-19,000	ESRI, Redlands, California	*www.esri.com*
ArcInfo®	7.2/1998	17,000-19,000		
GeoMedia Pro®	4.0/2000	7,000-8,000	Intergraph, Huntsville, Alabama	*www.intergraph.com*
MGE® (requires MicroStation)	8.0/2001	4,000-6,000 + 4,000-5,000 for MicroStation		
Smallworld®	3.1	50,000-60,000	Smallworld Systems, Cambridge, UK	*www.smallworld-us.com*

Table 2-2. GIS Software: Desktop / Small Systems

Software	*Version / Year*	*Price Range (US$)*	*Vendor*	*Web site*
ArcView®	8.1 / 2001 3.2 / 1998	1,400-1,600 1,100-1,300	ESRI, Redlands, California	*www.esri.com*
PC ArcInfo®	4.0 / 2000	3,000-4,000		
ArcCAD® (requires AutoCAD)	11.4.1	400-600 + 3,500-4,000 for AutoCAD		
Autodesk Map® (requires AutoCAD®)	2000i	4,000-5,000 (including AutoCAD)	Autodesk, San Rafael, California	*www.autodesk.com*
GeoGraphics® (requires MicroStation®)	8.0	1,500-1,8000 + 4,000-5,000 for MicroStation	Bentley Systems, Exton, Pennsylvania	*www.bentley.com*
GeoMedia®	4.0	1,400-1,600	Intergraph, Hunts-ville, Alabama	*www.intergraph.com*
MapInfo Professional®	6.0	1,400-1,600	MapInfo Corp., Troy, New York	*www.mapinfo.com*
Maptitude®	4.1	500-700	Caliper Corp., Newton, Massachusetts	*www.caliper.com*
Geo/SQL®	5.5	250-350	Geo/SQL Technologies, Calgary, Canada	*www.geosql.com*

Raster GIS Software

Tables 2-3 and 2-4 list raster GIS software packages for professional (large systems) and desktop (small systems) use, respectively. Some raster GIS packages, such as ERDAS, can also be classified as "geographic imaging" products. Additional information about geographic imaging software is provided in Chapter 3 (GIS Applications Software).

Table 2-3. Raster GIS Software: Professional / Large Systems

Software	*Version*	*Price Range (US$)*	*Vendor*	*Web site*
ArcGIS Spatial Analyst®	8.1/2001	2,000-3,000	ESRI, Redlands, California	*www.esri.com*
ARC GRID®	7.2/1998			
MGE® (requires MicroStation®)	8.0/2001	4,000-6,000 + 4,000-5,000 for MicroStation	Intergraph, Huntsville, Alabama	*www.intergraph.com*
IMAGINE®	8.5/2001	Depends on options. Starting price is 5,000.	ERDAS Atlanta, Georgia	*www.erdas.com*

Table 2-4. Raster GIS Software: Desktop / Small Systems

Software	*Version*	*Price Range (US$)*	*Vendor*	*Web site*
ArcGIS Spatial Analyst®	8.1/2001	2,000-3,000	ESRI, Redlands, California	*www.esri.com*
ArcView Spatial Analyst®	3.2/1998			
MFworks® for GeoMedia®	2.6	800-1,000	Intergraph, Huntsville, Alabama	*www.intergraph.com*
Idrisi32®	2.0	1,400-1,600	Clark Labs, Worcester, Massachusetts	*www.clarklabs.org*
ILWIS®	3.0	2,000-2,500	ITC, Netherlands	*www.itc.nl*
GRASS	5.0	Free	Baylor University, Waco, Texas	*www.baylor.edu/grass/*

GIS SOFTWARE BUSINESS

Figure 2-1 shows the percentage of users in the United States for various GIS software products in 1998. This chart is based on the Framework Data Survey conducted by the National States Geographic Information Council (NSGIC) and National Spatial Data Infrastructure (NSDI) in October 1998 (Somers, 1999).

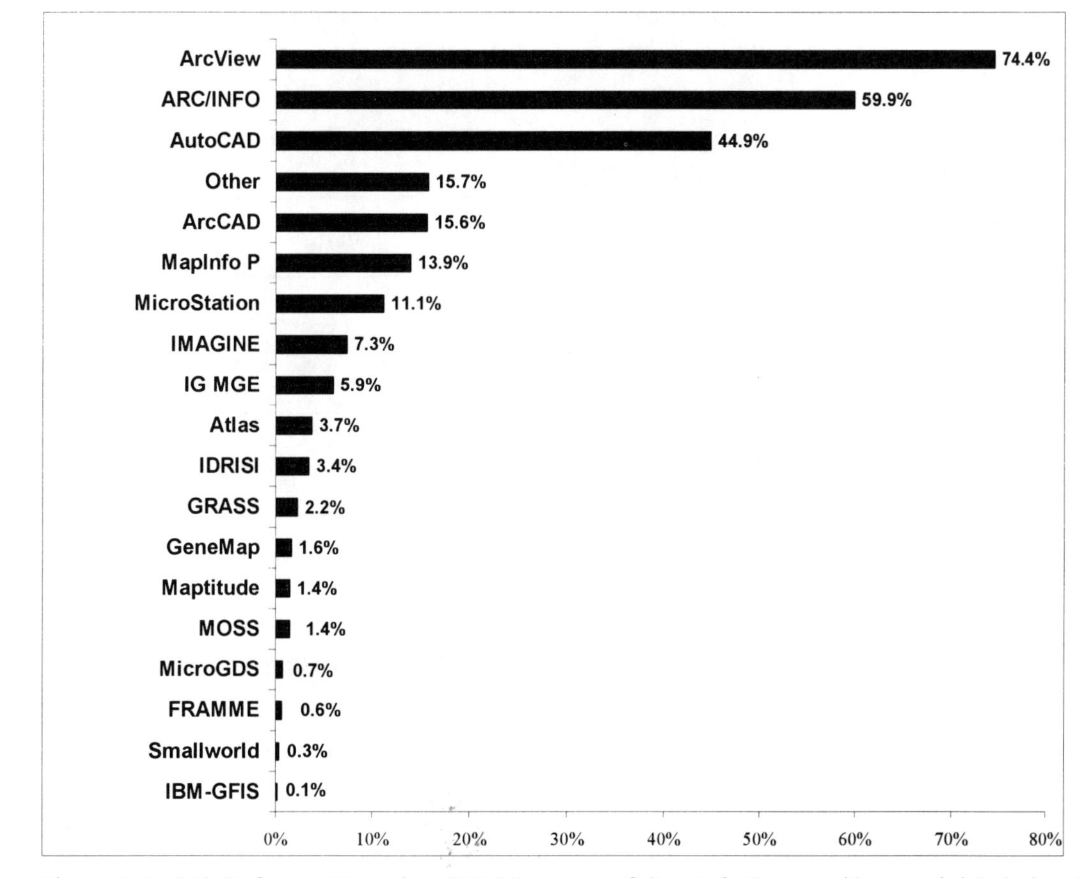

Figure 2-1. GIS Software Users in 1998 (Courtesy of Geo Info Systems/Geospatial Solutions)

The survey included 5,200 respondents from 1,683 counties in 14 states of the United States. The survey indicates that ESRI's ArcView and ArcInfo products top the list at 74.4% and 59.9%, respectively. Another survey conducted by Daratech Inc., an independent marketing research firm indicates that from 1998 to 1999 core worldwide GIS business (hardware, software, and services) revenue grew by 10.6% to an estimated $1.5 billion (Daratech, 2000). The GIS software business was estimated at $845 million in 1999, representing an increase of 12.8% from 1998. Table

2-5 summarizes the 1999 revenue and market share of the top revenue producing GIS software vendors (Barnes, 2000). ESRI maintained its first position in 1999 with its broad market presence among federal agencies, local governments, mapping organizations, petroleum and forestry companies, and retail businesses. Intergraph held the second position, driving revenue from its flagship MGE product suite and new-generation GeoMedia products in government, transportation, utilities, and communication markets. The Daratech report indicates that in 1999 the personal computer (PC) market (including Windows NT) was 69% of the core GIS software market. Capturing 36% of the business, Intergraph's GIS solutions accounted for the single largest share of the PC market.

Table 2-5. 1999 Worldwide GIS Software Revenue and Market Share by Vendor

Vendor	*Revenue (million US$)*	*Market Share (%)*
ESRI	296.6	35.1
Intergraph	238.7	28.3
MapInfo	46.7	5.5
Autodesk	39.6	4.7
Smallworld	36.7	4.3
MicroStation	23.5	2.8
ERDAS	18.0	2.1

Source: Daratech, 2000

In 1999, the worldwide revenue for GIS business for hardware, software, and services was 1.5 billion dollars.

Major software packages from all categories (professional, desktop, vector, and raster) are discussed below.

ESRI GIS Solutions

Environmental Systems Research Institute (ESRI), with headquarters in Redlands, California (USA), was established in 1969. With over a million users and 2,600 employees worldwide, ESRI is recognized as the leading supplier of topologically structured GIS packages. They have ten regional offices in the United States and 90 international offices and affiliates. ESRI's business includes development and maintenance of GIS software, software installation and support, database design, database automation, and programming. ESRI is expected to continue to lead in the innovation of future of GIS applications (Cheves, 2000).

ArcInfo®, ArcView®, ArcGIS®, MapObjects®, and IMS® are the leading GIS software products of ESRI. Federal agencies, state governments, and consulting firms are heavy users of ArcInfo and ArcView (Gilbrook, 1999). ESRI's software packages relevant to water, wastewater, and stormwater system applications are listed below.

- ArcGIS®
- ArcInfo®
- ArcView®
- Relevant Extensions
  - Spatial Analyst® (ArcView and ArcGIS)
  - 3D Analyst® (ArcView and ArcGIS)
  - Network Analyst® (ArcView)
  - Image Analysis® (ArcView)
- SDE® (Spatial Database Engine)
- MapObjects® and ArcObjects®
- Internet Map Server (IMS®)
- ArcLogistics® (vehicle routing and scheduling)
- PC ArcInfo®
- ArcCAD®

Some GIS users, such as most civil engineers, are users rather than creators of GIS data. Such casual GIS users will not have to use all the ESRI software listed above. As described below, ArcView is most appropriate for this type of GIS user.

ARCINFO

At the press time, two ArcInfo versions are being used most commonly: ArcInfo 7.x and ArcInfo 8.x.

ArcInfo 7.x

ArcInfo, previously known as ARC/INFO, is ESRI's flagship product. It is the leading GIS software in the world, with approximately 100,000 users worldwide. ArcInfo is a high-end GIS with tools for automation, modification, management, analysis, and display of geographic information. Up through Version 6, ArcInfo was UNIX workstation software with a command-driven shell. Version 7.1 was released both for UNIX and Windows NT operating systems. ArcInfo 7.x and older versions had steep learning curves and required extensive training. Up through versions 7.x, ArcInfo used its arc-node topological model to provide full topology.

Additional GIS functionality and applications are provided by optional add-on modules called extensions. Extensions are special scripts that are loaded into a GIS software to increase its capabilities. Priced at approximately $2,500 each, ArcInfo extensions are completely integrated with ArcInfo functions and data and add significant application-specific functions to ArcInfo. These extensions are listed below:

- ARC GRID: Raster geoprocessing (including support for hydrologic modeling)
- ARC TIN: Three-dimensional visualization and analysis
- ARC NETWORK: Utility modeling
- ArcSDE: Special release of Spatial Database Engine (SDE) for ArcInfo
- ARC COGO: Coordinate geometry for survey data
- ArcScan: Creates vector data from scanned raster images
- ArcPress: Prints high-quality maps on raster media
- ArcStorm: Database Storage Manager
- Geostatistical Analyst: A geostatistical package (for version 8.x)

ArcInfo 8.x

ArcInfo 8.1, released in 2000, has a user-friendly and menu-driven interface that runs on personal computers with Windows NT or Windows 2000 operating systems. Windows XP environment is supported for versions 8.2 and higher.

ArcInfo 8.x is an object-oriented software. The essence of enterprise GIS, ArcInfo 8.x serves as the core of an ArcGIS system that can include ArcView, ArcSDE, ArcIMS, and more. ArcInfo's development environment, ArcObjects, lets users easily build custom ArcInfo applications and interfaces using industry-standard development tools, such as Visual Basic for Applications (VBA). Additional ArcInfo 8.x information is provided below in the ArcGIS section.

ArcView

ArcView 3.x and ArcView 8.x are the two most common versions of ArcView at the present time. ArcView 3.2 has been the most popular version during 1999-2002. Major revisions of ArcView 3.2 are not expected. The next scheduled release (ArcView 3.3) is expected to provide functional improvements (e.g., a faster map projection utility) and bug fixes.

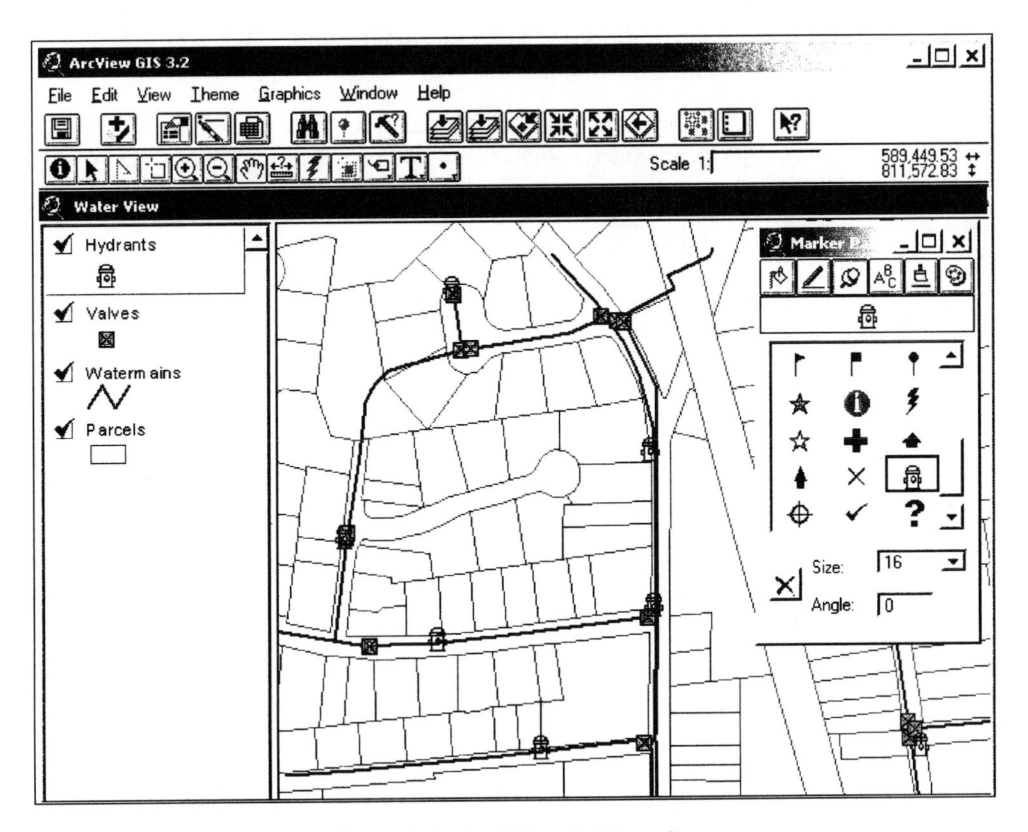

Figure 2-2. ArcView 3.2 Interface

ArcView 3.x

Through the 1990s, professional GIS packages designed for heavy production work were not user-friendly for casual GIS users. For example, older versions of ArcInfo and MGE were complex software systems that required substantial training and technical skills. Most GIS application users, such as civil and environmental engineers, do not have the technical expertise to develop, run, and maintain a complex GIS. ArcView 3.x, on the other hand, is a user-friendly desktop mapping and GIS tool that can be learned by most people without extensive training or GIS experience. Figure 2-2 shows the user-friendly interface of ArcView 3.2. ArcView 3.x is a sophisticated desktop mapping and GIS package that brings the power of GIS to the average PC user. With more than 500,000 users, ArcView may be considered the world's most popular desktop GIS package. The main features of ArcView 3.x are as follows:

- It is user-friendly and provides context-sensitive, on-line help.
- It is available for Windows, Mac, and UNIX operating systems.
- It can access and link documents, images, tables, text, graphics, spreadsheets, maps, multimedia, and CAD drawings in an integrated and comprehensive way.
- It provides routine display and plotting of water, wastewater, and stormwater system maps and provides tools to query the GIS database.
- It helps a user to quickly select and display different combinations of data for creative visualization of mapping data.
- It is compatible with the ArcInfo, PC ArcInfo, and ArcCAD file formats.
- It provides editing tools for creating new maps and modifying existing ones.
- It provides address geocoding and address matching capability, (i.e., tabular data containing street addresses can be displayed as points on a map).
- It can directly use AutoCAD drawings (DWG files) or AutoCAD interchange files (DXF files).
- It uses dBASE format files for data management and can access information from other applications such as FoxPro, Lotus 1-2-3, and Microsoft Excel.
- It can link map information to SQL databases such as Oracle, Ingres, Informix, and Sybase, which are common in local, state, and federal government.
- It supports Dynamic Link Libraries (DLL) and Dynamic Data Exchange (DDE), which can be used to provide efficient third party interfacing and integration capability.
- GIS data can be explored and retrieved by selecting features or formulating logical expressions.
- GIS data can be displayed as pie charts, bar charts, or tables for presentations and reports.

- Avenue (object-oriented development language for ArcView 3.x) or Microsoft Visual Basic can be used to create custom applications and user interfaces.
- Many ArcView extensions are available to supplement its standard capabilities, such as Spatial Analyst for performing hydrologic analyses in a raster GIS.
- Costing approximately $1,200, and running on off-the-shelf PCs, ArcView is affordable.

The above features make ArcView a cost-effective tool for developing applications. A limitation of ArcView is that it is not intended to support the data conversion activities of a GIS production shop. Although ArcView is capable of digitizing paper maps, there is limited feature editing capability. Therefore, ArcView in most appropriate for management, manipulation, and modification of existing GIS layers created using professional GIS products like ArcInfo. Organizations that do not have ArcInfo can hire consultants to convert existing maps to ArcInfo coverages by scanning and digitization. The widespread use of ArcInfo among government agencies, coupled with a wide range of exchange formats supported by ArcInfo, makes it highly probable that most third-party data providers will use ArcInfo and therefore be able to directly supply data in either ArcInfo coverage or ArcView's native shapefile format. If spatial data in any format are available for a project, they can be converted to an ArcView compatible format via ArcInfo (Hutchinson and Daniel, 1995).

ArcView 2.x and 3.x were designed using a client-server architecture. This feature allows the user to extract information from external DBMS programs and establish inter application communication (IAC) through standard protocols like DDE for Windows, RPC for UNIX, and Apple Events for Macintosh. Through IAC, ArcView 2.x could act as a client or a server in an integrated application. For instance, real-time global positioning system (GPS) data could be transferred to ArcView for real-time display through an interface program and DDE. On a multi-user computer network, ESRI's ArcStorm could be used as the data server, ArcInfo as the process server, and ArcView as the client application providing GIS services to end users (Dangermond, 1994).

ArcView 2.x and 3.x were developed using a new object-oriented programming language and development environment called "Avenue." Avenue provided a full-function, object-oriented, application-development environment with a scripting language designed specifically for GIS. The simplest application of Avenue is customization of the ArcView interface by adding or deleting various buttons or menus. Advanced Avenue applications consist of writing and compiling the scripts to extend

ArcView's basic capabilities for specific applications. Sophisticated user scripts can be linked to menu buttons and executed when a particular button is clicked. A sample Avenue script for displaying digital video files in ArcView is shown in Figure 2-3. This compact script uses ArcView's "HotLink" function and Windows Media Player software. It can be used to launch a video file by clicking on a feature that has been linked to the video file. For example, as shown in Figure 2-4 this script can be used to display AVI or MPEG files from sewer system TV inspection video tapes simply by clicking on a sewer pipe in ArcView.

```
MovieHotLinktoMediaPlayer
theVal = SELF
if (not (theVal.IsNull)) then
  if (File.Exists(theVal.AsFileName)) then
    System.Execute("C:\Program Files\Windows Media Player\MPLAYER2.EXE "+theVal)
  else
    MsgBox.Warning("File "+theVal+" not found.","Hot Link")
  end
end
```

Figure 2-3. Sample Avenue Script

Older versions of ArcView did not have the capability to reproject GIS data from one projection system to another. Version 3.2 provides a map projection wizard for vector data only. Image data should be projected in other programs. ArcView 8.1 provides on-the-fly coordinate and datum projection capability for both vector and raster data.

ArcView has three types of extensions: bundled, optional, and user. Bundled extensions are supplied with the software. Optional extensions should be purchased separately. User extensions are developed by the ESRI user community and can be downloaded, free of charge, from ESRI's ArcScript Web site (*arcscripts.esri.com*). The bundled extensions for ArcView 3.2 include the following:

- Projection Utility
- Report Writer
- CAD Reader
- Database Access
- Dialog Designer
- Digitizer
- GeoProcessing

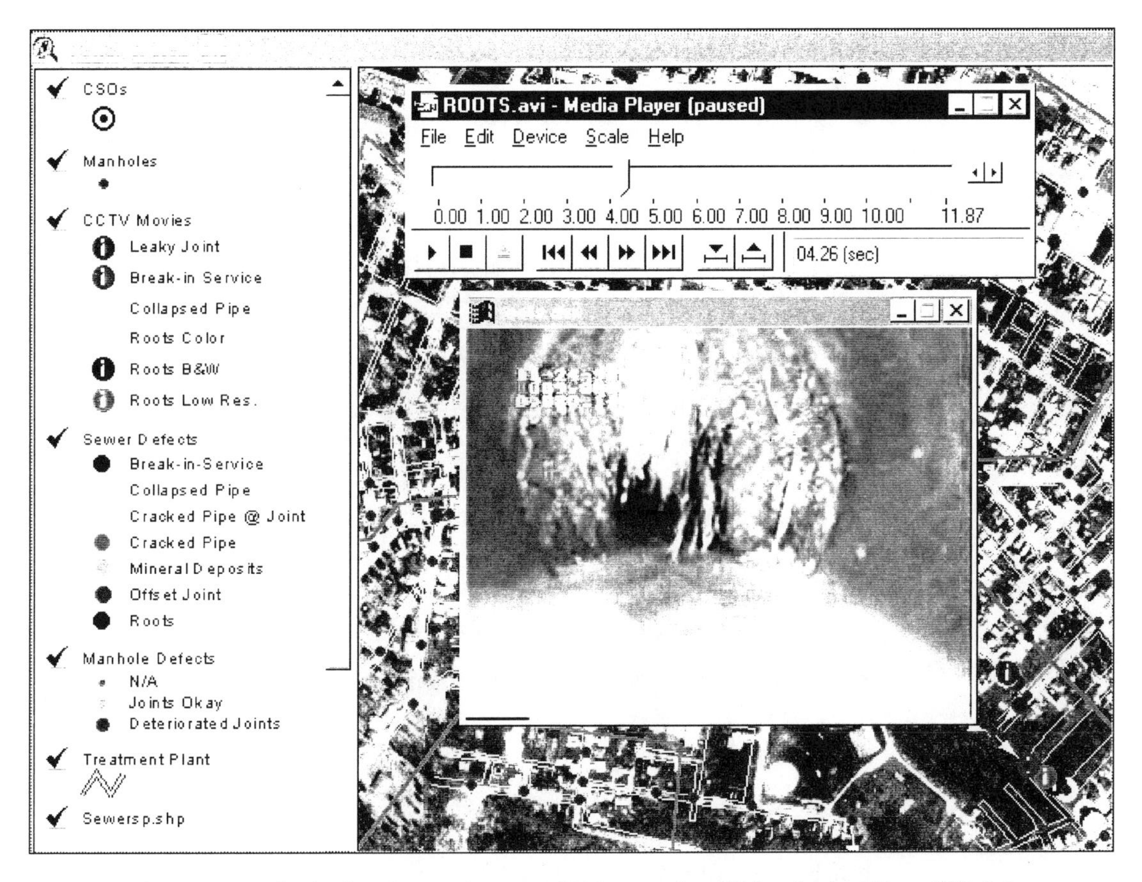

Figure 2-4. Displaying Sewer System TV Inspection Video in ArcView GIS 3.2

- Graticules and Measured Grids
- Image Readers
- Legend Tool
- Military Data Reader
- VPF Viewer
- DXF Export
- SDTS Data Import
- DIGEST (ASRP/USRP)
- MGRS Coordinate Display
- S-57 Data Converter
- RPF Indexer

The optional extensions are as follows:

- Spatial Analyst: Raster GIS with hydrologic modeling support ($2,000-3,000)
- 3D Analyst: 3D visualization and analysis ($2,500-3,500)
- Image Analysis: Image analysis (re-projection, mosaicking, etc.), ($2,000-3,000)
- Network Analyst: Routing analysis ($1,400-1,600)
- StreetMap 2000: Enhanced DYNAMAP 2000™ street database
- StreetMap: Geocoded US streets
- Tracking Analyst: Track real time data (GPS)
- ArcPress: Printing high-quality maps on raster media
- Business Analyst: Analyzing markets
- Internet Map Server: Publishing maps on the world wide web (Web)

ArcView 8.x

ArcView 8.1 is a new release after version 3.2a. The jump from version 3 to 8 signifies a major revision to the software as well as alignment with ESRI's new software suite called ArcGIS 8.1.

Though ArcView 8.x product suite is functionally equivalent to ArcView 3.x, it adds several new capabilities and notable improvements. For example, while the software features of the two versions are similar, the interfaces to access those features are substantially different. So, at a minimum migrating from ArcView 3.x to 8.x would require learning the new interface. ArcView 3.x has a limited support life and will be eventually phased out. Thus, in order to avoid learning both the versions, the new users should start with ArcView 8.x.

With release 8.1, ArcView is now a subset of ArcInfo and uses the same interface and satellite applications (e.g., ArcCatalog). ArcView 8.x is the entry point into the ArcGIS product suite. It is based on Microsoft Component Object Model (COM) technology for Windows and runs on Microsoft Windows 9X/NT/2000 platforms. This version permits the direct reading of more than 40 data formats and uses a new ArcCatalog application for browsing and managing data. One of the new features of ArcView 8.1 is on-the-fly projection both for vector and raster data. ArcView 8.1 includes new industry-specific symbology including symbol sets for water and wastewater systems. GIS customization and application development uses built-in VBA, which means that Avenue has become obsolete with this release. ESRI's abandonment of Avenue as the preferred programming language for customization and application development is perhaps the most significant change in ArcView 8.x (Jenkins, 2002). However, this change is good in the sense that it eliminates the need to learn a new vendor-specific programming language. In fact, development of new ArcView applications has never been easier thanks to the new generation of college graduates that are already familiar with VBA. ESRI has, nonetheless, provided technical white papers on Avenue-to-VBA migration and examples for converting existing Avenue scripts to VBA programs.

Additional ArcView 8.1 information is provided below in the ArcGIS section.

Useful ArcView Web sites

ArcView 3.x Home Page	*www.esri.com/software/arcview/index.html*
ArcView 8.x Home Page	*www.esri.com/software/arcgis/arcview/index.html*
ArcScript Home Page	*arcscripts.esri.com*

ARCGIS

Between 1997 and 2000, ESRI worked vigorously to rewrite its key software platform using Information Technology (IT) standards. This effort resulted in reengineered ArcInfo 8.x and ArcIMS based on a new object-component architecture using a host of industry and government standards, such as Structure Query Language (SQL), Common Object Request Broker Architecture (CORBA), COM, TCP/IP, HTTP, XML, JAVA, and C++. These developments produced ArcGIS 8.x, which represents ESRI's new software architecture rather than a new software product.

Designed for Microsoft Windows NT4 and 2000 operating systems, ArcGIS includes ArcView, ArcEditor, ArcInfo, ArcGIS extensions, ArcSDE, and ArcIMS. As shown in Table 2-6, ArcGIS 8.1 has three desktop clients jointly referred to as "ArcGIS Desktop":

1. ArcView GIS 8.1: ArcCatalog, ArcMap, and ArcToolbox LT
2. ArcEditor 8.1
3. ArcInfo 8.1

Table 2-6. ArcGIS Desktop Products

ArcView	*ArcEditor*	*ArcInfo*
Data access	ArcView	ArcEditor
Mapping	***Plus***	***Plus***
Customization	Coverage and geodatabase editing	Advanced geoprocessing
Spatial query		Data conversion
Simple feature editing		Workstation ArcInfo

All three clients have the same user interface, the same development environment using VBA, the same extensions, and the same data models (e.g. ArcFM Water). This architecture would allow users throughout an enterprise to share scripts, custom tools, applications, and extensions. ArcGIS 8.1 integrates the ArcView 8.1 and ArcInfo 8.1 environments and adds new software tools to bridge the gap between the two products.

ArcGIS desktop is a scalable and modular software suite with the same underlying executables and user interface. ArcGIS is scalable because it can be installed on an individual PC or shared across a distributed network. It is modular because users can install the system in pieces. Additional functionality is enabled as users move from ArcView to ArcEditor to ArcInfo (ESRI, 2001).

ArcView 8.x consists of ArcCatalog, ArcMap, and ArcToolbox LT. These applications are standalone and execute independently of each other. ArcCatalog is like Windows Explorer for ArcGIS. It allows the user to view GIS data, preview geographic information, view and edit metadata, work with tables, and define the schema structure for geographic data layers. ArcMap provides mapping and editing functions as well as multiple window displays for map-based analysis. Figure 2-5 shows an ArcMap screenshot for the same system shown in Figure 2-2 using ArcView 3.2. ArcToolbox is a collection of tools for working with data in ArcGIS 8.x. ArcToolbox for ArcInfo provides more than 170 tools for geoprocessing, data conversion, map sheet management, overlay analysis, map projection, and other functions. ArcToolbox for ArcView and ArcEditor (light or LT version) contains 36 commonly used tools for data conversion and management. ArcIMS browser-based viewer and stand-alone ArcExplorer viewer are lightweight Web clients for ArcGIS Desktop.

ArcEditor 8.1 is a new software application that includes all the functionality of ArcView and allows the editing of topologically integrated features in a geodatabase or coverage. It enables feature editing from a multi-user geodatabase or coverage and allows custom feature classes, feature-linked annotation, dimensioning, and raster editing. ArcEditor is a simple interface that serves as a “tool selector” for ArcGIS users. Indeed, the only difference between the full-strength flavor of ArcGIS called ArcInfo 8.x and the lighter-weight version of ArcGIS called ArcView 8.x is the number of specialized tools available to the ArcEditor toolbox. In terms of functionality ArcEditor lies in between ArcView and ArcInfo.

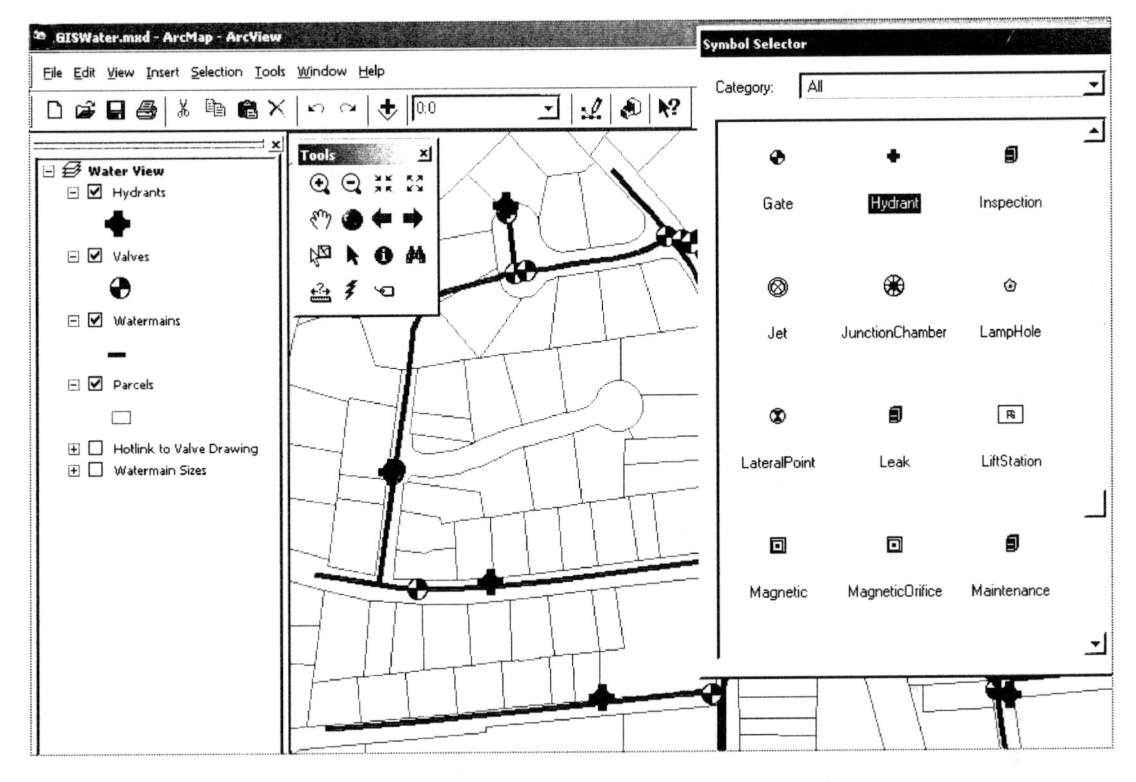

Figure 2-5. ArcGIS ArcMap Screenshot

ArcInfo 8.1 is the most powerful and functionally rich application in the ArcGIS product family. It includes all the ArcView and ArcEditor modules but adds advanced geoprocessing and data conversion capabilities. It also includes a complete ArcToolbox application and a full version of ArcInfo Workstation (ARC, ARCEDIT, ARCPLOT, ARC Macro Language).

ArcGIS 8.1 includes substantially improved raster GIS capabilities. For instance, the raster capabilities previously available in ARC GRID are accessible from the wizard-based ArcToolbox. ArcSDE 8.1 enables joint, multi-user editing and provides transacted views of a database. ArcIMS adds Internet mapping services to an ArcGIS system. A significant new capability is that all ArcGIS desktop clients can now dynamically stream GIS data across the Web from an ArcIMS server. These new layer types can be symbolized, mapped, queried, edited, and analyzed just like local data (ArcNews, 2001). They can also be saved locally for later use, although a key advantage of reading these layers directly from a remote server is ensuring that the most recently maintained iteration of each layer is used each time. GIS data maintenance then becomes completely transparent to remote ArcGIS users.

ArcGIS Web Site

ArcGIS Home Page *www.esri.com/arcgis*

Geodatabase Model

For data storage and manipulation, a DBMS uses a data model, such as hierarchical, network, and relational data models. Data models such as CAD, image, or TIN represent real-world objects in a digital format. Conventional georelational GIS data models, like ESRI's arc-node data model, represent geographic objects and their spatial relationships digitally and store attribute information in a DBMS. The GIS software manages the link between the features and the database. The georelational data model has limitations in modeling the details of geographic objects, and it cannot easily support user-specific or domain-specific features. To overcome these limitations, a new object-oriented (OO) data model was introduced in the late 1990s. This model which can store spatial data inside a relational database management system (RDBMS) is described in Chapter 6 (Database Design).

A geodatabase is a storage mechanism for both spatial and attribute data. It provides centralized storage for a wide variety of geographic information in a DBMS.

Concurrently with the release of ArcInfo 8, ESRI also introduced a new OO data model called a geodatabase model which is one of the new features of ArcGIS 8.x described above (Maguire, 1999). A geodatabase model is a framework for capturing key geographic and descriptive information about a class of landscape features, and for attaching behaviors to the features. A geodatabase model is a receptacle for storing all types of GIS data, including shapefiles, coverages, raster data, and tables. A geodatabase is implemented directly on commercial relational or object-relational database management systems (Zeiler, 1999). For example, a personal geodatabase can be stored in Microsoft Access MDB format.

The geodatabase is accessed using ArcGIS 8.x, in which objects interact through interfaces designed according to a common standard. A geodatabase allows the user to create custom objects (e.g., manholes, streams, and watersheds) rather than simple geometric features (points, lines, and polygons). Therefore, an implication of the geodatabase model is that it makes data more intelligent, which reduces the application development needs. ESRI is developing geodatabase models in collaboration with partners from industry and academia. ArcFM Water and ArcGIS Hydro are two such geodatabase models. ArcFM Water is a data model for water and sewer systems, whereas ArcGIS Hydro is applicable to surface water hydrology and hydrography. More information about these data models is provided in Chapter 6 (Database Design).

There are two types of geodatabases—personal and multi-user. Personal database support is built into ArcGIS implemented with Microsoft Jet®. A personal geodatabase is stored in a Microsoft Access® file and can be edited using ArcView 8.x. It is suitable for project-level GIS. A multi-user database is deployed using ArcSDE and requires a DBMS, such as Oracle® or Microsoft SQL Server®. It allows many users to read and write to the same database at the same time (subject to DBMS permissions) and is suitable for enterprise-level GIS (ESRI, 2001a). Multi-user geodatabase creation and editing requires ArcEditor 8.x.

ArcCAD

Like Autodesk Map, ESRI's ArcCAD brings GIS functionality to AutoCAD. Autodesk Map and ArcCAD should be attractive GIS solutions for experienced AutoCAD users. ArcCAD is inexpensive at a price of about $500. However, because ArcCAD runs from within AutoCAD, ArcCAD users must own AutoCAD, which sells for $3,500 to $4,000.

There are three ways to create ArcCAD GIS coverages:

1. From AutoCAD drawings,
2. From ESRI's ArcInfo coverages, or
3. From scratch using AutoCAD design tools and ArcCAD data automation tools.

In CAD drawings, entities are used to represent objects such as points, lines, and circles. In conventional GIS layers or coverages, features are used to represent objects such as points, lines, and polygons. Figure 2-6 shows the difference between AutoCAD entities and ArcCAD features. ArcCAD "links" (stored in .LNK directory) define one-to-one correspondence between entities and features. ArcCAD creates topology through its CLEAN and BUILD commands.

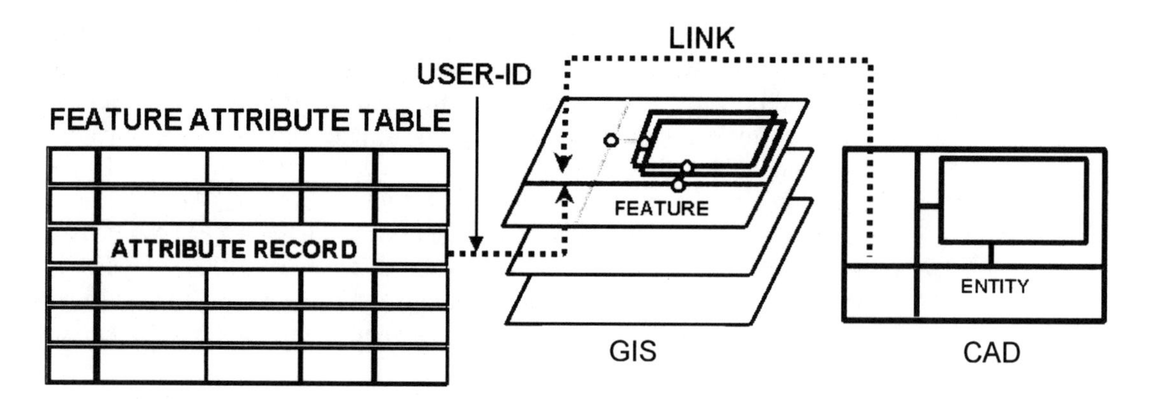

Figure 2-6. Difference Between GIS Features and CAD Entities

SDE

In the mid-1990s, GIS software vendors added an intermediate piece of software called "middleware" to the client-server model. For instance, ESRI introduced a software layer called SDE between the client and the server to facilitate data management and delivery (Goldstein, 1997). SDE allows multi-user query of the GIS database without conversion or replication in a DBMS. SDE is supported by enterprise RDBMS packages, such as Oracle's object-relational database called Spatial Data Option (SDO). Also referred to as Oracle Spatial Cartridge or Oracle 8*i* Spatial, SDO allows GIS data stored in Oracle to be distributed across the enterprise in several different ways. ArcSDE is an ArcInfo extension that integrates SDE client/server technology with ArcInfo. It includes the SDE for Coverages Server, which serves all supported ArcInfo vector data types without the use of a DBMS. It also includes the SDE for DBMS Server, which is available for Oracle, Informix, Microsoft SQL Server, Sybase, and IBM DB2.

An example of how ArcGIS products can be combined for an enterprise GIS implementation would be to have one ArcInfo 8.1 seat and ArcSDE 8.1, an RDBMS, a small staff of geodatabase editors using ArcEditor 8.1, and many staff querying and viewing the geodatabase using ArcView 8.1 (ESRI, 2001b).

ESRI Software Case Study

The Massachusetts Water Resources Authority (MWRA) provides water and wastewater services to 2.5 million people in 60 municipalities of the Greater Boston area. MWRA is best known for its $3.5 billion upgrade of the Boston Harbor wastewater treatment plant. The MWRA service area spans more than 800 square miles and contains several treatment plants, 780 miles of large-diameter pipelines, and dozens of pumping stations and tunnels. Initiated in 1989, the MWRA GIS has grown from a tool to support general planning-level applications to an information management system that supports automated mapping, H&H modeling, site-specific analysis, maintenance, and facilities management. Initially, MWRA operated ArcInfo on a Digital Equipment VAX platform with a VMS operating system and then migrated to a Digital Alpha 3000 operating system in a UNIX environment. MWRA began using ArcInfo LIBRARIAN in 1987 to manage data in a structured hierarchy that included production, test, and development directories. The system managed about 65 data layers throughout the service area that occupied approximately 11 gigabytes of disk storage. MWRA used the ARC GRID, ARC NETWORK, and ARC TIN modules of ArcInfo. ARCGRID was used to process 5-meter Digital Elevation Model (DEM) data provided by the Massachusetts Department of Environmental Protection into watershed slope and aspect. ARC NETWORK was used to perform valve-to-valve tracings within the transmission system to facilitate leak isolation and identify impacts to customers of temporary shutdowns required for scheduled maintenance and emergency repair. ARC TIN was used extensively for contouring water quality samples. MWRA's custom Oracle work was performed in Forms 4.0 and SQL Writer. An integration project was started in 1998 to develop user-friendly interfaces to Oracle data through both Oracle Browser and ArcView and to use ArcView to provide the front end for dBASE databases (Estes, 1998).

AUTODESK GIS SOLUTIONS

Autodesk Map (formerly AutoCAD Map) is Autodesk's entry into the desktop GIS market. It is Autodesk GIS Division's professional solution for automated mapping and GIS analysis in the CAD environment. Like ESRI's ArcCAD, Autodesk Map brings GIS functionality to AutoCAD. It allows the creation, maintenance, and production of maps and geographic data. The latest version is Autodesk Map 5 for AutoCAD 2002. The previous version was AutoCAD Map 2000i for AutoCAD 2000i.

As of 2000, there were approximately 2 million installations of AutoCAD worldwide with thousands of organizations relying on it for mapping and facilities management on a daily basis. AutoCAD-based GIS solutions allow people to use AutoCAD not just to produce maps and drawings, but to develop GIS applications as well. Autodesk Map runs from within AutoCAD and is accessible through a new menu. It appeals to those who already know AutoCAD and do not wish to learn a new system. AutoCAD is bundled with Autodesk Map at the time of purchase. Autodesk Map supports topology, georeferenced raster images, and the rubber sheeting of images. Its advantages include easy connection to relational databases (especially Microsoft Access), easy attachment to drawings, and improved raster capabilities. Path Trace and Flood Trace are examples of the topological capabilities of Autodesk Map. The Path

Trace feature finds the shortest travel time between two points. The Flood Trace feature can be used to identify property owners (parcels) within a given distance from a project and generate a mailing list to inform the residents of project issues (Bell, 1999). Autodesk Map 5 features a direct connection to the Oracle 8*i* spatial database, thematic mapping using wizards, the first redesign to the user interface since version 1, and an enhanced import/export engine. A client to the enterprise GIS, the product is built on the AutoCAD engine and reads and writes the latest versions of DWG files.

The Internet is simplifying how GIS applications and maps are created and maintained. The Web facilitates the sharing of GIS data between different organizations. Intranets are making it easy to share GIS data between different co-workers in the same organization. These possibilities are the result of a new generation of software from the leading GIS and CAD software companies. For example, built on AutoCAD 2000i, AutoCAD Map 2000i not only allows traditional CAD work but also enables users to collaborate with design professionals anywhere in the world and moves their designs and processes, simply and easily, to their company's intranet and the Web. Autodesk's CAD Overlay 2000i provides raster data and image analysis capability. The AutoCAD Map Today browser, which is built into AutoCAD Map and MapGuide, enables the development of Internet applications.

Autodesk OnSite Enterprise 2 delivers interactive digital design and mapping information from a company's central server to the user on a Microsoft CE-based mobile computing device, such as Compaq's iPAQ. The product is a platform for developing multi-user mobile solutions using Autodesk OnSite View 2 for the mobile clients.

INTERGRAPH GIS SOLUTIONS

Intergraph was the first GIS vendor to create products on the Windows NT platform in 1994. As the founding member and a strong supporter of the OpenGIS consortium, Intergraph's developers work with other GIS vendors to encourage adoption of non-proprietary standards for use and exchange of spatial data formats. Intergraph provides excellent GIS and CAD integration opportunities because of its products for designing water and sewer systems, such as InWater® and InSewer®. For example, in 1992 San Diego's Water Utilities Department integrated ArcInfo land-based data and InSewer's engineering CAD data in Intergraph's MGE GIS software to design a large trunk line (Robinson, 1993). Intergraph's GIS and mapping programs are described below:

- MGE® is Intergraph's professional strength GIS software with both the vector and raster GIS capabilities. Since moving to NT in 1994, MGE has gained production experience around the world. Interoperable with UNIX and accessible to all GIS users, MGE satisfies most GIS workflows right out of the box. The latest version, MGE 8, includes GeoMedia data server technology that allows the MGE users to easily capture, integrate, and maintain data from supported data sources into their MGE projects. This capability is very effective in expanding the use of MGE applications.

- GeoMedia® is Intergraph's universal information integrator. It serves as a visualization and analysis tool and as an open platform for custom GIS solutions. It is the core software program in Intergraph's scalable enterprise suite of GIS and Internet software, and is as close to the two ideals of "open GIS" and "enterprise solution" as users are likely to find with today's technology (Thrall, 1999a).

- GeoMedia Professional® is a product specifically designed to collect and manage spatial data using standard databases. It is an upgraded version of GeoMedia 2 that provides an integrated raster/vector environment, thematic mapping tools, data capture and editing, and buffer zone creation. It allows users to store geometric and attribute data in a single, unified database using standard database tables. It also represents parcels, streets, and other geographic components as feature objects with both spatial and feature attributes. GeoMedia Pro is compatible with a wide variety of GISs and RDBMSs, such as MGE, FRAMME, MicroStation, ArcInfo, ArcView, GE Smallworld, Oracle SDO, Microsoft SQL Server, and Microsoft Access.

- ActiveFRAMME® (formerly FRAMME) is an automated mapping/ facilities management/GIS (AM/FM/GIS) software. Successful operations in today's utilities require that decision-makers throughout an organization have access to up-to-the-minute facility information. ActiveFRAMME allows creation of a facility model showing all work in progress across the organization. Because this model is continuously updated, multiple users in different departments can refer to the most recent information. ActiveFRAMME allows users to model and access facility data as objects defined by all related components and attributes such as, maps, photos, installation date, manufacturer, etc. These attributes are updated during day-to-day operations, so data are kept current and everyone makes decisions based on the most recent information. Using Field View, which displays facility data in both office (online) and field (offline) environments, anyone in the enterprise can access the facility model from any location.

- GeoMedia WebMap® is Intergraph's Web-based map visualization tool with real-time links to one or more GIS data warehouses. It is an integration tool for different data formats that converts data into a common projection system on-the-fly (Lowe, 2000).

- GeoMedia WebEnterprise® creates dynamic, custom, Web-mapping applications that can analyze and manipulate geographic data.
- MFworks® for GeoMedia is a raster-based GIS package. It provides grid-based software visualization, mapping, and analysis capabilities. It can be used either as a GeoMedia 3 add-on to add raster GIS functionality to GeoMedia or as a stand-alone software program.
- SMMS® for GeoMedia is a desktop tool for the creation of geographic metadata and the management of geographic data.
- Digital Cartographic Studio® products for GeoMedia provide a complete cartographic platform for mapping and GIS professionals.

GE Smallworld GIS Solutions

Smallworld was established in Cambridge, England, in 1988 by eight engineers who had a vision "to make the world smaller" by creating a way to bring people closer together through the use of geographic and spatial information. Calling their new company Smallworld Systems, they rejected the industry standard of transferring paper maps to a computer screen and instead developed a whole new way to work with spatial data. In October 2000, General Electric (GE) acquired Smallworld and renamed it GE Smallworld. GE Smallworld's products include a wide range of application products that automate processes for designing physical facilities, managing operations, analyzing networks, improving responses to service outages, analyzing market opportunities, and addressing a host of other business needs. Smallworld specializes in managing utility networks such as communication, water, and wastewater systems. GE Smallworld Water Distribution Manager and Wastewater Distribution Manager are the company's products for water and wastewater systems. This product line provides an asset management system for water and wastewater utilities and supports all aspects of water and wastewater networks, from pumping stations to customer connection. The model simplifies the collection, maintenance, and analysis of water and wastewater network data.

Bentley Systems GIS Solutions

Bentley Systems GIS solutions are built around their MicroStation CAD package. MicroStation GIS products include the following:

- MicroStation GeoGraphics®
- MicroStation GeoOutlook®
- MicroStation GeoTerrain®

- MicroStation GeoCoordinator®

MicroStation GeoGraphics is a fully integrated CAD/GIS solution with raster and vector functionality that was created for engineers and application developers. It extends MicroStation's industry-standard data capture and editing tools with a database interface and powerful spatial analysis capabilities. MicroStation GeoGraphics integrates seamlessly with MicroStation/J and provides tools to input, manage, analyze, and visualize geographical information within MicroStation. Because most mapping projects require some level of customization, MicroStation GeoGraphics functions as both an end-user product and an application development platform. Its open design provides a wide range of mapping functions that enable users to extend functionality and create custom applications on a wide range of hardware and operating systems.

MicroStation GeoOutlook is a stand-alone data access and decision support tool for mapping and GIS applications. It provides an easy-to-use environment designed to complement MicroStation-based mapping projects that are built in MicroStation 95 and MicroStation GeoGraphics.

GeoTerrain by GEOPAK is full-function 3D software for digital terrain modeling (DTM) and the analysis of surface terrain. The software handles the broad scope of requirements for terrain modeling, contouring, thematic model generation, surface visualization, data editing, and sophisticated volume calculations between DTM surfaces.

With MicroStation GeoCoordinator, MicroStation and MicroStation GeoGraphics, users can integrate data from different map coordinate systems and projections.

MicroStation Software Case Study

Arizona Water Company (Phoenix, USA) is the largest investor-owned water utility in the state of Arizona. It manages the distribution of water to the dry areas of the state, installing new water lines to ranches, private dwellings and communities, and maintaining them with mylar atlas maps and manual data management techniques. Faced with booming real estate development accompanied by an increase in the demand for water resources, Arizona Water realized it needed GIS software to facilitate a more efficient way of maintaining it distribution system atlas maps. As an existing user of MicroStation's engineering software, Arizona Water wanted a complete automated mapping and facilities management system based on MicroStation. The final solution consisted of an integrated system which included Oracle, MicroStation, MicroStation GeoGraphics and GEOTELEC's GeoGraphic Utility System (GUS). This new system runs on an NT 4.0 server and NT 4.0 workstations. The system allows users to transfer their rendering data directly into the hydraulic modeling software and GeoGraphics allows them to display the results of the model graphically. Arizona Water's engineers feel that this combination of software gives them the ability to be more effective when choosing location and size of new water line installations (Bentley Systems, 2001).

MapInfo GIS Solutions

MapInfo GIS and mapping products include the following:

- MapInfo Professional®: Software for mapping and geographic analysis.
- MapInfo MapXtreme® Java Edition: MapInfo's Java-based Internet mapping server for the deployment of mapping applications.
- MapInfo MapXtreme® NT Edition: Windows NT Internet mapping server for the deployment of mapping applications.
- MapInfo ProViewer®: An easy-to-use tool for viewing maps and tables created in MapInfo Professional.
- MapBasic®: An application-development environment for MapInfo Professional.
- MapInfo MapX®: A tool for embedding mapping functionality into new and existing applications.
- MapInfo SpatialWare®: An information management tool for storing, managing, and manipulating location-based data.
- MapInfo DriveTime®: A software module for creating drive-time polygons around any location.
- MapInfo GIS Extension®: A spatial data cartridge for Oracle SDO.
- MapInfo MapXtend®: A developer tool for creating location-based wireless handheld applications.

MapInfo GIS Extension®

The MapInfo GIS Extension data cartridge adds a rich set of spatial functions to Oracle SDO for enhanced spatial analysis capabilities. MapInfo GIS Extension works together seamlessly with Oracle SDO, providing an enterprise-wide solution to better leverage spatial information. It operates on Oracle's object-relational geometry data and is accessible through all programmatic interfaces supported by Oracle, such as OCI, ODBC, JDBC, PL/SQL, and SQLJ. Once installed, it provides access to a set of SQL functions to develop various GIS applications.

CLARK LABS GIS SOLUTIONS

With approximately 30,000 users worldwide, Idrisi32® is a product of Clark Labs at Clark University in Worcester (Massachusetts, USA). Idrisi is not an acronym; the product is named after a cartographer born in 1099 AD in Morocco, North Africa. Clark Labs, with 12 years of experience in the development and distribution of GIS and image processing tools, continues its tradition of providing affordable access to the frontiers of spatial analysis.

Idrisi32 can provide both raster and (with appropriate companion software) vector GIS capabilities, but it is primarily a raster software. It offers raster analytical functionality for GIS and remote sensing database query, spatial modeling, and image enhancement and classification. The software includes built-in functionality for environmental modeling and natural resource management, including change and time-series analysis, multi-criteria decision support, uncertainty analysis, and simulation modeling (Simonovic, 2002). The most recent 32-bit Release 2 uses the latest object-oriented development tools to work as a powerful research tool on NT workstations and desktops. Idrisi's interface is not as user-friendly as MapInfo and ArcView. For example, data import and raster and vector integration tasks are not very intuitive (Binford, 2000). However, this very reason plus the low academic price of $75 to $150 make Idrisi32 ideally suited for academic use and GIS tutorials. In fact, more than 30 universities use this software in their GIS and remote-sensing courses.

Idrisi32 images have an open data structure and can be easily manipulated by user programs (FORTRAN, Visual Basic, etc.). This customization capability makes Idrisi32 a valuable tool for such applications as the following:

- Developing GIS-based distributed surface runoff models
- Manipulating DEM data for extracting streams and watershed boundaries
- Estimating SCS runoff curve numbers
- Estimating runoff using Manning's equation

Clark Lab's software for vector operations is CartaLinx which sells for $300 to $500. It has a relational database engine and is intended as a companion to a variety of GIS and desktop-mapping software products such as Idrisi32, ArcInfo, ArcView, and MapInfo. CartaLinx is used to create layers that consist of the spatial definitions of features in vector format and associated attribute value files. These data are then typically exported to a GIS, either as entire coverages or as a series of map layers.

The strengths of Idrisi32 software are its raster data analysis, help files, object orientation, animation (creating and playing audio and video files), and geostatistics and interpolation algorithms.

Other GIS Solutions

Representative GIS products from other software vendors are described below.

Geo/SQL

Geo/SQL, produced by Geo/SQL Technologies, is based on the latest distributed database and SQL technology. It supports complete topology, spatial analysis, and thematic mapping at a reasonable price. Geo/SQL is available both in standalone Windows and AutoCAD add-on versions. On the desktop, Geo/SQL works with major GIS formats, including ESRI, MapInfo and Autodesk. Geo/SQL not only uses popular GIS formats but also provides seamless spatial data using SQL database technology, such as Oracle SDO and Geo/SQL Spatial SQL. Reportedly, any SQL database that supports ODBC may be used to manage the large volumes of geographic and descriptive data required by the most demanding applications. Its AutoCAD-based, seamless, object oriented, topological database is capable of linking objects with multiple, on-line, dedicated, or distributed databases via modem or network. Geo/SQL's AutoCAD interface and its structured nonprocedural command language allow the development of GIS applications. Application developers can modify Geo/SQL menus to build turnkey applications. According to Geo/SQL Technologies, links to existing analytical and operations-management software can be prototyped in a matter of hours, not weeks or months. Its open architecture allows the use of many specialized third-party applications, such as automated mapping/facilities management (AM/FM), maintenance programming, and water analysis.

U.S. Army's GRASS

The Geographic Resources Analysis Support System (GRASS) was developed by the Environmental Division of the U.S. Army Construction Engineering Research Laboratory (USACERL). It is a raster GIS, although it can use both raster and vector data formats. USACERL's last release of GRASS was version 4.1 in 1993. After 1993, a "GRASS Research Group" located at Baylor University (Waco, Texas, USA) took control of GRASS development. This group's first release was GRASS 4.2 in 1997. The latest (2001) release of the program is GRASS 5, which represents the first major change in GRASS functionality in several years, with the most notable change being support for floating point and null values. GRASS also has a new and much easier-to-use Windows interface. GRASS is a public domain software and can be downloaded free of cost.

GRASS can digitize, edit, store, and overlay vector data from traditional maps, while using the greater analytical power of raster technology. GRASS has several design features that are helpful in hydrologic modeling. Its "Watershed" function maps streams and watershed boundaries from digital elevation models. Its "Gdrain" function finds the drainage path from a designated starting point. For example, GRASS was used to create a hydrologic model of Pinon Canyon (Colorado, USA) using data for elevation, vegetation, soils, streams, seasonal variations of the water table, and rainfall versus stream runoff, among other data. The GRASS information can be converted to hydrologic submodels, which in turn can be converted to thematic models.

SCS-GRASS is a special version of GRASS that is nationally supported by the Natural Resources Conservation Service or NRCS (formerly Soil Conservation Service or SCS). The SCS-GRASS application interface program provides an interactive, menu-driven shell that calls other "C" programs to create application products in a GRASS GIS environment. An SCS-GRASS interface is anticipated for SSURGO and STATSGO soil databases and other RDBMS packages. Detailed SSURGO and STATSGO information is presented in Chapter 5 (Internet GIS).

MICROSOFT'S MAPPOINT®

MapPoint® 2000 marked Microsoft's debut in the desktop mapping business. Designed for seamless integration with other Microsoft Office applications, MapPoint brought basic mapping capability to Microsoft Office users in an easy-to-learn format. With a price tag of just under $250, MapPoint 2002 is an example of GIS technology as a broad-based consumer product. MapPoint is a basic thematic mapping tool for business applications and should not be considered a full-featured desktop GIS package. MapPoint is very easy to learn although it has limited GIS capabilities (Thrall, 1999).

MapPoint mapping software for business use allows the creation of maps in the form of Microsoft Office documents. It also allows users to import their own data (sales figures, demographics, etc.) into maps to reflect themes and comparative relationships significant to their business. For example, users can import contact information from Microsoft Outlook and display it as points on a map. MapPoint will take care of address matching and geocoding these contacts. It contains many of the capabilities of Microsoft Trip Planner and Expedia Streets software.

MapPoint 2002 supports GPS data for waypoints, territory management, and drive-time analysis. Additional MapPoint information is available at the software Web site.

MapPoint Web Site

MapPoint Home Page *www.microsoft.com/mappoint/*

Kodak's ENVI®

Research Systems, Inc., a Kodak company, is a technical software company that develops visualization and data analysis tools. Its main GIS product is called Environment for Visualizing Images (ENVI), which is an advanced, yet easy-to-use, remote-sensing software. Additional ENVI information is available at the software Web site.

ENVI Web Site

ENVI Home Page *www.rsinc.com/envi/*

GIS Software Selection

The selection of GIS software should be dictated by the intended GIS application. The purchaser should review the technical capabilities of alternative products, ease of installation, hardware demands, learning curve, training requirements, and cost. The cost must include maintenance, upgrade, and technical support fees. The purchaser should research the details of the software, test the system using demonstration versions, and question the vendors about project-specific needs (Lee, 1998).

Because it is natural for people to stay with what they are already comfortable with, existing hardware, software, and personnel often influence GIS software selection. For example, AutoCAD users are naturally inclined to use Autodesk Map for their GIS needs. Based on this logic, Intergraph users may prefer GeoMedia, MicroStation users may prefer GeoGraphics, and ArcInfo users may prefer other ESRI products and extensions.

Needs Analysis

For large GIS implementation projects, a needs analysis (or needs assessment) study should be conducted to select appropriate GIS software. The analysis will identify and quantify the GIS needs of an organization and its stakeholders, such as customers, departments, elected officials, regulators, and clients. It will also define how a GIS will benefit an organization by relating specific organizational resources and needs to specific GIS capabilities (Wells, 1991).

A careful needs analysis is critical to a successful GIS implementation and should be the first task of any GIS project. Needs should be quantified by inventorying GIS-related resources (maps, software, data, technical staff, etc.) and interviewing policy makers, senior managers, users, and technical staff.

Most needs analyses utilize a decision matrix of user needs (applications) and available software capabilities to help select the most appropriate software. For example, Table 2-7 shows a sample decision matrix for various water, wastewater, and stormwater applications of a buyer who wants to use ESRI software products. This table indicates that ArcView, ArcInfo, and SDE should be given a high priority because they are required to implement all the GIS applications of the organization. Additional needs analysis information is presented in Chapter 9 (Wastewater System Applications).

Table 2-7. Sample Software Selection Matrix

Application	*AV*	*SA*	*NA*	*3DA*	*AI*	*AC*	*SDE*	*MO*
Planning and engineering	🖫			🖫	🖫	🖫	🖫	🖫
Operations & maintenance	🖫		🖫		🖫	🖫	🖫	🖫
Infrastructure management & construction	🖫		🖫		🖫	🖫	🖫	🖫
Water resources & hydrology	🖫	🖫			🖫		🖫	
Finance and administration	🖫				🖫		🖫	

AV = ArcView, SA = Spatial Analyst, NA = Network Analyst, 3DA = 3D Analyst, AI = ArcInfo, AC = ArcCAD, SDE = Spatial Database Engine, MO = MapObjects

SUMMARY

This chapter presented representative GIS development software products from several companies. Based on 1998-99 user surveys, ESRI, Autodesk, and Intergraph appear to provide the most widely used products. A side-by-side comparison of their capabilities is difficult because of the large number of features and options that each one offers. Anyone trying to select a GIS software product should begin with an assessment of the organization's application needs and compare the various GIS software products to those needs, rather than comparing the products to one another. The major GIS implementation cost is usually data conversion. Hardware and software are relatively inexpensive—usually less than 20% of total GIS cost. This dictates that one should avoid excessive focus on software selection and devote an appropriate amount of effort on other GIS implementation issues, such as GIS database design, user training, and data maintenance.

SELF EVALUATION

1. What are different types of GIS software? What type of GIS software do you have?
2. Why is the selection of GIS development software important for implementing a GIS?
3. Use the matrix in Table 2-8 to compare the features of three GIS development software products you are most familiar with. Fill in the blank cells of the table. Answer Yes or No where applicable.

Table 2-8. Software Comparison Matrix

Features	*Software No. 1*	*Software No. 2*	*Software No. 3*
General			
Software Name			
Vendor			
Latest Version			
Year			
Number of installations			
Price			
Platform			
Workstation			
PC			
Operating System			
Windows NT			
Windows XP			
Windows 2000			
Windows 98			
UNIX			
Linux			
Hardware Demand			
RAM			
Disk Space			
Processor			
Other			
CAD Packages Required			
AutoCAD			
MicroStation			
Other			
Capabilities			
Vector			
Raster			
Topology			
Georeferencing			
Image processing			
Street geocoding			
Customization			
Support for H&H modeling			
Support for facilities management			
Scripting language name			
Other			

Chapter 3

GIS Applications Software

GIS applications are developed by extending the core capabilities of a GIS software. GIS software selection is dictated by the intended GIS application.

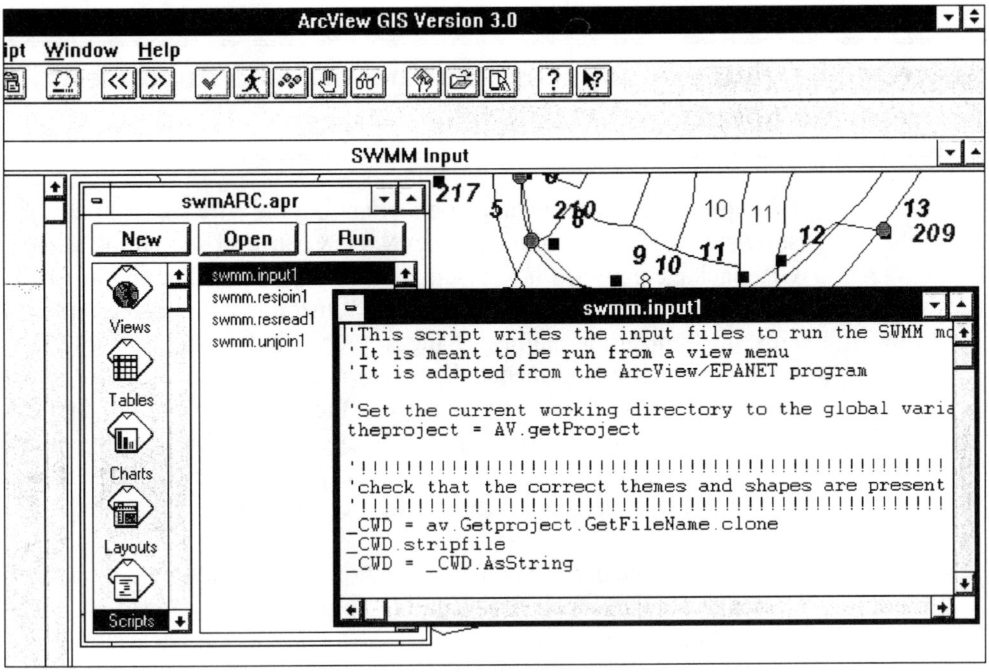

ArcView GIS Version 3.0

ipt Window Help

SWMM Input

swmARC.apr

New Open Run

Views Tables Charts Layouts Scripts

swmm.input1
swmm.resjoin1
swmm.resread1
swmm.unjoin1

217 5 210 8 10 11 9 10 11 12 13 209

swmm.input1

```
'This script writes the input files to run the SWMM mo
'It is meant to be run from a view menu
'It is adapted from the ArcView/EPANET program

'Set the current working directory to the global varia
theproject = AV.getProject

'!!!!!!!!!!!!!!!!!!!!!!!!!!!!!!!!!!!!!!!!!!!!!!!!!!!!!
'check that the correct themes and shapes are present
'!!!!!!!!!!!!!!!!!!!!!!!!!!!!!!!!!!!!!!!!!!!!!!!!!!!!!
_CWD = av.Getproject.GetFileName.clone
_CWD.stripfile
_CWD = _CWD.AsString
```

ArcView® GIS Application Development Environment

LEARNING OBJECTIVE

The learning objective of this chapter is to become familiar with major application development software products. Major topics discussed in this chapter include

- Application development software
- Image processing software
- Internet GIS software
- Database management software
- Statistical analysis software
- Document management software
- AM/FM/GIS software
- Computer modeling software

APPLICATION DEVELOPMENT SOFTWARE

GIS application software is used to run GIS applications, such as hydraulic modeling or work order management. Like GIS production software, there are many types of GIS applications software, not all of which can be listed here. Some representative examples are given in this chapter.

> **Creative application developers are the people who dream with their eyes wide open.**

GIS applications are developed by extending the core capabilities of a GIS software. Creative application developers can find something useful in a GIS package and turn it into something innovative. The two methods for developing GIS applications are GIS customization and programming. GIS customization mainly changes the default GIS user interface by adding new tools and menus that perform news tasks. The customization capability of a GIS software package is the key to developing applications using this method. Customization is appropriate for small applications. For larger applications, new computer programs must be written and linked to GIS. Many applications use a combination of both methods.

Basic GIS applications and customization, such as adding a new button or a menu, may be achieved without programming. However, advanced applications, such as creating a link to a computer model, almost always require some programming using a scripting language. A scripting language is a programming language that is (usually) embedded in another product, such as Microsoft's Visual Basic for Applications (VBA) or Autodesk's AutoLISP. Scripts are small computer programs written in a scripting language. Scripts must be compiled and linked to GIS software before they can be executed, which is cumbersome. A set of scripts can be converted to an "extension" for faster and user-friendly installation and execution.

Older (pre-ArcGIS) versions of ArcInfo and ArcView are customized using Arc Macro Language (AML) and Avenue scripts, respectively. PC ArcInfo is customized using Simple Macro Language (SML). The latest versions of ArcInfo and ArcView (versions 8 and higher) can be customized using VBA. Because VBA does not compile code, all customizations must be saved to a map document or template. A stand-alone programming environment, such as Visual Basic (VB), C++, or Delphi is required to create full-blown custom applications (Fitzpatrick, 2002).

Avenue is the native object-oriented scripting language for ArcView 2.x and 3.x built (integrated) into ArcView. To the extent that these earlier versions of ArcView continue to be used, application developers and programmers can use Avenue to modify the user interface, build custom tools, and develop solutions for specific applications. Avenue allows simple customization, such as the addition of a new tool to do a new task or the creation of complete turnkey applications. Avenue's full integration with ArcView 3.x benefits the user in two ways: first, by eliminating the need to learn a new interface; second, by letting the user work with Avenue without exiting ArcView (ESRI, 1995). The figure on the first page of this chapter shows the integrated Avenue programming interface of ArcView 3.x. A sample Avenue script for playing digital video files of sewer system TV inspection tapes is described in Chapter 2 and shown in Figure 2-3. Another Avenue script for calculating node elevations of a water distribution system model from digital elevation model (DEM) data is shown in Chapter 8 (Water System Applications) (Hsu, 1999). Avenue developers can also customize the ArcView 3.x interface using ESRI's "Dialog Designer" extension. Like VBA, Dialog Designer allows programmers to call up Avenue scripts by clicking on buttons (Bell, 1999a).

Applications for ArcInfo 7.x and earlier are developed using its scripting language, AML. Designed for these versions of ArcInfo, ArcTools is an AML-based graphical user interface (GUI) composed of sets of tools that represent individual ArcInfo commands or sets of commands. It provides generic messaging and file browsing tools useful for general application development. Initially released with ArcInfo version 6.1.1, ArcTools allows ArcInfo customization and makes ArcInfo easier to use.

For software developers who want to include mapping capabilities in their applications, MapObjects and ArcObjects are appropriate. MapObjects

Version 2 is an ActiveX control (OCX) with more than 45 programmable ActiveX Automation objects that can be plugged into many standard Windows development environments such as VBA, Visual C++, Delphi, and PowerBuilder. MapObjects supports a wide variety of data formats (e.g., shapefiles, ArcInfo coverages, CADD formats, MrSID, TIFF, JPEG). ArcObjects is the COM-based framework that allows developers to enhance the ArcGIS Desktop user interface and extend ArcGIS data models.

The preceding paragraphs discuss how to develop GIS applications using ESRI products. ESRI products were used for illustration purpose only. Other GIS packages that allow interface customization and have a scripting language can be similarly used for developing GIS applications.

GEOGRAPHIC IMAGING AND IMAGE PROCESSING SOFTWARE

The recent image data proliferation has increased the need for desktop tools to manage images. In the past, image analysis software was reserved for geospatial professionals using high-performance workstations (Hurlbut, 1999). These days, user interfaces are becoming friendlier, wizards are replacing obscure command lines, and use of raster GIS by semiskilled end users is growing. For instance, it is now possible to orthorectify and mosaic several aerial photographs and extract 3D objects using wizard-based software on a desktop computer. The gap between the raster GIS and image processing is closing. The recent trends indicate a convergence of image processing and raster GIS into single systems (Limp, 2001).

Representative examples of geographic imaging and image processing products are listed in Table 3-1.

Table 3-1. Geographic Imaging and Image Processing Software

Software	*Version*	*Price Range (US$)*	*Vendor*	*Web site*
IMAGINE®	8.5	Starting at 5,000	ERDAS	*www.erdas.com*
Image Analysis®	1.1	2,000-3,000	ERDAS and ESRI	*www.erdas.com*
MrSID®	1.4	1,000–3,500	LizardTech	*www.lizardtech.com*
Geomatica®	N/A	N/A	PCI Geomatics	*www.pcigeomatics.com*
ENVI®	3.4	3,500-4,500	Research Systems	*www.rsinc.com*
ER Mapper®	6.1	4,000-6,000	Earth Resources Mapping	*www.ermapper.com*
Image Analyst® for MicroStation	N/A	N/A	Z/I Imaging Corporation	*www.ziimaging.com*

Geographic imaging and image processing products help visualize, manipulate, analyze, measure, and integrate geographic imagery and geospatial information. Some of these programs, such as Multi-resolution Seamless Image Database (MrSID), help to preprocess raster data for GIS applications. Others add image processing capability to a GIS. For example, Image Analysis is an ArcView GIS extension that adds image processing capability to ArcView.

IMAGINE

ERDAS Inc.'s flagship product, IMAGINE 8.x, provides remote sensing capabilities and a broad range of geographic imaging tools. It contains tools to make production faster and easier, such as on-the-fly reprojection, a batch wizard to automate routine procedures, the ability to create and edit ESRI shapefiles, and faster and easier map production capabilities. Three components make up the ERDAS IMAGINE product suite, providing a scaleable solution for a project's specific needs.

- IMAGINE Essentials: A mapping and visualization tool that allows different types of geographic data to be combined with imagery and quickly organized for a mapping project.
- IMAGINE Advantage: Builds upon the geographic imaging capabilities of IMAGINE Essentials by adding more precise mapping and image processing capabilities. It analyzes data from imagery via image mosaicking, surface interpolation, and advanced image interpretation and orthorectification tools.
- IMAGINE Professional: A suite of sophisticated tools for remote sensing and complex image analysis. IMAGINE Professional contains all of the capabilities of IMAGINE Essentials and IMAGINE Advantage, plus radar analysis and advanced classification tools like the IMAGINE Expert Classifier. It also includes graphical spatial data modeling, which is an advance capability to analyze geographic data.

ArcView Image Analysis extension was developed as a collaborative effort between ESRI and ERDAS. It allows georeferencing of imagery to shapefiles, coverages, global positioning system (GPS) points, or reference images; image enhancement; automatic mapping of feature boundaries; change detections for continuous and thematic imagery; multispectral categorizations for land cover mapping and data extraction; vegetation greenness mapping; and mosaicking imagery from different sources and different resolutions. Image Analysis provides user-friendly dialog boxes that make complex operation, such as vegetation index analysis, easy to perform. "Image Difference" and "Thematic Change" functions can be used for change detection analysis using continuous or thematic (classified) data, respectively (Hurlbut, 1999).

ERDAS Stereo Analyst extension for ArcView allows users to collect and visualize spatial data in true stereoscopic viewing and to roam with real-time pan and zoom. Software features include the ability to collect and edit 3D shapefiles and visualization of terrain information and watersheds.

ERDAS MapSheets is an easy-to-use mapping and geographic presentation software package for Windows 95, Windows NT 4.0, and Digital Equipment Corporation's NT/Alpha. Because of its compatibility with Microsoft Office, MapSheets allows use of OLE technology to easily incorporate maps and images into reports, presentations, and spreadsheets. It is reported to be as easy to use as a word processor or a spreadsheet, because it works directly with Microsoft Office software. It allows adding a map to a Microsoft Word report, using Excel to query database attributes, using corporate data with Access, and making presentations in PowerPoint. MapSheets allows reshaping images and drawings that have different projections. Its Change Detection feature allows viewing changes from one image or drawing to another.

MrSID

Water and sewer system GISs generally have digital orthophoto basemaps. Orthophotos are stored in extremely large files that take a long time to display. For example, the City of Loveland, Colorado, had four aerial photo images of 1.3 gigabytes (GB) each (Murphy, 2000a). Raster images are storage hungry and compressed images loose resolution; there is a tradeoff between image size and resolution. MrSID is a new image file type (.SID) from Lizardtech, Inc. It encodes a large, high-resolution image to a fraction of its original file size while maintaining the original image quality. Images become scalable and can be reduced, enlarged, zoomed, panned, or printed without compromising integrity. MrSID's selective decompression and bandwidth-optimization capabilities also increase file transfer speeds.

MrSID provides the world's highest compression ratios, averaging about 40:1 but as high as 100:1. For example, Mecklenburg County, North Carolina, had 708 sheets of 1 in. = 1,000 ft black and white digital orthophotos covering 538 square miles of their jurisdiction. Each 9 in. × 9 in. photo was scanned at a resolution of one-foot to one-pixel, resulting in approximately 23 MB georeferenced TIFF files. This procedure created 16 GB of imagery stored on 27 CD-ROMs. Delivering a compression ratio of 1:28, MrSID took 14 hours on a Pentium PC with 512 MB of RAM, to compress 16 GB of imagery to a single 608 MB MrSID Portable Image Format that could be stored on a single 650 MB CD-ROM (Kuppe, 1999). Similarly, MrSID was able to compress 18 GB of Washington DC Orthophotos on one CD-ROM.

The MrSID ArcView GIS extension that is included with ArcView allows MrSID images to be instantly decompressed and displayed within ArcView GIS. It takes advantage of MrSID's image compression and retrieval capability and offsets the problems of working with large images in ArcView GIS. MrSID's ArcView GIS extension gives users the ability to work with any size raster image while providing instantaneous, seamless, multi-resolution browsing of large raster images.

GEOMATICA

Geomatica represents the most aggressive movement toward the integration of GIS and image processing functions in one software package (Limp, 2001). Geomatica from PCI Geomatics, unites previously separate technologies that were dedicated to remote sensing, image processing, GIS, cartography, and desktop photogrammetry into a single integrated environment. Although Geomatica is available in several configurations, all have a consistent user interface and data structure.

INTERNET GIS SOFTWARE

Internet GIS software combines the Internet and GIS technologies to display spatial data across the Internet. Distributing GIS data via the Internet allows for real-time integration of data from around the world. With Internet GIS, the users do not need GIS data and software installed on their local computers (Peng, 1998).

Web browsers now provide experts and novices with a common, powerful, inexpensive, and intuitive interface for accessing a GIS. Because casual public users can now access a GIS through a Web browser, the GIS staff of a water or sewer utility is freed up to spend more time on improving the database for providing better customer service (Irrinki, 2000).

Internet GIS uses a client-server architecture. Therefore, two types of software are required to Web-enable a GIS: client-side and server-side. Client-side software is very easy to use but does not provide data creation or editing capability. It costs less than the server-side software or is available for free. Server-side software, accessible through browsers, makes GIS usable over the Web. Based on the capabilities, prices range from $5,000 to $25,000. Representative Internet GIS software examples include

- ArcIMS, RouteMap and ArcExplorer from ESRI
- GeoMedia Web Map and GeoMedia Viewer from Intergraph
- MapXtreme (NT or Java) and MapXsite from MapInfo
- MapGuide from Autodesk

RDBMS Software

A database management system (DBMS) is a computer program for organizing the information in a database. A relational database allows accessing information from different tables without joining them together physically. A relational database management system (RDBMS) is a DBMS with the ability to access data organized in tabular files that can be related to each other by a common attribute.

In the past, data retrieval using the sequential search method was slow. Therefore, geographic data were stored in map (graphic) files rather than in databases. Today, faster computers have eliminated the speed problem and new spatial indexing techniques are available to expedite the searches. Suppose you want to located a manhole by its ID number. Today's databases do not start with manhole number 0001 and keep looking until the required manhole is found. A variety of indexing measures make the search process much faster than a sequential search. Now suppose you want to locate all the manholes within a given sewershed. To accomplish this, old systems performed a point-in-polygon geometric calculation to determine if the manhole coordinates were contained within the coordinates defining the sewershed boundary. The new spatial indexing system first selects the manholes that have an index number similar to the sewershed. This set is then passed to a traditional point-in-polygon routine to determine which manholes are in the sewershed (Limp, 2001a).

As user databases have become larger, with more concurrent users, it has been a natural transition to use database management system technology to store geographic data. To have open access to geographic information in the GIS database, the DBMS should be "open." Initially, most GIS database systems were proprietary. Fortunately, they are now embracing industry standards and using popular systems such as Oracle and Microsoft. For example, Intergraph released a new product in 2000 that integrates GeoMedia, Oracle Spatial, and the Oracle Workspace Manager feature of the Oracle database to allow multiple users to work on projects in the same geographic area (Murphy, 2000a). Open access to data in databases allows users to take advantage of DBMS technology to store and manage data, to support multiple users and applications concurrently on the same database, and to integrate heterogeneous data at the desktop. Using DBMS to store and manage data provides a superior solution for backup/recovery, replication, failover remote synchronization, and multi-user access (ESRI, 2000).

Suppose your maintenance department is using your water system layer to enter field inspection results. If the engineering department tries to access the water layer while the maintenance department is using it, they will get "file already in use" error. This problem is solved by middleware software like ESRI's Spatial Database Engine (SDE) described in Chapter

2 (GIS Development Software). SDE runs on top of a conventional RDBMS, permits storing geographic shapes together with their attributes in an RDBMS, and allows multi-user query of GIS database without conversion or replication in a DBMS. Spatial tables can coexist with other attribute tables in the database. Consider storing your data in an RDBMS and accessing it using SDE if your GIS either has an extremely large number of features or if you have a dynamic database where many people need to update and view the same data at the same time.

A side benefit of storing spatial data into database records is the ability to define and store relationships among specific records in different tables. For example, suppose that a valve is attached to a water main. In earlier GIS data structures, the valve would be in a point theme and the water main would be in a line theme, making it possible to move the main and forget to move the valve. With relationships, it is possible to build formal database rules that attach objects together. Thus, moving one automatically moves the other or issues a warning (Limp, 2001a).

The most common GIS data storage approach is a "hybrid" or "file system" approach in which attributes and spatial information are stored in separate files. The latest object RDBMS models from Oracle, IBM, Informix, Ingres, and so on provide an "integrated" approach to GIS data storage. Their spatial extensions (add-ons) enable their main RDBMS products to store and retrieve both attribute and spatial data. These extensions allow storage of spatial data as just another column in the database tables, alongside the attributes. This approach is suited especially to water and wastewater utilities that already have most of their facilities data in an RDBMS and want to add seamless connection to their spatial data (Lowe, 2000a). These extensions also facilitate linking data warehouses with GIS software packages. Representative examples include

- Spatial Data Option from Oracle
- SQL Server from Microsoft
- DB2 Universal Database and GeoManager from IBM
- Spatial DataBlade from Informix

Oracle Spatial Data Option®

In March 1997, Oracle released version 7.3.3 of its flagship RDBMS, the Universal Server, which uses an extension called Spatial Data Option (SDO). SDO also supports middleware products, such as ESRI's SDE, that allow GIS data stored in Oracle to be distributed across the enterprise in several different ways. SDO provides native support of geographically referenced data and facilitates the storage of geometry, points, lines, polygons, and topologies within the RDBMS (Goldstein, 1997). Once the spatial data are loaded into the universal data servers, common spatial calculations and queries can be performed. For example, SDO has structured query language (SQL) operators and functions for buffering, intersecting, unioning, computing area and length, and searching within a specified distance (Lowe, 2000a). The 2001 release called Oracle 9*i* provides basic spatial functionality into its core database software via a free integrated feature known as "Locator." The Locator module provides data management functions for location information, including spatial R-tree and quadtree indexing and simple location analysis. It supports Oracle native spatial data types (SDO_GEOMETRY) for points, lines, and polygons plus geometric elements like arc and compound-line strings, compound polygons, circles, and rectangles. Location data can be manipulated using SQL. Because Locator does not support the high-end spatial analysis capabilities needed by most GIS users, complex GIS applications will still require Oracle 9*i* Spatial, a priced option for the enterprise edition. Oracle 9*i* Spatial supports additional spatial functions (area, buffer, and centroid calculations) and advanced coordinate and linear referencing systems (Engelhardt, 2001).

MICROSOFT SQL SERVER®

SQL Server is the premier enterprise RDBMS from Microsoft Corporation. SQL Server 2000 provides important new ways to handle image data types and database views and introduces support for user-defined functions. With ESRI's ArcSDE 8.1 for SQL Server, applications that take advantage of spatial queries and data visualization can be developed. ArcSDE and the Geodatabase model enable SQL Server to be a repository for an organization's vector and raster data, supporting advanced, spatial data types like the ability to store three-dimensional coordinates, true curves, and complex networks and associating these with relationships, rules, behavior, and other object properties.

IBM DB2 SPATIAL EXTENDER®

In 2001, ESRI and IBM jointly developed an integrated software solution for the multi-user enterprise GIS system. IBM DB2 Universal Database (UDB) with DB2 Spatial Extender is designed to take advantage of ESRI's geodatabase model. The Spatial Extender is a modern object-related DBMS that integrates spatial data management into core RDBMS technology and enables full integration of spatial and attribute data. It enables more meaningful representations of spatial data and behavior rules in applications. This development has resulted in ArcGIS being a fully integrated application of DB2. This product has been implemented in Cook County (Illinois, USA) and Metropolitan Government of Nashville, Davidson County (Tennessee, USA).

INFORMIX SPATIAL DATABLADE®

Informix provides a place for spatial data alongside all the other types of information typically found in conventional databases. Informix's Spatial DataBlade spatially enables their core line of database servers, Informix Dynamic Server™ and Internet Foundation™, with full support for built-in spatial types and functions at no additional cost.

STATISTICAL ANALYSIS SOFTWARE

Most GIS packages compute only simple statistics. Users who need sophisticated statistical functionality in their GIS must obtain a statistical extension. For example, S+GISLINK is a product from Statistical Sciences (Seattle, Washington, USA) that allows GIS users to move GIS spatial data to and from statistical packages. SAS/GIS, developed by the SAS Institute Inc. (Cary, North Carolina, USA, *www.sas.com*), is an add-on to the SAS enterprise software that allows SAS users to include spatial variables. It allows several basic tools to merge spatial and attribute data. It also allows basic thematic mapping and robust address matching/geocoding capabilities. SAS/GIS Version 6.0 costs $500-1,000 if added to an existing SAS installation or $3,000-$4,000 if purchased with the base SAS modules.

The generation of predictive surfaces, their accuracy, and their estimation of error are critical to modeling and analysis. ESRI's Geostatistical Analyst, an extension to ArcGIS 8.x, provides tools for surface generation using geostatistical tools and analyzes the error of the resulting estimation (surface). The Geostatistical Analyst can help spatial scientists understand and use Kriging and other advanced mathematical methods used for surface generation. Potential fields in which the Geostatistical Analyst can be applied include soils science, agriculture, epidemiology, exploration geology (petroleum engineering, mining, and so on), hydrology, environmental science, and any discipline that samples point locations, such as rain gauges and water quality samplers.

DOCUMENT MANAGEMENT SOFTWARE

Document management systems are computer-based tools for storing, accessing, and managing utility data efficiently. Most utility records that contain critical information for maintaining, repairing, and upgrading the water and sewer systems have existed in paper archives. These records deteriorate with age and are expensive to store and retrieve. These drawbacks can be eliminated by scanning the paper documents and adding them to an electronic document management system (Irrinki, 2000). These systems are referred to as Document Image Management Systems (DIMS). As described above, the GIS data can now be accessed using simple web browsers. Thus, once the GIS links have been established, the utility personnel can use the Internet to access the utility documents from any location 24 hours a day.

Until recently, DIMS were monolithic systems that were available from a few suppliers who sold complete but proprietary systems. Around 1996, people started to mix and match various DIMS components (capture, store, retrieve, view/print, and workflow) to suit their specific needs. Thanks to such experiments, desktop GIS programs like ArcView, can be used to integrate these components in a single "knowledge management system" (White and McConnell, 1998).

Generally, the documents maintained in a document management system are inaccessible from outside the document environment. Conversely, the managed documents have no spatial relations. Thus, it is impossible to dynamically link documents and map features. This limitation is overcome by linking the GIS with the document management software. The GIS can link asset inventories such as pumps, hydrants, and valves to source documents, such as maps and drawings. For example, Rancho California Water District implemented an Engineering Document Management System by integrating ArcView GIS with the AutoManager WorkFlow software (Cyco International). An application called GeoDoc was developed to link features in ArcView with documents in AutoManager. Using GeoDOC, users can browse through drawings and documents in AutoManager and select all associated features in ArcView or vice–versa (ESRI, 1998).

The town of New Hartford in New York, comprising approximately 11,000 tax parcels, utilized the ESRI software family linked with a powerful document imaging software. A user interface was developed using the MapObjects software. By providing spatial point and click access to relevant GIS data and document information, what used to require a visit to the records room and a call to the Assessor's office can now be obtained and printed from the engineer's desk within minutes. The New Hartford DIMS includes property owner information and photos from the Assessor's office, as-built drawings for existing sewers, elevation and flow data for upstream/downstream manholes, sewer connection cards from the Facilities Maintenance office, and maintenance records and field notes (Cleveland and Clair, 2001).

GIS Technology Inc. has developed an ArcView extension that provides a seamless link between ArcView and FileNET's Panagon Integrated Document Management (IDM) system. With IDM for ArcView, users can select features in a map theme (e.g., parcels) and find documents related to those features (e.g., tax records, business licenses, map book pages).

Latest image compression and Internet technology is helping governments and organizations to post or transfer documents such as plat maps and engineering plans, legal and public records, historical documents, and land use maps online. Thanks to this approach, GIS users worldwide can access and use a wealth of information without going to government offices. For example, as described above, LizardTech's MrSID software can reduce image files, such as aerial photographs, to truly portable sizes without compromising visual quality. LizardTech's DjVu software can make the scanned documents Web-ready in extremely small file sizes. DjVu files download faster, take up less storage space, and maintain the clarity of original documents. Reportedly, DjVu preserves the most detailed full-color maps in perfect clarity, easily accessible through standard hyperlinking in many GIS applications. DjVu offers pan and zoom of source-quality documents in real time at TIFF-to-DjVu reduction ratios of more than 1,000:1. The town of Greene County (Ohio, USA) reduced their average map size from 8 MB in other file formats to less than 400 KB in DjVu for efficient distribution via the Internet.

Useful Document Management Web Site

GIS Technology Inc. *www.gistech.com*

AUTOMATED MAPPING / FACILITIES MANAGEMENT / GIS SOFTWARE

Automated Mapping / Facilities Management (AM/FM) is CADD technology used for managing utility system data. AM/FM is a digital infrastructure management database. GIS and AM/FM are different, with their own advantages and applications. For many years, people have used both systems separately. Developing and maintaining two different systems is expensive and inefficient. Thanks to the latest advances in computer hardware and software, integrated AM/FM and GIS systems called "AM/FM/GIS" systems are now available. AM/FM/GIS systems are especially useful for asset inventory, inspection and maintenance, and work management. Representative AM/FM/GIS software is listed in Table 3-2.

Cityworks (formerly known as Pipeworks) is an ArcView Extension developed by Azteca Systems that helps users integrate their GIS and work management. It works with ESRI coverages and shapefiles and stores project data in any SQL database, such as SQL Anywhere, Oracle, and Sybase. Cityworks capabilities include data inventory, data editing, work order management, work order scheduling, network tracing, maintenance histories, inspections, and condition ratings. Network tracing is a valuable tool for sewer system networks. For example, it can identify all the storm sewer pipes contributing flow to a given stormwater outfall or identify valves that must be closed to isolate and repair a broken water main. Cityworks can also be used for managing and recording a TV inspection program for sewers.

Table 3-2. AM/FM/GIS Software

Software	*Vendor*	*Web site*
ArcFM	ESRI and Miner & Miner	*www.esri.com*
CASS WORKS	RJN Group, Inc	*www.rjn.com*
Cityworks	Azteca Systems	*www.azteca.com*
FRAMME	Intergraph	*www.intergraph.com*
GBA Sewer Master and Sewer Master	George Butler Associates	*www.gbutler.com*
GeoPlan	Regional Planning Technologies	*www.rpt.com*
GeoWater and GeoWasteWater	MicroStation	*www.bentley.com*
GIRIS	Generale d'Infographie, Vivendi	*www.seureca.com*
IMS-AV	Hansen Information Technologies	*www.hansen.com*
Infrastructure Analysts	BaySys Technologies	*www.baysys-gis.com*
Water and Sanitary /Storm Sewer Management Systems	Stantec	*www.stantec.com*
WATERview and SEWERview	CarteGraph Systems	*www.cartegraph.com*

H&H Modeling Software

The hydrologic and hydraulic (H&H) computer modeling software can run both inside and outside the GIS. When modeling software is run inside a GIS, it is considered "seamlessly integrated" in GIS. Most modeling programs run in stand-alone mode outside the GIS, in which case the application software simply shares GIS data. This method of running applications is called a "GIS interface." Additional information about GIS integration methods and H&H modeling software is provided in Chapter 7 (Modeling Integration).

Tables 3-3 and 3-4 list representative software packages with GIS integration or interface capabilities for modeling water distribution and sewer systems, respectively. Table 3-5 lists similar software packages for hydrologic modeling of watersheds.

Table 3-3. Water Distribution System Modeling Software

Software	*Vendor*	*Web site*
CEDRA AVWater	CEDRA Corporation	*www.cedra.com*
Cybernet and WaterCAD	Haestad Methods	*www.haestad.com*
H_2ONET and H_2OMAP Water	MWH Soft	*www.mwhsoft.com*
MIKE NET	DHI Water & Environment	*www.dhi.dk*
InfoWorks and HydroWorks	Wallingford Software	*www.wallingfordsoftware.com*
PIPE2000	University of Kentucky	*www.kypipe.com*
SynerGEE	Stoner Associates	*www.stoner.com*

Table 3-4. Sewer System Modeling Software

Software	*Vendor*	*Web site*
CEDRA AVSand	CEDRA Corporation	*www.cedra.com*
H_2ONET and H_2OMAP Sewer	MWH Soft	*www.mwhsoft.com*
InfoWorks and HydroWorks	Wallingford Software	*www.wallingfordsoftware.com*
Mouse GIS	DHI Water & Environment	*www.dhi.dk*
PCSWMM GIS	Computational Hydraulics Int.	*www.chi.on.ca*
SewerCAD and StormCAD	Haestad Methods	*www.haestad.com*
SWMENU	Portland Bureau of Environmental Science	*N/A*
SWMM DUET	Delaware Department of Natural Resources	*N/A*
XP-SWMM	XP-Software	*www.xpsoftware.com.au*

Table 3-5. Watershed and Hydrologic Modeling Software

Software	*Vendor*	*Web site*
Hydro Extension	ESRI	*www.esri.com*
Mike SHE Mike 11 GIS Mike Basin	DHI Water & Environment	*www.dhi.dk*
HEC-Geo HMS & HEC-Geo RAS	Hydrologic Engineering Center (HEC)	*www.hec.usace.army.mil/software/*
BASINS	U.S. EPA	*www.epa.gov/OST/BASINS/*
GIS Hydro	University of Texas at Austin	*www.ce.utexas.edu/prof/maidment/*
Watershed Modeling System	Brigham Young University	*www.ems-i.com*
River Modeling System (RMS) and RiverCAD	Boss International	*www.bossintl.com*

Representative Examples

Some representative examples of leading H&H modeling software with GIS capability are discussed below.

HEC Software

The Hydrologic Engineering Center (HEC) is an office of the U.S. Army Corps of Engineers (COE) established to support the nation's hydrologic engineering and water resources planning and management needs. To accomplish this goal, HEC develops state-of-the-art comprehensive computer programs that are also available to the public. HEC-1 and HEC-2 are COE's legacy DOS programs for hydrologic and hydraulic modeling, respectively. Recently HEC-1 and HEC-2 have been replaced with Windows programs called Hydrologic Modeling System (HEC-HMS) and River Analysis System (HEC-RAS), respectively. HEC Geo-HMS and HEC Geo-RAS have been developed as ArcView GIS extensions for HEC-HMS and HEC-RAS users, respectively. They allow users to expediently create hydrologic input data for HEC-HMS and HEC-RAS models. Free downloads of these programs are available from the HEC software Web site listed in Table 3-5.

HEC-GeoHMS is an ArcView GIS extension that allows users to visualize spatial information, delineate watersheds and streams, extract physical watershed and stream characteristics, perform spatial analyses, and create

HEC-HMS model input files. HEC-GeoHMS requires ArcView Spatial Analyst Extension.

HEC-GeoRAS for ArcView is an ArcView extension that allows users to create a HEC-RAS import file containing geometric attribute data from an existing digital terrain model and complementary data sets. Results exported from HEC-RAS may also be processed in HEC-GeoRAS. ArcView's 3D Analyst extension is required to use HEC-GeoRAS. ArcView's Spatial Analyst extension is also recommended. An ArcInfo version of HEC-GeoRAS is also available.

The U.S. Army Engineer Research and Development Center (ERDC) (formerly Waterways Experiment Station) has also been responsible in part for the development of modeling systems for watersheds, surface water, and groundwater. More information about the ERDC products and projects is available at ERDC Web site *www.erdc.usace.army.mil.*

Haestad Methods Software

Haestad Methods Inc. (Waterbury, Connecticut, USA) provides a wide range of H&H computer models, publishes textbooks, and offers continuing education for the civil engineering community. Haestad Methods is a growing company with more than 100,000 users in more than 160 countries around the world. Haestad Methods provides the following H&H modeling software for water, wastewater, and stormwater systems:

- WaterCAD and Cybernet for water distribution systems
- SewerCAD for sanitary sewer systems
- StormCAD for storm sewer systems

GIS interface software for these programs is sold separately. These programs provide import and export wizards to transfer data between GIS and computer models. For example, WaterCAD is the Haestad Method's water distribution analysis and design tool. It can analyze water quality, determine fire flow requirements, and calibrate large distribution networks. WaterCAD's "Import Shapefile" wizard is used to import WaterCAD model input parameters from ArcView GIS shapefiles. Similarly, the "Export Shapefile" wizard exports hydraulic model networks to ArcView. This feature lets users build and maintain their water and sewer networks directly inside a GIS. Haestad Methods has also developed a geodatabase, Water Data Model, for integrated WaterCAD, SewerCAD, and StormCAD modeling in ArcGIS.

Geographic Engineering Modeling Systems (GEMS) is the next generation of Haestad Methods' hydraulic network modeling products. GEMS is a new product family that combines water, wastewater, and stormwater system modeling with an open database architecture. It has been reported that the GEMS technology will eliminate the need for a separate model database (Shamsi, 2001). It will streamline the process of accessing information on, for example, facilities management, zoning, topography, and gauging stations for use with the model. GEMS-based H&H models can be run within ESRI's ArcGIS software using the geodatabase structure.

DHI Software

DHI Water and Environment (formerly Danish Hydraulic Institute, Hørsholm, Denmark, *www.dhi.dk*) is a global provider of specialized consulting and numerical modeling software products for water, wastewater, river and coastal engineering, and water resources development. DHI has developed a large number of hydraulic, hydrologic, and hydrodynamic computer models for water, wastewater, and stormwater systems. They also specialize in integrating their computer models with GIS so that modelers can use both modeling and GIS technologies within a single product.

Since 1998, DHI has embarked on an ambitious program to link its models with the ESRI family of GIS products. Many of their modeling systems now support GIS data transfer. For example, DHI has developed ArcView GIS extensions and interface programs for a number of their products. DHI's main products and their price range are listed in Table 3-6.

Table 3-6. DHI Software List

Software	*Applications*	*Price Range (US$)*
MOUSE	Modeling of urban storm/sanitary sewer	15,000-16,000 for both programs
MOUSE GIS	GIS interface for MOUSE	
Mike SWMM	EPA SWMM modeling	3,500-4,500
Mike SHE	Integrated and distributed modeling of surface and groundwater flow	6,000-8,000
Mike SHE GIS	ArcView based pre- and post-processor	400-600
MIKE 11	Floodplain hydraulics	5,000-7,000
MIKE 11 GIS	ArcView extension for floodplain mapping	4,500-5,500
MIKE BASIN	River basin modeling in ArcView	3,500-4,500

MIKE 11 GIS is a spatial decision support system for river and floodplain management. MIKE BASIN, which provides a versatile decision support system for integrated water resources planning and management, runs inside the ArcView GIS. MOUSE GIS links sewer system hydraulic modeling with GIS and other database management systems. Mike SWMM is DHI's graphical user interface for U.S. EPA's Storm Water Management Model (SWMM) program. Mike SWMM provides data editing and querying, facilitates animation of results and exporting of graphics to other programs like Excel and Word, and provides an ArcView GIS link for importing and exporting between the computer model and the GIS.

SUMMARY

In this chapter we learned that GIS applications, such as H&H modeling and AM/FM/GIS, can be developed by extending the core capabilities of GIS development software. Application software can be obtained from government agencies or purchased from software companies. This chapter provided software lists for various water, wastewater, and stormwater applications. Although some representative software examples were discussed, readers are cautioned not to believe that those are the only available products. Readers are encouraged to obtain more information about application software by visiting various Web sites listed in this chapter.

SELF EVALUATION

1. What is the difference between the GIS development and GIS application software?
2. How are GIS applications developed?
3. What is the difference between a script and an extension?
4. Prepare a list of GIS application software in your organization grouped by the following categories: image processing, Internet GIS, RDBMS, AM/FM/GIS, and H&H modeling.
5. What is the best way to select GIS application software?

Chapter

4 GIS DATA

You should be aware of the intended use and accuracy of your GIS data. Data quality and accuracy should be evaluated in the context of the GIS application in which the data will be used.

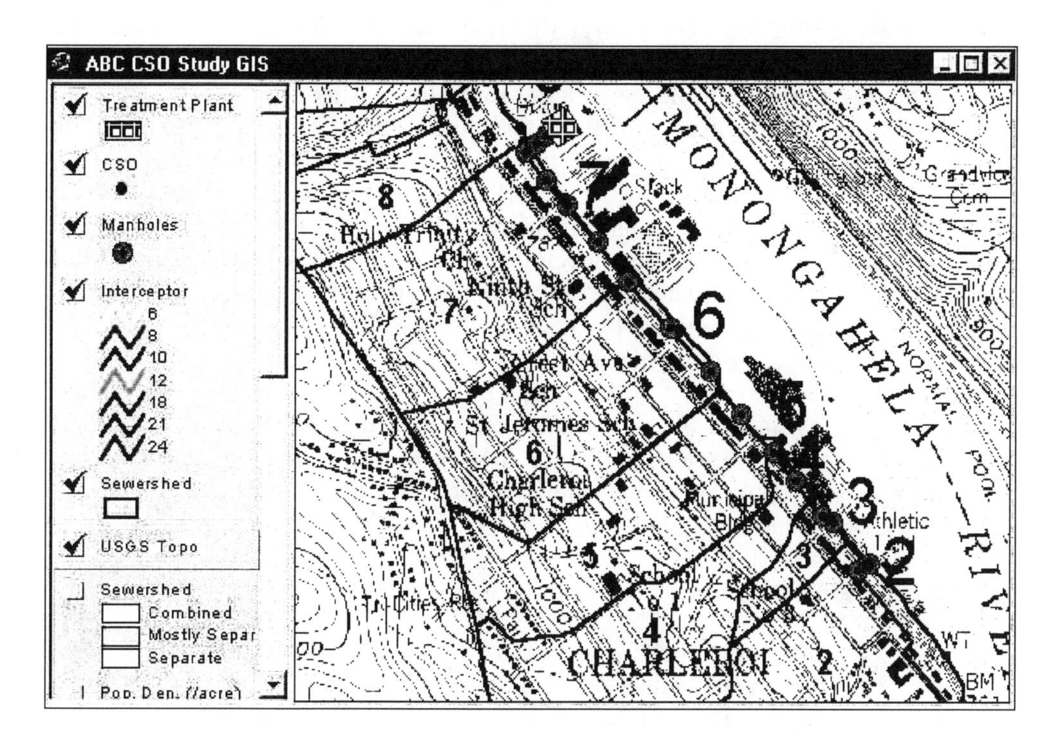

USGS Digital Raster Graphic (DRG) Data as a GIS Basemap
Charleroi, Pennsylvania

LEARNING OBJECTIVES

The learning objectives of this chapter are to understand various types of GIS data, know their limitations, and appreciate their benefits. This chapter provides information about the public domain, commercial, and custom GIS data that are generally useful in developing GIS applications for water, wastewater, and stormwater systems. Major topics discussed in this chapter include

- Data types
- Data quantity and quality
- Data limitations
- Data formats
- Data sources
- Metadata

GIS DATA FOR SAN DIEGO'S SEWER SYSTEM

The GIS data users are often perplexed by many different GIS packages, many different data formats, and little interoperability. For example, in 1992 San Diego's Water Utilities Department integrated ArcInfo land base data and InSewer's engineering CAD data in Intergraph's MGE GIS software to design a large trunk line. This process required translating ArcInfo data into MGE format and moving it to InSewer. ArcInfo-to-MGE conversion required converting ArcInfo export files to ASCII text files for use in MGE ASCII Loader. A training set was created to map ArcInfo sewer main database columns into the generic MGE relational database. Because ArcInfo data were not available in a single file, SQL commands and shell scripts were written to manipulate and combine them for easy retrieval by InSewer. Finally, DBAccesses was used to create an InSewer input file and the trunk main sewer network (Robinson, 1993).

Thanks to the recent advances in GIS data and software, data exchange among various GIS and CADD programs is now much easier. For example, users can employ off-the-shelf data conversion utilities described in this chapter.

IMPORTANCE OF GIS DATA

The most challenging part of a GIS application project is to obtain the right kind of data in the right format at the right time. Therefore, data is the most important component of a GIS. Without data, you simply have a computer program, not a GIS. A GIS accepts any kind of data that have a spatial component. These data types are quite diverse—from aerial photographs to satellite imagery, contour maps to parcel maps, and sewer drawings to Global Positioning System (GPS) data. GIS data can also be found in some unexpected places, such as customer billing records. With the right tools and by using a technique called geocoding, GIS can convert the billing records, or any postal address, to a point on a map.

Data conversion is a laborious and expensive task. In some projects, data collection and entry cost can be up to 80% of the total project cost. Thus, a thorough understanding of GIS data requirements is essential.

DATA TYPES

GIS data (also referred to as geographic, spatial, or geospatial data) describe real-world objects using three characteristics (Gorokhovich, 2000)

1. Their position with respect to a known coordinate system
2. Their attributes related to position, such as pipe size or manhole depth
3. Their spatial (topological) interrelationship describing how they are related to each other

Because spatial is special, GIS applications require special kind of spatial data. There are two types of data required for GIS applications:

1. Assets Data
2. Applications data

Assets data define the physical infrastructure, such as water mains and manholes. Assets data form the backbone of a GIS and therefore should be in place before developing any GIS applications. Applications data are required for creating GIS applications, such as work order management and H&H modeling. Leak detection surveys, manhole inspections, and watershed boundaries are examples of applications data. Applications data are needed at the time of developing GIS applications. Detailed information about the assets and application data for water, wastewater, and stormwater systems is provided in Chapter 6 (Database Design).

Assets data are created by data conversion (scanning and digitization) and data collection (field survey and GPS). Required data are usually scattered among a multitude of different organizations and agencies. For instance, it is claimed that 800 worker years would be needed to convert England's sewer system to digital format including field verification work (Bernhardsen, 1999).

Applications data can be created by data conversion and data collection, obtaining public domain data, and purchasing commercial data. This chapter provides information about public domain and commercial GIS data that are useful in developing GIS applications for water, wastewater, and stormwater systems.

DATA QUANTITY AND QUALITY

Construction of an appropriate GIS database is the most difficult and expensive part of developing GIS applications. A successful GIS application requires careful consideration of both the quantity and the quality of data.

DATA QUANTITY

The question of quantity should be evaluated carefully before the data conversion phase of a GIS project is started. Too little data will limit the application. Too much data will be wasteful. Sometimes, more data are captured than required by the application. This approach is called a "data-driven" or "bottom-up" approach. For example, to support the development of early nonpoint source analysis applications and erosion modeling, U.S. Natural Resources Conservation Service spent substantial time and effort to capture detailed soils series survey data. However, the final application ended up using the simple Universal Soil Loss Equation, which required the simpler soil associations data instead of the detailed soils series data (EPA, 2000). Considerable time and money could have been saved if an appropriate model and its data requirements were identified before starting the data conversion process.

DATA QUALITY

The famous computer industry proverb "garbage in, garbage out" conveys very well the importance of GIS data quality. A GIS is only as good as the data used to create it. Data quality roughly means how "good" the data are for a given application. Data quality is important because it determines the maximum potential reliability of the GIS application results. Use of inappropriate data in a GIS application may lead to misleading results and erroneous decisions, which may erode public confidence or create liability.

There are two types of data errors—inherent errors embedded in the source of data and operational errors introduced by the users during data input, storage, analysis, and output. Inherent errors can be avoided by using the right kind of data. Operational errors can be prevented by quality control and training.

Application developers should be aware of data quality issues, including the sources and magnitude of error. For example, spatial information in U.S. Geological Survey (USGS) 1:24,000 scale (7.5-minute) topographic maps is certified to have 90% of its features within 50 ft (15 meter) of their correct location. Fifty feet is large enough to incorrectly place an NPDES outfall inside or outside your sewershed boundary. Such an error can result in an inaccurate sewershed model, which can lead to improper design and inadequate abatement measures for controlling sewershed wet weather overflows.

Once map data are converted into a GIS environment, the data are no longer scaled, since the data can be scaled as desired to create any output map scale. However, the spatial data can never be any more accurate than the original source from which the data were acquired. GIS data are typically less accurate than the source, depending on the method of data conversion. Therefore, if data were digitized from a source map scale of 1 in. = 2,000 ft, and a map was created at 1 in. = 100 ft, the map accuracy of features shown is still 1 in. = 2,000 ft (PaMAGIC, 2001).

The data must be of appropriate scale and highly documented (see the section on metadata, later in this chapter) to be useful in GIS applications. Users must be extremely conscious of the nature of the source information to avoid abusive extrapolations and generalizations. For example, a Massachusetts water utility, puzzled by abnormally high lead and copper levels in their distribution system, found that interpretations of water quality problems based on aggregate monitoring data can be misleading unless analysis is performed at the appropriate scale. Incorporation of system and monitoring site physical characteristics data, plus monitoring results, into GIS could have saved them considerable investigatory effort (Schock and Clement 1995).

Data quality and accuracy should be evaluated in the context of the GIS application in which the data will be used. For some applications, supposedly low resolution data, such as USGS digital orthophoto quadrangles (DOQs), may be acceptable. In other applications, even the presumably high-resolution data, such as 1-meter IKONOS imagery, may not be accurate enough.

Precision and accuracy are two entirely different measures of data quality and should not be confused. A GIS can determine the location of a point feature precisely as coordinates with several significant decimal places. However, lots of decimal places in coordinates do not necessarily mean that the feature location is accurate to a hundredth or thousandth of a distance unit. Simply stated, high coordinate precision does not always represent high positional accuracy. Always review the metadata to understand the accuracy of your data.

GIS users work with the database as a surrogate for the real objects, just as an architect works with the drawings and models of a building. Uncertainty is defined as the difference between the database and the real-world values. As a rule of thumb, a database built from a map will have positional inaccuracies of about 0.5 mm at the scale of the map because this is the typical line width of the drawing instrument. This can cause inaccuracies of up to 12 meters in a database built from 1:24,000 mapping, such as USGS DRGs (Goodchild, 1998).

A detailed discussion of data quality and errors is outside the scope of this book, but the reader is encouraged to become familiar with this subject. More information on data errors can be found in Longley et al. (1999), Griffin (1995), and most GIS textbooks.

DATA FORMATS

As defined in Chapter 1 (GIS Basics), there are two kinds of geographic data formats—vector and raster. Both kinds of data are stored in various file formats. Table 4-1 provides a list of geographic data formats supported in ArcInfo 8 (ESRI, 2000).

The number of data formats has increased exponentially with the growth in the GIS industry (Goodchild, 2002). According to some estimates, there might be more than 80 proprietary geographic data formats (Lowe, 2002). Why are there so many geographic data formats? One reason is that a single format is not appropriate for all applications. For example, a single format cannot support both fast rendering in a command and control system and sophisticated topological analysis in a natural resource information system. Different data formats have evolved in response to diverse user requirements.

DATA TRANSLATION

A GIS software cannot read all the data formats simply because there are so many formats. Disparate data formats should be converted to one of the formats compatible with your GIS software. Although many GIS packages provide data conversion for the most common data formats, no GIS can support all possible conversions. Historically, transferring digital data from one GIS format to another has been very difficult, if not impossible. Typically, data would be exported to an ASCII file by the originating GIS then imported into the target GIS. The target GIS would then need to have built-in tools for reading the many possible ways to define spatial data in ASCII. Luckily many GIS packages support data conversion with some of the more common ASCII formats, including DXF (Autocad), Arc Ungenerate (ArcInfo), TIGER (U.S. Census Bureau), and DLG (USGS roads, hypsography, and hydrography) (Huse, 1995).

Data translation from one format to another can potentially lead to the loss or alteration of data. For instance some platforms can support better numeric precision than others. Depending on the mapping scale and the coordinate system, this could seriously affect data quality. In one system, curves representing a river might be defined mathematically, but in another described as a series of straight line segments. Or data brought from one topological model to another may require extensive "cleaning" to make it compliant with the target GIS.

Table 4-1. GIS Data Formats Supported by ArcInfo 8

Vector	*Raster*
Automated Mapping System (AMS)	Arc Digitized Raster Graphics (ADRG)
Coverage	Band Interleaved by Line (ESRI BIL)
Computer Graphics Metafile (CGM)	Band Interleaved by Pixel (ESRI BIP)
Digital Feature Analysis Data (DFAD)	Band SeQuential (ESRI BSQ)
Encapsulated PostScript (EPS)	Windows Bitmap (BMP)
Native MicroStation Drawing File (DGN)	Device Independent Bitmap Format (DIB)
Dual Independent Map Encoding (DIME)	Compressed ARC Digitized Raster Graphics (CADRG)
Digital Line Graph (DLG)	Controlled Image Base (CIB)
Drawing Exchange Format (DXF)	Digital Terrain Elevation Data (DTED) Levels 1 and 2
AutoCAD Drawing (DWG)	ER Mapper (Basic Image Format)
MapBase File (ETAK)	Graphics Interchange File (GIF)
ESRI GeoDatabase	ERDAS Imagine
Land Use and land Cover Data (GIRAS)	ERDAS 7.5 GIS
Interactive Graphic Design Software (IGDS)	ESRI GRID File Format (GRID)
Initial Graphics Exchange Standard (IGES)	JPEG File Interchange Format (JFIF)
Map Information Assembly Display System (MIADS)	Multiresolution Seamless Image Database (MrSID)
MOSS Export File	Spatial Database Engine Raster File Format (ArcSDE Raster)
TIGER/Line file	Tag Image File Format (TIFF and GeoTIFF)
Spatial Data Transfer Standard (SDTS)	
Topological Vector Profile (TVP)	
ArcView GIS Native Format (Shapefile)	
Vector Product Format (VPF)	

After ESRI, 2000.

Users must possess considerable expertise to overlay, combine, or analyze different map layers or images. Converting from one format or type of data to another is cumbersome, time consuming, and error-prone. That is why efforts are currently underway in the GIS industry to standardize data formats and database management systems by promoting open platforms, data, and database management systems. For example, the Open GIS Consortium (OGC) (*www.opengis.org*) has been created with the vision of "the complete integration of geospatial data and geoprocessing resources into mainstream computing." OGC was formed in 1994 to facilitate access and geoprocessing of data held in systems or networks. OGC is a 150-member standards body that is trying to replace the proprietary interfaces found in some GIS software with a common set of interfaces based on SQL, CORBA, and OLE/COM standards. OGC defines an open GIS as "Open and interoperable geoprocessing" or "the ability to share heterogeneous geodata and geoprocessing resources transparently in a networked environment." OGC also promotes the ability to browse various data formats over the Internet, which will be very beneficial for the Internet applications of GIS. What do Autodesk, Intergraph, MapInfo, Microsoft, Mitsubishi, Oracle, and Sun Microsystems have in common? They are not all GIS vendors, but all of them are represented in the OGC (McKee, 1998). OGC has made considerable progress in promoting common data formats, but faces a legacy of innumerable proprietary formats. Clearly, we have a long way to go to achieve the goal of full GIS interoperability (Goodchild, 2002).

There are two approaches to transferring different data formats between different systems.

1. Develop software that translates data by a "direct read" into memory.
2. Develop intermediate data translators.

Permanent data conversion to a common format might be sometimes too costly, time consuming, or even impossible for users, data providers, and software companies. In the first approach, no permanent conversions of format or projection are involved; instead, the data are read in native formats and projections. This approach provides an easy way to access, visualize, share, and disseminate GIS data without the need of complex preprocessing or translation. MapFusion from Global Geomatics (*www.globalgeo.com*) shares more than 80 GIS data formats. Being referred to as the "Mapster" of the GIS industry, MapFusion standardizes various data formats on-the-fly using the scalable Peer to Peer (P2P) approach first pioneered by Napster in the music industry. MapFusion software includes a set of adapters that recognize common GIS formats and allow the user to access them transparently. In addition to map display, MapFusion also allows GIS analysis capabilities by adding GIS software specific extensions for ArcView 3.x, MapInfo Professional, etc. This approach can bring a standard GIS much closer to the goal of full interoperability (Lowe, 2002; Goodchild, 2002). ArcView's direct reading of AutoCAD DXF and DWG file formats is another example of this approach. Similarly, PCI's Geogateway software provides read-write access to a wide variety of raster data formats without first converting them.

Data translators convert one data format to another format that can be read directly by your GIS. Some sample data conversion resources are discussed below.

- USGS provides various free data conversion utilities. For example, free Digital Line Graphics (DLG) translators are available that convert DLG formats to CAD and GIS formats, such as DXF, ArcView shapefile, or MapInfo MIF format. DLG32 is a free downloadable software for Windows 95 that allows preview and evaluation of USGS data in DLG Optional (DLG-O), DRG, and SDTS formats. It does not have editing capabilities, and is not a substitute for commercial GIS software. Such utilities can be downloaded from the USGS or GeoCommunity Web sites listed below.

- GIS Tools Web site (given below) also provides data conversion utilities for converting USGS DLG and TIGER files to ArcView shapefile and MapInfo MIF file formats.

- Safe Software Inc. provides an ArcView Extension called "FME Themes" that can be used to directly read and view more than 50 diverse data types including DGN, DTED, DXF, DWG, GenaMap, MIF, MID, NTF, NTX, Oracle 8i Spatial, OGDI, SDTS (TVP and DEM), TAB, TIGER, and so on. Imported data can be edited or analyzed using ArcView GIS, and the results can be saved as shapefiles.

Useful GIS Data Conversion Web Sites

GeoCommunity data conversion utilities	*software.geocomm.com/translators/dlg/*
GIS Tools data conversion utilities	*www.gistools.com*
MapFusion	*www.globalgeo.com*
Safe Software	*www.safe.com*
USGS	*mcmcweb.er.usgs.gov/viewers/dlg_view.html*

DATA SOURCES

Based on the source of their availability, there are three main types of GIS data that can be used to develop GIS applications for water, wastewater, and stormwater systems

1. Public domain data
2. Commercial data
3. Custom data.

Different GIS data are available in different formats, projections, and scales. It is difficult and sometimes frustrating to get the two different data sets to align or match up. Therefore, the user should consider the following factors carefully when obtaining public domain data or purchasing the commercial data.

- Date of data (i.e., how current they are)
- Scale
- Resolution
- Projection
- Datum
- Source
- Method (i.e., how they were produced)
- Format
- GIS compatibility (i.e., are they GIS-ready?)
- Copyright limitations (e.g., can you distribute the data to other people?)
- Cost

PUBLIC DOMAIN DATA

Public domain data are nonproprietary data developed by government agencies for public use. A large number of federal, state, and local governments are willing and eager to share their existing GIS data. These data are often free when available online or for the cost of media when shipping is necessary. At times, data sharing agreements or waivers must be signed or acknowledged before the data can be used by others. The most popular public domain data include the following:

- USGS Digital Elevation Model (DEM)
- USGS Digital Raster Graphics (DRG)
- USGS Digital Orthophoto Quadrangles (DOQ)
- USGS Digital Line Graphs (DLG)
- USGS Land Use/Land Cover (LULC)
- USGS Watersheds (HUC)
- EPA Reach Files
- EPA Land Use/Land Cover (LULC)
- National Hydrography Dataset (NHD)
- Census Bureau TIGER/Line files
- NRCS SSURGO soils data
- NRCS STATSGO soils data
- FEMA flood data

Detailed information about these data and downloading instructions are provided in Chapter 5 (Internet GIS). All of the above data are available in digital formats, which can be used in a GIS project. Some data formats require minor pre-processing to convert the public domain data in a GIS compatible (GIS-ready) format. Public domain data that are available only as paper maps can be scanned and optionally vectorized for use in a GIS.

As described above, a successful GIS application requires the right level of data quality, accuracy, and integrity. Thus, users should be aware of the scale, resolution, accuracy, quality, and intended use of public domain data before using them in their GIS application projects. Although much of the public domain GIS data are not detailed enough to be used as site maps, they can be used for preliminary studies or vicinity maps. For example, USGS topographic maps (or DRGs, as described in Chapter 5) at scales of 1:24,000 show an area in detail that may be useful for some engineering and local area planning departments. This level of mapping is designed for use by landowners, townships, and county natural resource planning and management departments. Less detail is shown at scales of 1:63,360 and 1:100,000. These maps cover large areas and are used in land management and planning. The 1:100,000 scale data may not be suitable for high-precision measurement applications, such as engineering problems, property transfers, or other uses that might require highly accurate measurements of the Earth's surface. Maps at a scale of 1:250,000 cover very large areas on each sheet and, therefore, are suitable for broad regional planning and management uses covering state, regional, and multi-state areas.

The accuracy of most public domain data depends on the application for which those data were originally collected. However, some data are used outside their intended applications. In these situations, the users should be aware of the intended use and accuracy of such data. For example, the USDA soil maps were originally developed to aid farmers in planning their crops. Soil maps, however, see wide use for very different applications, such as hydrologic modeling. Similarly, US EPA's STORET data represent point water quality data, which are typically extrapolated to represent water quality of an entire stream reach (Griffin 1995). This is an example of point features data being used to describe line features. Such extrapolations are not necessarily wrong, as long as users remember the limitations of their data when interpreting the GIS maps and using them for decision making.

Another potential drawback of public domain data is that they may not be current. For example, much of the USGS data are considered dated. Some USGS cartographic data date back to the 1950s in areas where "nothing is going on" or where states do not require regular updates.

Commercial GIS Data

Although commercial data costs money, they are usually much more usable because they have already been reformatted to a GIS-ready format and in many instances, they are more current. It can be "penny wise and pound foolish" to automatically reject commercial data because they are not free. Free data are not always free. Users can spend a lot of time downloading, converting, cleaning, optimizing, and enhancing data before using them. Commercial data are often immediately usable and often include enhancements not available from public domain sources (Thoen, 2001). Moreover, the cost of spatial data is falling rapidly because of competition in data acquisition, processing, and distribution.

When opting for free public domain GIS data, users should be aware of data quality, file format, GIS compatibility, map projection, Internet download times (if applicable), and data storage issues. Evaluation of these factors may sometimes favor purchasing the commercial GIS data. For example, the cost of geocoding the public domain streets data in-house may be more than the cost of buying the commercial geocoded streets data. In this case, purchase of commercial data can be more cost effective than the free Internet download. Similarly, although commercial TIGER data are more expensive than the free Census Bureau TIGER data, the data enhancements (e.g., double-line roads), data projections, and GIS compatible file formats may justify the up-front cost. Other benefits of commercial data are technical support, product updates, and free data upgrades.

Geospatial Solutions Magazine's Web site maintains an alphabetical list of GIS data vendors in their "Buyers Guide," which can be accessed from the URL below. GEOWorld Magazine's home page (www.geoplace.com) also provides a list of geospatial data vendors and their products in a matrix format. This list can be accessed from the URL below or from the "data vendors" link on the "Geo Resources" tab of the GEOWorld home page.

Web Sites For GIS Data Vendor Lists

Geospatial Solutions Buyers Guide	*www.geospatial-online.com/geospatialsolutions/*
GIS Data Depot	*www.gisdatadepot.com/catalog/*
GEOWorld Magazine	*www.geoplace.com/bg/2000/0300/data_vendors.asp*

There are a lot of commercial data vendors, not all of which can be listed here. Some representative examples of commercial data vendors and their main products are given below.

AMERICAN DIGITAL CARTOGRAPHY

Web site: *www.adci.com*

Products and Cost

- EtakMap: $1,000–$1,500/county
- TIGER: $2,000–$2,500/state
- Digital Section Lines: $1,800–$2,200/state
- DEM-based contours: $25–$100/quad, based on scale
- Scanned and vectorized contours: $25–$800/quad, based on scale

DELORME

Web site: *www.delorme.com*

Products and Cost

- Topo USA 2.0: Seamless topographic maps for the entire United States for $100

- 3D Topo Quads: USGS 7.5-minute quadrangle maps plus DeLorme's own updated topographic and street-level data with 3D mapping capability ($150 per state)

- The data includes free standalone desktop mapping software without a GIS interface

- For GIS query and analysis of DeLorme topo data, users may purchase XMapGeographic GIS software. This inexpensive program provides geographic querying, geocoding, image registration, 3D draping, GPS tracking, and trip routing functions (Bell, 2001).

ESRI

Web sites: *www.esri.com/data/online*

data.esri.com

ESRI's ArcData Online provides both free and commercial data that may be licensed and downloaded. ArcData Online is a collection of digital information products developed cooperatively by ESRI and many other data providers. Included are data sets for applications such as urban and transportation planning, natural resources, environmental, agriculture and demographics. Some specific examples of ArcData are DLGs (roads, railroads, rivers, lakes, streams, wetlands, contours, states, counties, cities, etc.), TIGER/Line data (census blocks, census block groups, major roads, railroads, hydrography, political boundaries, etc.), and Climate-data (monthly summaries of meteorological data).

A substantial amount of sample data are also distributed with ESRI GIS software. The 1999 ESRI Data & Maps is a five CD-ROM set of ready-to-use data that is bundled with ArcView GIS 3.2 and MapObjects. This dataset includes many types of map data at many scales of geography provided in shapefile format. These data are useful for general purpose basemaps. Examples of United States data include

- Boundaries and attributes for Census Bureau's TIGER data census tracts and census block groups and centroids and population for census blocks

- Major water features and permanent streams in a single file for the entire United States. These data are based on the TIGER/Line File data at scale of 1:100,000.

- Major roads from Geographic Data Technologies (GDT) and federal, state, and county highways and railroads from the National Transportation Atlas

- Polygon data for parks, large area landmarks, airports, congressional districts, and urbanized areas (cities of 50,000 people or more)

ETAK

Web site: *www.etak.com*

Etak is the premier publisher of highly-accurate map data, real-time traffic information, and advanced mapping technology in the United States. Reportedly, Etak doesn't simply repackage government data. Their digital databases are developed using a unique and meticulous integration of a wide variety of sources, including aerial photographs, ZIP+4 files, field data capture, and TIGER. For example, ETAK data shows highway width, ramps, and one-way streets. EtakMap Premium Digital Map Data contains the complete nationwide network of roadways, extensive points-of-interest listings, political boundaries, and more than 100 additional attributes. These data are available in MapAccess, shapefile, MapBase, and MapInfo formats. EtakGuide is a route guidance and travel guide CD-ROM for in-car navigation systems.

GEOGRAPHIC DATA TECHNOLOGY (GDT)

Web site: *www.geographic.com*

GDT is a digital map database compiler that provides street center line, postal address, and boundary files for location-based applications. The company also supplies data to insurance companies for defining appropriate rate territories and to telecommunication firms for locating new service areas. Their data time span is annual, semiannual, or quarterly. Geographic scale is county, state, or national. Their United States tiles supply data at the ZIP code and census boundary level (Thrall, 2001).

Products and Cost

GDT provides a myriad of data resources to create spatial products and services to understand and analyze geographic relationships. Their products include Dynamap street networks and many geocoding, postal, boundary, and international products. They also provide Wessex Streets 7.0 which is based on the TIGER 98 data and cost $400–$600 for the complete nationwide set. In addition to enhanced TIGER/Line streets, the Wessex data also include the following layers:

- Elevation contours
- Water and boundaries
- Water ports
- Census block group boundaries
- Railroads
- Voting districts
- City boundaries
- Airport runways

GeoLytics

Web site: *www.censuscd.com*

Demographic data and basic census data on housing are key components to residential market analysis. CensusCD products from GeoLytics, Inc. provide low-cost demographic data, housing data, TIGER boundary files, projections, and consumer expenditures for all fields in the U.S. Census of Population and Housing, as well as historical data and expenditure data by category. The data time spans are census year, current year, and five-year projections. Geographic scale is U.S. census block groups, counties, and states (Thrall, 2001).

Products and Cost

- CensusCD Blocks (census blocks GIS data and attributes) with ArcView shapefile and MapInfo export
- All states for $500–1,000
- Single state CD-ROM for $300–$600

Horizons Technology

Web site*:* *www.suremaps.com*

Sure!Maps from Horizon Technology, Inc. are geo-referenced and seamless USGS topographic raster maps plus free map projection and extraction software for $2.5/quadrangle. Border and legend information is clipped away to create a seamless mosaic that enables scrolling across a study area without gaps, borders, or overlaps. SureMAPS can also be downloaded from the Internet via ArcExplorer and ArcData Online, ESRI's innovative Internet data publishing programs. Reportedly, once users have displayed a map product, they can select their specific area of interest and download the map image in the projection of their choice.

LAND INFO INTERNATIONAL

Web site: *www.landinfo.com*

LAND INFO International offers vector map layers (e.g., contours, roads, hydrology) in 10 different formats for more than 125 countries. They also provide georeferenced topographic maps, DEMs, parcel tax maps, satellite imagery, aerial photos, and FEMA flood maps.

LANDVIEW

Web site: *landview.census.gov*

LandView reflects the collaborative efforts of the U.S. Environmental Protection Agency (EPA), the U.S. Bureau of Census (USBC), the U.S. Geological Survey (USGS), and the National Oceanic and Atmospheric Administration (NOAA) to produce a "Federal Geographic Data Viewer" that provides the public with ready access to published federal spatial and related data.

LandView is a Windows-based desktop mapping system that includes database extracts from the EPA, USBC, USGS, and NOAA. These databases are presented in a geographic context on maps that show jurisdictional boundaries; detailed networks of roads, rivers, and railroads; census block group and tract polygons; schools; hospitals; churches; cemeteries; airports; dams; and other landmark features. Unfortunately, LandView is a standalone mapping system and its database and maps cannot be imported to other mapping and GIS programs.

The latest version, LandView IV is available for purchase on a single DVD for approximately $100 and covers the entire United States. LandView IV replaces LandView III and contains both database management and mapping software. It has the following features:

- It presents detailed information for EPA-regulated sites and maps their locations.
- It maps point features from the USGS Geographic Names Information System (GNIS). The GNIS contains geographic names for all known populated places, features, and areas in the United States that are identified by a proper name.

- It maps a detailed network of major and minor roads, rivers, and railroads for the entire United States based on TIGER/Line files.
- It presents census demographic and socioeconomic data from the Census Bureau along with census maps of legal entities (states, counties, cities and towns, and congressional districts), and statistical entities (such as census tracts and block groups).

NATIONAL GEOGRAPHIC

Web site: *www.nationalgeographic.com/topo/*

National Geographic provides a software and data package called TOPO! TOPO! Image Support software is an extension for ArcGIS 8.1 or ArcView 3.x that creates high quality images of USGS 1:24,000 and 1:100,000 maps of the United States when used with TOPO! state series CD-ROMs. These CD-ROMs feature seamless statewide USGS coverage enhanced by 3D digital shaded relief. The software adds a TOPO! Import Tool to ArcView or ArcGIS toolbar that can be used to import a seamlessly mosaicked georeferenced image of a user defined area from the CD-ROM. TOPO! creates an image catalog of all the maps needed to fill the user defined area, copies them to hard drive, and imports them as a single layer or theme. According to National Geographic, TOPO! state series data are GPS ready. They can be used to create routes and upload waypoints to a GPS. Users can draw a freehand route and TOPO! displays its distance and generates an elevation profile. The 2002 price for software extension plus one data CD-ROM is $600. Additional CD-ROMs covering one state or multiple states (in the Northeast and Mid-Atlantic region) are $100 each. Figure 4-1 shows a sample TOPO! image in ArcMap 8.1.

USGS EARTH SCIENCE INFORMATION CENTER (ESIC)

Web site: *ask.usgs.gov/products.html*

Products and Cost

- USGS GEO-DATA for $30–$50 per CD-ROM
- 1:100,000 scale Digital Line Graph (DLG) data
- 1:2,000,000 scale DLG Data
- Digital Orthophoto Quadrangles (DOQ)
- Digital Raster Graphics (DRG) data
- VMap (supersedes Digital Chart of the World)

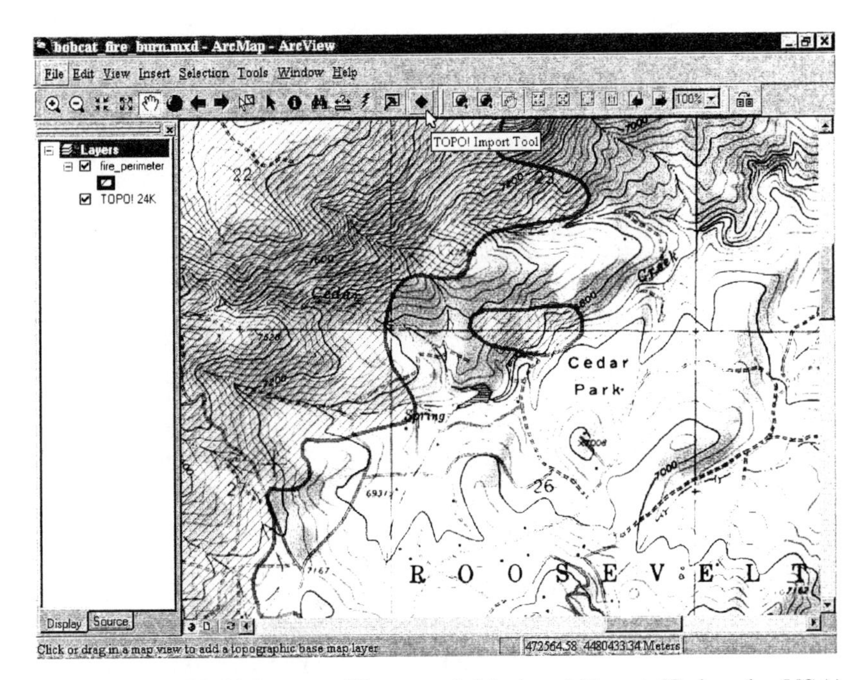

Figure 4-1. TOPO! Image of Roosevelt National Forest (Colorado, USA) Imported in ArcMap 8.1 (Map Courtesy of National Geographic)

Other Commercial Data Sources

ChartTiff: collarless / seamless USGS DRGs, DEMs, and DOQs	*www.charttiff.com*
Geowarehouse: GIS/GPS data and products	*www.geowarehouse.com*
Map Mart: USGS DEMs, DRGS, and DOQs, (free 100K and 250K collarless DRGs) and TIGER data in MapInfo and ArcView formats	*www.mapmart.com*
MapFactory: AlphaMap products for selected U.S. cities	*www.alphamap.com*
Maptech: export full, cropped, or seamless USGS topographic maps at user-specified datum and projection	*www.maptech.com*
Micropath Corporation: Full/collarless DRGs and DEMs	*www.micropath.com*
Natural Systems Analyst, Inc.: DOQQ, IKONOS, and Landsat 7 imagery	*www.gis1.com*
Sanborn Map Company, Inc.: detailed city and tax parcel maps in GIS- and CAD-compatible data formats	*www.sanbornmap.com*

TopoDepot: CAD and GIS vector 7.5-minute topographic quadrangles of the continental United States and Hawaii	*www.topodepot.com*
TopoZone: a la carte (custom) topographic maps	*www.topozone.com*

CUSTOM DATA

Standard public domain or commercial datasets at scales of 1:100,000 and 1:24,000 are inherently inferior to datasets at scales of 1:4,800 and 1:2,400. A roads layer created by digitizing the 1:24,000 paper maps and containing only the line work is much inferior to one created by photogrammetric means using 1:4,800 quality source documents and augmented with road names and maintenance information. Custom data must be created when public or commercial data are not accurate enough to meet the project specifications.

Custom data are those data that are neither available in the public domain nor from commercial data vendors; these data must be created in-house or by a GIS consultant. Custom data creation requires data conversion using scanning, digitization, and data collection using field survey and GPS. Digital orthophotos, LIDAR data, and parcels are examples of custom data.

VECTORIZATION

Scanned images can be automatically converted to vector data without cursor digitization using a process called "vectorization." Vectorization is a cost-effective data conversion method but it is based on new technology that is still improving. Vectorization may not always produce reliable results, especially in the unsupervised (unmanned) mode. Complex, faded, or unclear maps may require post vectorization manual touch-up.

For example, Lockheed Martin Tactical Defense System's AUTOGRAPHICS software can be used to vectorize scanned maps and drawings. AUTOGRAPHICS combines a trainable neural network with a user-friendly graphical interface to enable the efficient processing of color or grayscale hardcopy maps directly at an interactive workstation, without having to digitize the information via a cursor. The user selects specific examples of objects to be automatically identified. Six to eight example pixels are required for each type of object. AUTOGRAPHICS maintains a database of predictors that transforms a raw image scan into a thematic classification of the map. The process has been tested with a variety of scanners, as well as CD-ROM digital maps available from the Defense Mapping Agency.

Vidar Systems Corporation (*www.vidar.com/autographic.html*), a manufacturer of large format color scanners, is the licensed reseller of AUTOGRAPHICS. Reportedly, AUTOGRAPHICS extracts data 10 times faster than other products; scans, imports and separates most typical paper maps in just one hour; and cuts costs by 90% over the conventional data conversion methods. AUTOGRAPHICS is an ESRI-certified solution and produces output in ARC GRID or ArcInfo format.

DIGITAL ORTHOPHOTOS

Digital orthophotos are orthorectified raster images of aerial photographs. They are used as basemaps or landbases that lie beneath other GIS layers

High-resolution orthophotos with sub-meter accuracy can guide the public works crews directly to a manhole hidden behind vegetation.

and provide real-life perspectives of terrain and surroundings that are not available in the vector GIS layers.

Water, wastewater, and stormwater system applications of high resolution digital orthophotos are as follow:

- Utility mapping, network development and expansion, planning of collection and distribution system networks
- Improving service crews efficiency. An estimated 15% of all property marking visits could be eliminated if a one-call system technician could view a property on screen while talking on the telephone with a property owner (Corbley, 2000).
- Serving as backdrop in automatic vehicle location (AVL) systems
- Identifying vegetation encroachments

Typical vector data do not show vegetation. The vector layers can show the manhole location but may not include the vegetation hiding the manhole. High-resolution orthophotos with submeter accuracy can guide the public works crews directly to a manhole hidden behind bushes. Knowing the land cover characteristics before leaving for an emergency repair of a broken water main will allow the crews to bring the appropriate tools and equipment. Knowing whether the job will be on a busy intersection or in somebody's backyard will determine the kind of equipment, material, and personnel for the job.

LIDAR DATA

Light Imaging Detection and Ranging (LIDAR) is a new system for measuring ground surface elevation from an airplane. LIDAR sensors provide some of the most accurate elevation data in the shortest amount of time by bouncing laser beams off the ground. LIDAR technology was developed in the mid-1990s and combines GPS, precision aircraft guidance, laser range-finding, and high-speed computer processing to collect ground elevation data. Mounted on an aircraft, a high-accuracy scanner sweeps the laser pulses across the flight path and collects reflected light. A laser range-finder measures the time between sending and receiving each laser pulse to determine the ground elevation below. LIDAR system can survey up to 10,000 acres a day and provide horizontal and vertical accuracies up to 12 and 6 inches, respectively.

PARCELS

A parcel is defined as a contiguous single ownership interest in a legally described real property. A parcel can be composed of one or more adjacent lots or portions of lots in a subdivision owned by the same person. The parcel or cadastral data are the key to land ownership, parcel size, configuration, land use, improvement values, and other related information contained in federal, state, and local government or public and private agencies. Parcel maps are the foundation of most municipal GIS data because approximately 90%of all municipal governments depend on them. GIS facilitates the integration of parcel data with other layers, such as soil type, slope, and infrastructure (Zimmer, 2000).

Water, wastewater, and stormwater applications of parcel GIS data include planning and engineering. The parcel layer can be used to delineate land use classes to study potential for system expansion. It can identify the landowners along a potential water or sewer line construction route. The ownership information of the parcel layer can be used to create mailing lists or form letters for construction notification or smoke testing or to generate a list of residents to be notified in emergency situations, such as contaminated water supply or hazardous spill into the sewer system.

Parcel maps can be digitally converted from existing paper maps by digitization, scanning, or vectorization. Although this is the fastest and cheapest method, it is also the least accurate. A coordinate geometry or COGO-based method that recompiles the cadastral map from scratch using deeds and plats produce the most accurate map short of performing a new survey. Parcel maps were originally designed to aid the tax assessor in visually identifying a given parcel. We should avoid misuse of digital cadastral mapping data by using it in ways that exceed its intended purpose (Adams and Johnson, 1994). For example, parcel maps should not be used as construction drawings.

METADATA

Have you ever obtained GIS data and wondered about their production date, scale, units, and projection? Do you know the datums of your GIS data? Are they NAD27, NAD83 (1986), NAD83 (1992), NGVD29, or NAVD88? Do you know whether the height was orthometric or ellipsoid? Were the Northing/Easting State Plane coordinates or UTMs? Are the units in feet or meters? If feet, are they U.S. Survey or International feet? You probably do not have this information if your data supplier did not give you a metadata file for your GIS data. Without this information, you may not be able to use this data in a GIS or create an accurate GIS database.

Metadata is defined as "data about data." Metadata documents descriptive and practical information about a particular dataset, database, or GIS layer. The Federal Geospatial Data Committee's (FGDC) has defined a Content Standard for Geospatial Metadata. The Content Standard was derived as a part of the FGDC's creation of the National Spatial Data Infrastructure (NSDI) to promote the sharing of geospatial data throughout federal, state, and local governments. FGDC-compliant metadata allows for the inclusion of GIS data in the FGDC's national clearinghouse. The metadata standard outlines a formal and structured document that contains information that identifies a dataset and provides vital information pertaining to its use in GIS, such as projection information, file format, and the sources of the data. It also includes information about how to acquire the data and who to contact for more information about the data.

Metadata information is stored in a text (*.txt) or HTML (*.htm or *.html) file. Most USGS metadata files made before 1999 do not conform to any formal standard for physical format. The files are simple ASCII text, and the information is organized to mirror the FGDC content standard. DRGs made since 1999 use a similar but more stable format based on conformance with the "metadata parser" (mp) software tools. This format is also ASCII text but has more rigorous rules for text arrangement. Various sections of a typical metadata file are described below.

IDENTIFICATION INFORMATION

Identification information includes the title, creator or originator of the data, an abstract describing the content of the dataset, keywords for search engines, and contact information for a person or organization for questions pertinent to the content or technical details about the data itself. If the data are available directly through download, a link is provided to the dataset here. For example, the identification information for the 7.5-minute USGS DEMs consists of the following text:

```
Identification_Information:
  Citation:
    Citation_Information:
      Originator: U.S. Geological Survey
      Publication_Date: 19790701
      Title: 7.5 minute Digital Elevation Models
      Geospatial_Data_Presentation_Form: map
      Publication_Information:
        Publication_Place: Reston, VA
        Publisher: U. S. Geological Survey
```

DATA QUALITY INFORMATION

The data quality information contains information about the resolution or "scale" of the data, accuracy of the assignment of attributes, and georeferencing of the data.

Spatial Data Organization Information

Spatial data organization tells the user whether data are vector or raster and provides relevant details about the vector objects or pixel dimensions of raster data.

Spatial Reference Information

Spatial reference information is the most important section for GIS applications. It details the projection or coordinate system and associated modeling specifications necessary for using the data or determining the usefulness of the data.

Entity and Attribute Information

Entity and attribute information provides an overview of detailed information on the attributes in the tables or fields contained in a database.

Distribution Information

Distribution information discusses in what format the data are available. For data downloaded from the Pennsylvania Spatial Data Access (PASDA) Web site, the most relevant information is the format of the data (e.g., ArcInfo Export, shapefile) and what compression schemes, if any, have been applied to the data (see Chapter 5, "Internet GIS," for more information). For datasets that are not available online, this section will also include the physical formats in which the data are available (e.g., CD-ROM, tape cartridge) and contact information describing where and how to acquire the data. For example, the distribution information for the 7.5-minute USGS DEMs consists of the following text:

```
Distribution_Information:
  Distributor:
    Contact_Information:
      Contact_Organization_Primary:
        Contact_Organization: Earth Science Information
        Center, U.S. Geological Survey
      Contact_Address:
        Address_Type: mailing address
        Address: 507 National Center
        City: Reston
        State_or_Province: Virginia
        Postal_Code: 20192
      Contact_Voice_Telephone: 1-888-ASK-USGS
      Hours_of_Service: 0800-1600
      Contact_Instructions:
        In addition to the address above there are  other
        ESIC offices throughout the country.  A full list
        of these offices is at:
        URL: http://mapping.usgs.gov/esic/esic_index.html
```

SUMMARY

This chapter provided information about various GIS data types and formats. Three types of GIS data were discussed: public domain, commercial, and custom. Limitations and benefits of each data type were also presented. We also learned the importance of data quality for developing accurate and reliable GIS applications. It was emphasized that although errors can be introduced at many points along a GIS project, it all starts with data quality. Finally, importance of data documentation using metadata was described.

SELF EVALUATION

1. What are different types of GIS data? What type of GIS data do you use?
2. How are GIS data created?
3. Is it good or bad to have a lot of GIS data formats? Provide justification for your answer.
4. What is the best way to select GIS data for your application?
5. How is GIS data quality defined? Why is data quality important?
6. Why is it important to document GIS data? What is the best way to document your GIS data?

Chapter

5 INTERNET GIS

Can you download a free basemap for your GIS mapping project from the Internet? Will it be accurate enough for your GIS project? Read on to find the answer!

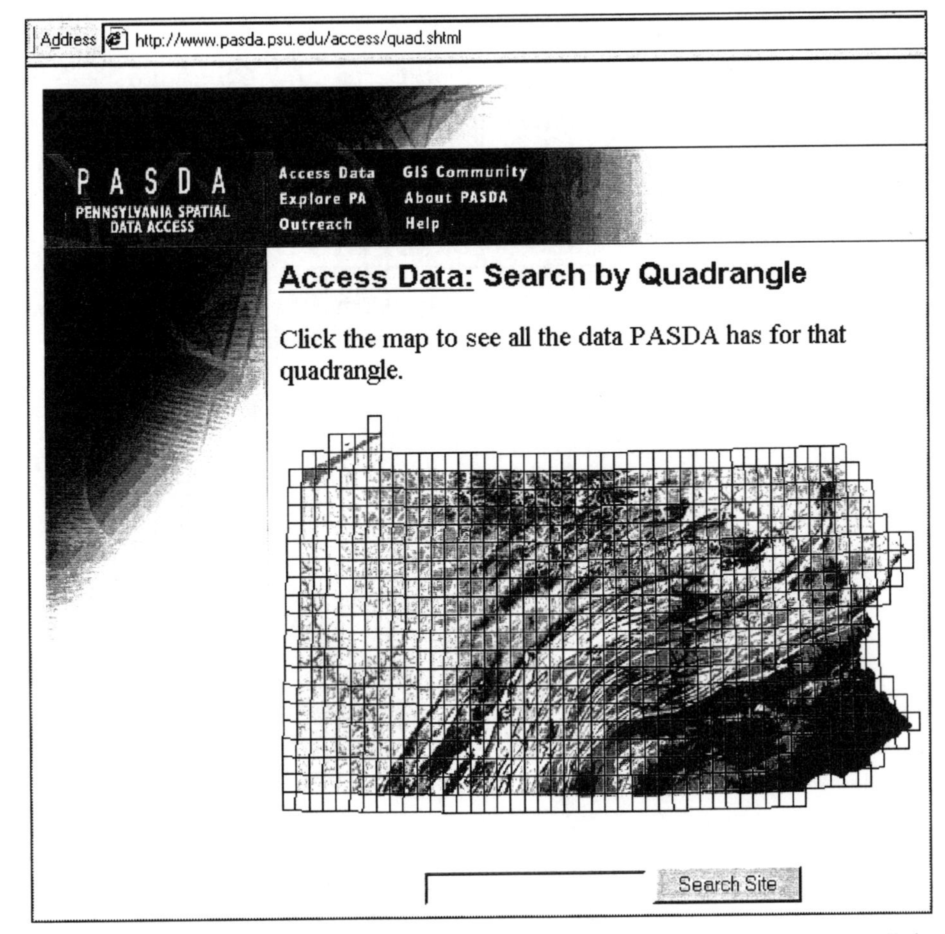

Pennsylvania Spatial Data Access (PASDA) Web Site Allows Users to Click on a Map of USGS Quadrangle Boundaries to Retrieve GIS Data.

LEARNING OBJECTIVE

Though the Internet has made public domain and commercial data more readily available, you can still encounter problems finding appropriate data to fill your specific needs. The learning objective of this chapter is to disseminate information about public domain GIS data that are generally useful in developing GIS applications for water, wastewater, and stormwater systems. Major topics discussed in this chapter include

- *Sources of free and public domain GIS data*
- *GIS data clearinghouses*
- *U.S. Geological Survey (USGS) GIS data*
- *U.S. Environmental Protection Agency (EPA) GIS data*
- *U.S. Bureau of Census (USBC) TIGER GIS data*
- *U.S. Natural Resources Conservation Service (NRCS) digital soils data*
- *Federal Emergency Management Agency (FEMA) flood data*
- *National Wetlands Inventory GIS data*

MISSISSIPPI RIVER BASIN USES INTERNET GIS DATA

Application	Development of a surface water quality GIS database using Internet GIS data
Year Published	2000
Reference	Swalm et al., 2000
GIS software	ArcInfo and ArcView
Other software	ERDAS, Oracle 8*i*, SAS
Hardware	SUN Microsystems Sparc 1000 and Sparc 20 workstations, Dell Precision 400 workstations, Gateway 200 personal computers, Hewlett-Packard 1000c printer and 755cm plotter
Public Domain GIS data	USGS digital raster graphics (DRG), digital line graphs (DLG), digital elevation models (DEM) and Natural Resources Conservation Service (NRCS) state soil geographic (STATSGO) data
Commercial GIS data	USGS 60-m resolution satellite imagery and NASA 5-m resolution Calibrated Airborne Multispectral Scanner (CAMS) imagery
Study area	Mississippi River Basin, United States
Organization	Tulane University Medical Center, New Orleans, Louisiana, U.S.A.

A large amount of fertilizers, herbicides, and pesticides applied yearly to the lands in the Mississippi River basin contribute to the algal booms, as some residual chemical breakdown of the compounds contained in agricultural runoff function as nutrient components. Amid concerns of algal boom and water quality, a research consortium integrated millions of data records from hundreds of sources to construct a massive GIS for the Mississippi River Basin. The main objective of the research study was to develop a water-quality parametric database to characterize the river using historical and contemporary water quality data and a GIS. The project team developed a relational database of water quality parameters and integrated the parametric data into ArcView GIS. GIS data for the basemap were obtained from multiple public-access Internet sites. Much of the data were downloaded as DLG files that were converted to vector coverages and shapefiles. Visual evaluation of GIS maps answered many research questions. For example, are the high dissolved nitrogen locations located near a land use change area? Or are they near an industrial complex? If yes, then what type of industry is located there? Highlights of the GIS database are as follows.

- Projection system: Universal Transverse Mercator (UTM) zone 15
- Basemap: 1:2,000,000-scale layers of road, railroad, political boundary, and hydrography
- Parametric database: 1.3 million records on 700 parameters
- 1:100,000-scale USGS DLGs
- 1:250,000-scale USGS land use/land cover (LULC)
- USGS DEMs
- NRCS STATSGO soils
- 60-m multispectral satellite imagery from USGS from the 1970s, 1980s, and 1990s

PUBLIC DOMAIN GIS DATA

Public domain data are nonproprietary data developed by government agencies for public use. A large number of federal, state, and local governments are willing and eager to share their existing GIS data. These data are often free when available online or for the cost of media when shipping is necessary.

GIS data have never been as easily available as they are today. The Internet has become the most popular medium for the electronic dissemination of GIS data to the public. The emergence of Internet technologies in the past few years has made it much easier for geographic data publishers to distribute their data to GIS users. For example, the United States Federal Government was used to distribute data on clumsy and expensive reel-to-reel tapes. Now most data are available on the Internet (Thrall, 2001a). According to many GIS industry experts, the Internet will also be the data delivery medium of choice in the future.

Although large volumes of public domain and commercial data for elevation, hydrography, soils, land use, and so on exist in digital format as described in this chapter, there may be some problems with these digital databases

- Inadequate resolution (scale)
- Inadequate accuracy
- Outdated data
- Different data formats
- Different projection (coordinate) systems

Most federal and state agency maps have small scales (1:24,000–1:250,000) that may not be suitable for small study areas. These maps may not be highly accurate because the National Map Accuracy Standards (NMAS) allow up to 40 ft (12 m) of positional error in 1:24,000-scale maps. Government agency maps of rapidly growing areas may quickly become outdated because they are updated on multiyear cycles. As discussed in Chapter 4 (GIS Data), there are approximately 80 different GIS data formats, not all of which can be read directly in a GIS. For example, U.S. Census Bureau TIGER/Line data format cannot be read in most GIS packages unless it is converted to one of the standard GIS formats. Finally, government data may be projected using a projection system appropriate for large areas that may not be suitable for small areas.

A successful GIS application requires data accuracy, data integrity (also called referential integrity), and multiple attributes capability (Singh, 1995). Thus, as described in Chapter 4, users should be aware of the scale, resolution, accuracy, quality, and intended use of public domain data before using them in their GIS application projects. For example, widely available and popular USGS digital orthophoto quadrangles (DOQs) have a resolution of 1 m, which may not be suitable for all types of GIS applications. At a minimum, municipalities wishing to use digital orthophoto images as a working basemap should acquire panchromatic images with a resolution of 2-ft pixels. Because land use density impacts the usability of the images for various tasks, it would be advantageous for rural areas to use the 2-ft pixel recommendation and urban areas to acquire image data at 1-ft or greater pixel resolution (PaMAGIC, 2001).

"Free" GIS data may not always be free because they may require a lot of pre-processing time to make them ready for your GIS. In fact free GIS data are sometimes worthless. Quite often, a significant effort is required to make free Internet data useful and usable. Most public domain GIS data are available in different formats, scales, projections, and datums. The users must do the projection and datum conversions so that the public domain data can be used with the private data. As described in Chapter 4, the users must also convert, the public domain data in a format that is compatible with their GIS.

Table 5-1 provides a summary of major public domain data sites that provide free Internet downloads of GIS data or ship most types of data on CD-ROM for a fee of about $50. This price and other data prices in the book are based on the average 2000-2002 prices.

Table 5-1. Main Public Domain GIS Data Sites

Organization	*Data*	*Web Site*
USGS	Digital Elevation Model (DEM)	*Earthexplorer.usgs.gov*
	Digital Raster Graphics (DRG)	*nsdi.usgs.gov/*
	Digital Orthophoto Quadrangles (DOQ)	*edcwww.cr.usgs.gov/doc/ edchome/ndcdb/ndcdb.html*
	Digital Line Graphs (DLG)	
	Land Use/Land Cover (LULC)	*edcftp.cr.usgs.gov*
	National Elevation Dataset (NED)	*dss1.er.usgs.gov*
	National Land Cover Database (NLCD)	*www-nmd.usgs.gov*
	National Hydrography Dataset (NHD)	
	Watersheds (Hydrologic Unit Codes)	
U.S. EPA	Streams (Reach Files)	*www.epa.gov/enviro/ index_java.html.*
	Land Use/Land Cover (LULC)	
U.S. Census Bureau	TIGER/Line data for census tracts and blocks and geographic features	*www.census.gov/geo/www/ tiger/index.html*
FEMA	DFIRM	*www.fema.gov/MSC/index.htm*
	Digital Q3 flood data	
NRCS	Soil Survey Geographic (SSURGO) soils data	*www.ftw.nrcs.usda.gov/ ssur_data.html*
	State Soil Geographic (STATSGO) soils data	*www.ftw.nrcs.usda.gov/ stat_data.html*

Web site URLs are current at the time of publication and may change in the future. The latest information (e.g., version, cost, URL) about various data and software products discussed in this book may be obtained from the author's Web site (*www.GISApplications.com*). If a URL does not work, go to the Web site's root address (home page) to search the revised link. For example, if *edcwww.cr.usgs.gov/doc/edchome/ndcdb/ndcdb.html* does not work, try *edcwww.cr.usgs.gov* or go to *www.usgs.gov* and search a revised link for edchome or ndcdb. Alternatively, try the FirstGov site (*www.firstgov.gov*), U.S. government's search engine. FirstGov is the first-ever government website to provide the public with easy, one-stop access to all online U.S. Federal Government resources.

When opting for free Internet download of public domain, GIS data users should be aware of long data download times. Download times can span several hours for large files (e.g., USGS DOQ images) when using slow modem connections. CD-ROM orders may be more appropriate for users who do not have high-speed Internet connections, such as T1, cable, or DSL.

GIS Data Clearinghouses And Geoportals

If you cannot find the GIS data you are looking for using your favorite search engine, do not assume that it does not exist. Try a different search engine, such as "Google" (*www.google.com*), which can search approximately 1.4 billion Web sites or use the so-called "Deep Web" search engines that focus on specialty areas, like mapping. Once you have mastered the search techniques, the Internet will overwhelm you with information. However, too much information can bog you down, and though the Internet "Geoportals" bring a world of geographic information to your desktop, it is still up to you to select the right data. GIS data clearinghouses help you sort through the mess of information you may experience when looking for GIS data on the Internet.

One of the best ways to discover GIS data is through geographic clearinghouses and other geographic data sites on the Internet. Some regional organizations act as a clearinghouse for local government data or are responsible for unified GIS data for a region. Clearinghouses promote a more open market in which GIS data with superior characteristics can be selected over inferior data. State, regional, nonprofit, and professional organization mapping and spatial data clearinghouse Web sites are the most valuable source of free local spatial data.

Many clearinghouse Web sites that provide free GIS data are available, but they cannot all be listed here. Some sample U.S. and international Web sites are listed in Tables 5-2 and 5-3, respectively. Representative examples are discussed in detail.

Table 5-2. Examples of U.S. Data Clearinghouses and Geoportals

Clearinghouse	*Web Site*
National Geospatial Data Clearinghouse	*www.fgdc.gov/data/data.html* *www.fgdc.gov/clearinghouse/clearinghouse.html*
FGDC Clearinghouse Gateway	*fgdclearhs.er.usgs.gov/*
Geography Network	*www.geographynetwork.com*
Geography Network list of clearinghouses	*www.geographynetwork.com/data/clearinghouses.cfm*
GIS Data Depot	*www.gisdatadepot.com*
GISLinx (categorized GIS links)	*www.gislinx.com*
Directions Magazine Data Center	*www.directionsmag.com/datacenter*
Pennsylvania Mapping and Geographic Information Consortium (PaMAGIC)	*www.pamagic.org*
Georgia Spatial Data Infrastructure	*gis.state.ga.us*
Iowa Department of Natural Resources Natural Resources GIS (NRGIS) Library	*www.igsb.uiowa.edu/nrgis/gishome.htm*
New Jersey Spatial Data Clearinghouse from Office of GIS (NJOGIS)	*njgeodata.state.nj.us*
New York State GIS Clearinghouse	*www.nysgis.state.ny.us/index.html*
Spatial Hydrology	*www.spatialhydrology.com*
Texas Natural Resources Information System – Texas State GIS Clearinghouse	*www.tnris.state.tx.us*
U.S. EPA Region 5, Office of Information Services	*www.epa.gov/reg5ogis/*

Table 5-3. Examples of Global / International Data Clearinghouses and Geoportals

Clearinghouse	*Web Site*
Center for International Earth Science Information Network (CIESIN)	*www.ciesin.org/data.html*
Canadian Geospatial Data Infrastructure	*ceonet.ccrs.nrcan.gc.ca/*
Global Spatial Data Infrastructure (GSDI)	*www.gsdi.org*

ESRI's Geography Network provides a collection of links to some of the more useful clearinghouse sites offering data that works with GIS software. The Geography Network is discussed below in detail. GIS Data Depot provides free downloads and custom CD-ROM creation. Help yourself to data from their on-line inventory of more than 2 terabytes of geospatial data including more than 46,000 of 24k-, 25k-, 63k-, 100k-, and 250k-USGS DRGs. Many DOQQ, DEM, NWI, DTED, and DCW data are also available.

PASDA: CLEARINGHOUSE EXAMPLE 1

Pennsylvania Spatial Data Access (PASDA) (*www.pasda.psu.edu*) is an excellent example of a GIS data clearinghouse. PASDA is Pennsylvania's official geospatial information clearinghouse and Pennsylvania' node on the National Spatial Data Infrastructure (NSDI). It provides free downloads of DRG, DEM, DOQ, roads, streams, watersheds, river conservation plans, and satellite imagery data for the entire state of Pennsylvania. PASDA supports search, display, and retrieval of GIS data, satellite images, aerial photographs, and metadata related to Pennsylvania.

PASDA has made searching for GIS data easy. There are many ways to search for data on PASDA. Data Download will take you to their data catalog which is a list of PASDA data by provider. Keyword search is a default Boolean "and" search that allows you to combine words and narrow your search. For example, you may search for watershed-related data simply by typing in the word "watershed" in the keyword search box. To narrow your search, you may combine "watershed" with "monitoring" to retrieve only those data sets containing watershed monitoring data. Browse by Topic capability allows you to browse the data catalog by subject such as transportation, demography, and environment. In addition, PASDA has created three map-based searches—county, USGS quadrangle, and watershed. These searches retrieve all data relevant to the county, quad, or watershed you select. The figure on the first page of this chapter shows how the users can point-and-click on the USGS quad boundaries overlayed on the state map to retrieve all the available GIS data for that quad. PASDA standard data are available in decimal degrees and UTM NAD 83.

ARC: CLEARINGHOUSE EXAMPLE 2

Atlanta Regional Commission (ARC) is another example of a well-coordinated effort for disseminating regional GIS data. ARC publishes a set of four CD-ROMs called Atlanta Region Information System (ARIS). ARIS contains the best available GIS data for its 10-member counties and the City of Atlanta. Available for $300-$400, ARIS blends GIS software, data (including Digital Orthophoto Quarter Quadrangle or DOQQ images), and visualization and multimedia presentation tools into an easy-to-use PC environment. It is designed to work with most GIS software, particularly ESRI's popular desktop GIS, ArcView.

NGDC: Clearinghouse Example 3

The Federal Geographic Data Committee (FGDC) coordinates the development of the NSDI. The NSDI encompasses policies, standards, and procedures for organizations to cooperatively produce and share geographic data. The 17 federal agencies that make up the FGDC are developing the NSDI in cooperation with organizations from state, local, and tribal governments; the academic community; and the private sector. FGDC has created the National Geospatial Data Clearinghouse (NGDC), which is a collection of more than 250 spatial data servers that have digital geographic data primarily for use in GIS, image processing systems, and other modeling software. NGDC provides search tools to locate GIS data on servers throughout the world. NGDC includes more than 200 sites and thousands of data sets (Lanfear, 2000). These data collections can be searched through a single interface based on their descriptions, or "metadata."

Some free GIS data may be outdated. For example, some USGS data date back to the 1980s. Local GIS data is generally more accurate. Check with local governments in your study area to determine if they belong to any regional GIS consortium or clearinghouse like PASDA or ARC.

If the Web is like the public library system, then ESRI's Geography Network is like a really nice librarian who keeps the library open 24 hours a day and makes it possible for everyone to find exactly what they need (Geospatial Solutions, December 2000).

Geography Network

ESRI's Geography Network provides both free and commercial GIS data on the Internet. The Geography Network is a global network of geographic information users and providers. It provides the infrastructure needed to facilitate the sharing of geographic information between data providers, service providers, and users around the world. The Internet is used to deliver geographic content to the user's browser and desktop. Through the Geography Network, one can access both live maps and downloadable data.

The site features a mini browser and software that can combine data from maps of different sources. The network allows users to access GIS data from numerous sources. The USGS and U.S. EPA are two of the government contributors to the Geography Network. Among its available GIS data are satellite imagery, land use, vegetation, flood risk zones, elevation, socioeconomic, and political data layers.

Figure 5-1 shows a Flood Risk Map for the City of Pittsburgh (Pennsylvania, USA) created using the Geographic Network. Created from FEMA Q3 Flood Data and GDT Dynamap/2000 street data, such maps can be used to help determine the relative flood risk of a specific location. The Q3 Flood Data are currently available for approximately 1,200 counties across the United States. ArcGIS 8.1 is designed to accept streaming data from the Geographic Network.

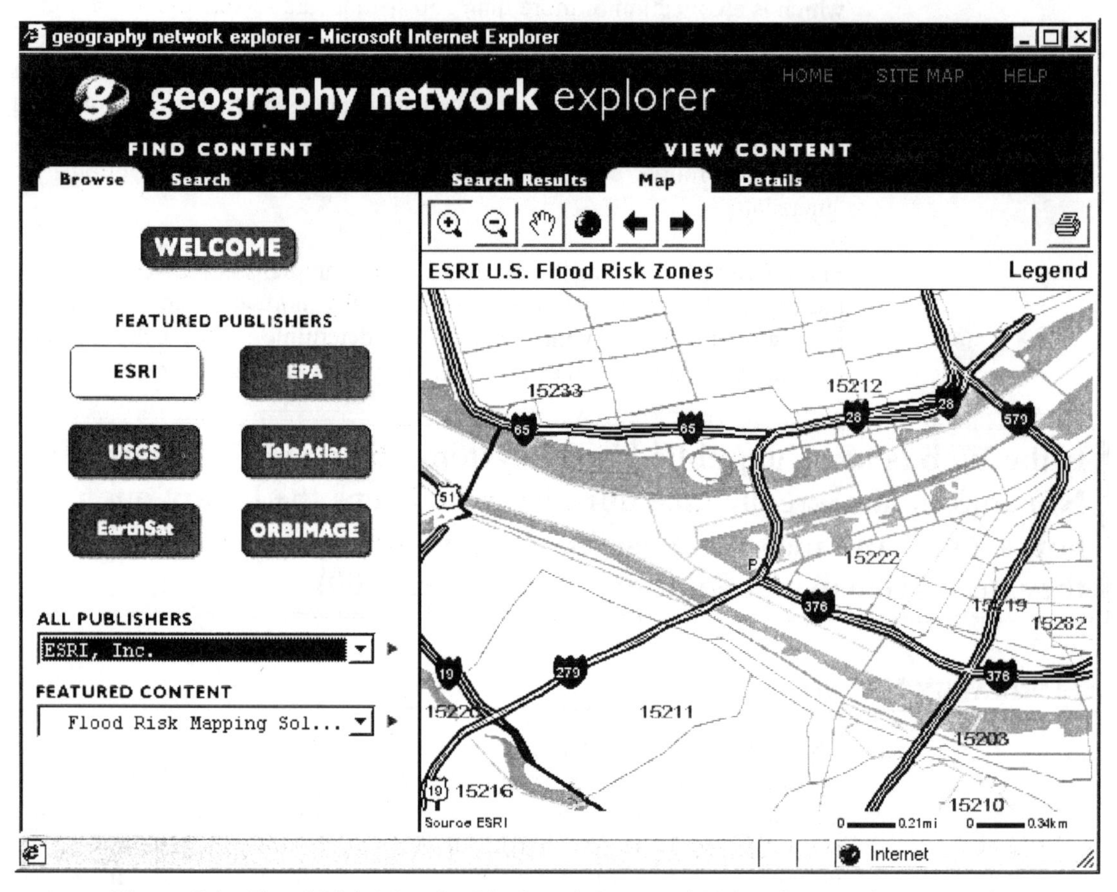

Figure 5-1. Flood Risk Map for Pittsburgh Prepared Using Geography Network

Geography Network Web Site

The Geography Network	*www.geographynetwork.com*

UNITED STATES GEOLOGICAL SURVEY GIS DATA

The USGS, established in 1879, has Federal responsibility for preparing and making multipurpose maps and fundamental cartographic data to meet the requirements of users throughout the United States.

The National Mapping Division of USGS has scanned all its paper maps into digital files, and all 1:24,000-scale quadrangle maps now have DEMs (Limp, 2001). USGS GIS data can be downloaded from many Web sites that continue to change their URL addresses. The most likely Web site for downloading USGS data is the Earth Resources Observation Systems (EROS) Data Center (EDC) Web site (*edc.usgs.gov*) shown in Figure 5-2.

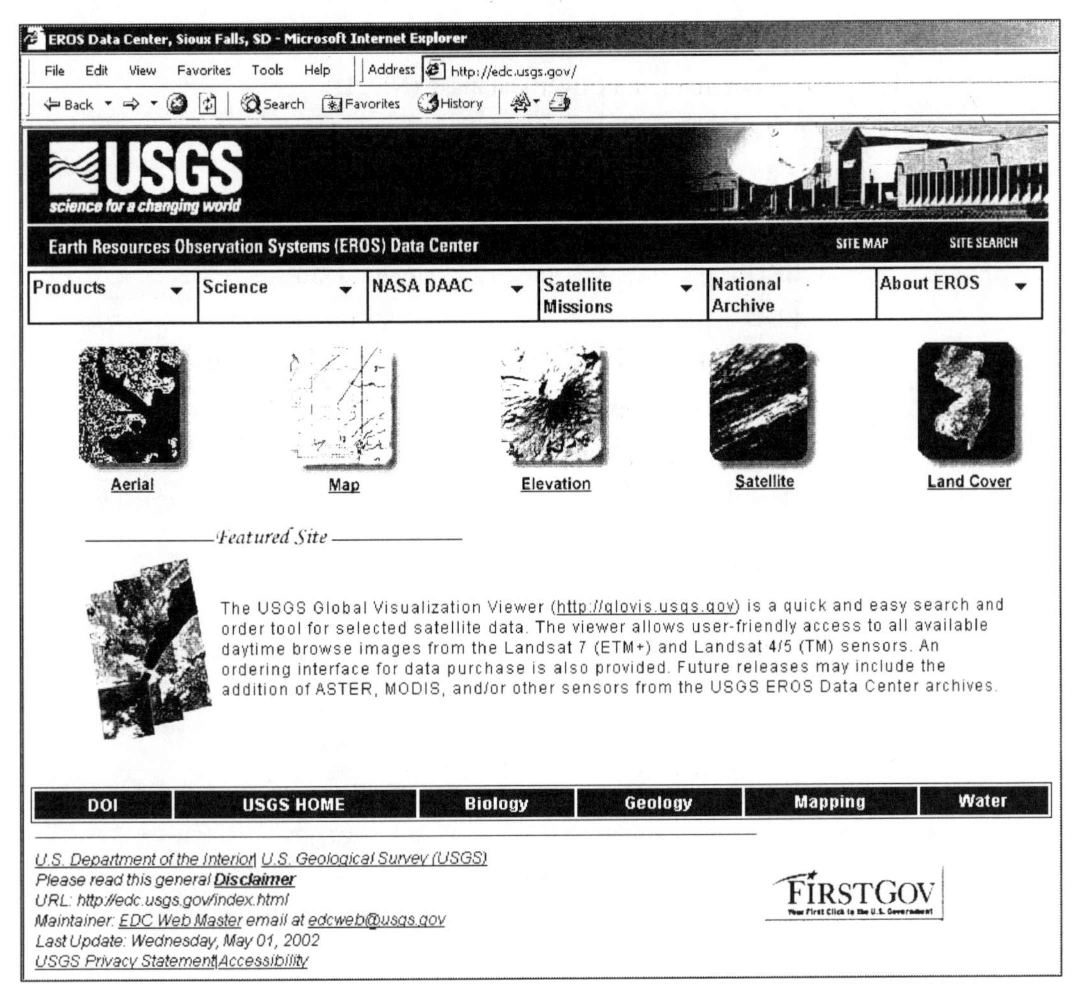

Figure 5-2. USGS Web Site for Downloading GIS Data

USGS provides the following spatial data that are useful in developing GIS applications for water, wastewater, and stormwater systems:

- Digital Elevation Models (DEM)
- Digital Raster Graphics (DRG)
- Digital Orthophoto Quadrangles (DOQ)
- Digital Orthophoto Quarter Quadrangles (DOQQ)
- Digital Line Graphs (DLG)
- Land Use/Land Cover (LULC)
- Hydrologic Unit Code (HUC) Watersheds
- National Elevation Dataset (NED)
- National Land Cover Database (NLCD)
- National Hydrography Dataset (NHD)

USGS Web Sites

EROS	*edc.usgs.gov*
EDC	*edcwww.cr.usgs.gov*
Geographic data download	*edc.usgs.gov/geodata/*
Digital maps and data	*ask.usgs.gov/digidata.html*
USGS National Mapping	*www-nmd.usgs.gov*
Data download	*edcwww.cr.usgs.gov/doc/edchome/ndcdb/ndcdb.html*
FTP download	*edc.usgs.gov/doc/edchome/ndcdb/ndcdb.html*
GIS data for water resources	*water.usgs.gov/GIS/*
Geo-Data Explorer	*geode.usgs.gov* *dss1.er.usgs.gov*
EarthExplorer	*earthexplorer.usgs.gov*
USGS node of the NGDC	*nsdi.usgs.gov*
Spatial datasets for water	*water.usgs.gov/lookup/getgisfoot?type=gis&value=erf1*
Global Land Information System (to be retired)	*edc.usgs.gov/webglis*

USGS data can also be purchased from USGS or its business partners. USGS CD-ROMs can be ordered from Customer Services at the USGS

EROS Data Center. For example, DEMs can be purchased on CD-ROM from the USGS EarthExplorer Web site for an entire county or state for a nominal fee. The price of most USGS CD-ROMs is $40–$60. More USGS data information can be found on the Web site of USGS node of the NGDC (*nsdi.usgs.gov*).

Digital Elevation Models

Topography is an important element for modeling surface runoff from watersheds and sewersheds. The quantity of runoff depends on the slope of land surface. The shape of the surface determines where the runoff goes. Both the surface slope and shape can be determined from a DEM.

DEM data represent spatially distributed topography in the form of regular grid of elevation values. DEM data files contain information for the digital representation of elevation in a raster form. Regardless of the underlying data structure, most DEMs can be defined in terms of (x,y,z) data values, where (x,y) represents the coordinates at a regularly spaced interval or grid, and z represents the elevation values. This data structure creates a square grid matrix with the elevation of each grid square, called a pixel, stored in a matrix format. Each pixel has a single value representing the elevation over the extent of that grid cell.

DEM data can be obtained from traditional land survey methods or by using remote sensing techniques (Chapter 8), such as aerial photography, airborne lasers or LIDAR (Chapter 4), radars, and spaceborne satellites.

Early DEMs were derived from USGS quadrangles, and mismatches at boundaries continued to plague the use of derived drainage networks for larger areas (Lanfear, 2000). The National Elevation Dataset (NED) produced by USGS in 1999 is the new generation of seamless DEM that largely eliminates problems of quadrangle boundaries and other artifacts. Users can now select DEM data for their area of interest.

USGS provides 90-m resolution DEM data for the entire United States and 30-m resolution DEM data for almost all locations in the United States. Ten meter resolution data are also available for some locations. The 30-m data are generally adequate for most watershed and sewershed studies. By applying the basic rule of "water must flow in the direction of the steepest downhill slope" to DEMs, it is possible to determine flow direction, flow length, flow slope, and so on by processing the DEM data in a raster GIS. DEMs can be used as source data for digital orthophotos and as layers in a GIS. GIS provides an integrated platform for DEMs, automatic watershed and stream delineation, and computation of watershed hydrologic parameters. DEMs are especially useful in distributed hydrologic modeling. Topography, elevation, and slope information can also be extracted from digital contour data or hypsography data. Contour data are available from USGS Digital Line Graphs (DLGs) described later in this chapter.

Digital Elevation Model Applications

DEM applications include the following:

- Automatic watershed and stream delineation
- Sewershed boundary delineation
- Estimation of watershed hydrologic parameters
- Hydrologic modeling
- Calculating node elevation in water distribution system models
- 3D visualization

Shamsi (2000) provides a review of DEM applications in hydrologic modeling and examples of automatic watershed and stream delineation.

Useful DEM Web Site

DEM Data	*edcwww.cr.usgs.gov/glis/hyper/guide/usgs_dem*

DIGITAL RASTER GRAPHICS

USGS topographic quadrangle maps are used by almost all civil engineers and surveyors on a daily basis. USGS topographic maps are distributed as both paper maps and as scanned raster files referred to as digital raster graphics or DRGs. A DRG is a scanned raster image of a USGS standard series topographic map, including all map collar information. Simply stated, DRGs are digital versions of topographic paper maps. The image inside the map neatline is georeferenced to the surface of the Earth using the UTM projection. The horizontal positional accuracy and datum of the DRG matches the accuracy and datum of the source map. The map is scanned at a minimum resolution of 250 dots per inch (dpi). The USGS produces DRGs of the 1:24,000-,1:24,000/1:25,000-, 1:63,360- (Alaska), 1:100,000-, and 1:250,000-scale topographic map series.

A DRG can be used as a background layer in a GIS to perform quality assurance on other digital products, and as a source for collecting and revising vector data. Figure 5-3 shows a basemap created by making a seamless mosaic of four 7.5-min (1:24,000) individual DRG images for a combined sewer overflow (CSO) study for the Authority of the Borough of Charleroi, Pennsylvania located near Pittsburgh. The individual DRGs correspond to USGS quadrangles for Monongahela, Donora, California, and Fayette.

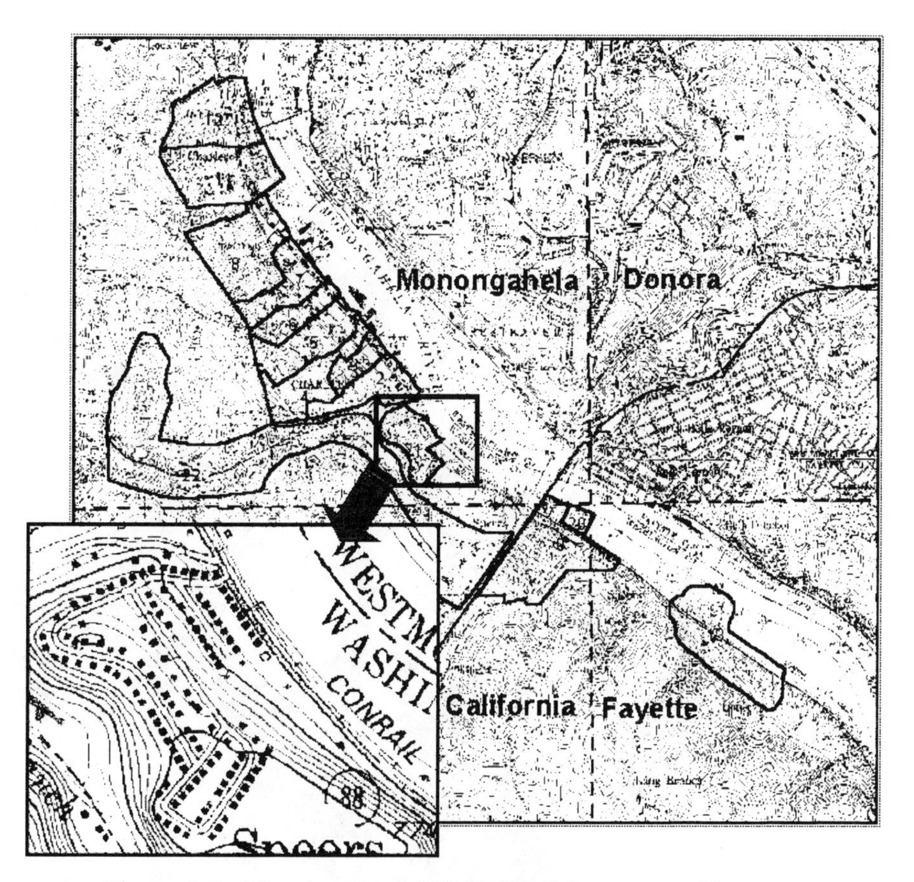

Figure 5-3. Mosaic of Four USGS DRG Images as a Basemap

By merging and combining DRGs, which have attributes inherent with their symbology, with other digital data, such as DOQs, which have updated photographic qualities, land use changes, such as new roads and houses, can be easily identified (Murphy, 2000). DRGs or DOQs can also be merged with DEMs to produce a hybrid digital file. This process is also called 3D draping. Figure 5-4 shows a 3D topographic map created by draping a DRG over a DEM for a study area in southwestern Pennsylvania. This map was created using data and software from DeLorme, Inc. (*www.DeLorme.com*).

The USGS DRGs are also available on CD-ROM by individual files. Each CD-ROM includes three files for a pre selected map scale, which are

1. One tagged image file format (TIFF) file
2. One georeferenced associated world file or TFW file
3. One federal geographic data or FGDC metadata file

Figure 5-4. A DRG Draped Over a DEM Showing Kiskiminetas River, Pennsylvania

The most common 7.5-min (1:24,000) DRGs are created using the following steps:

1. Paper topographic maps are scanned on a high-resolution scanner. Scanning resolutions range from 500 to 1,000 dpi with an output file of 160–300 MB.

2. The number of colors is reduced during the scanning phase by removal of screens and color quantization.

3. The raw scan file is then transformed and georeferenced using UTM coordinates of the sixteen 2.5-min grid ticks, which are interactively visited and assigned their respective UTM coordinates. The USGS program, XSHAPES4, then performs a piecewise linear rubber sheet transformation.

4. An output resolution of 2.4 m (8.2 ft) is chosen to resample the file to 250 dpi.

5. The image file is converted to a TIFF 6.0 image and further reduced by converting the file to a run length encoded PackBits compression (type 32773).

6. The color palette of the compressed DRG is then standardized by replacing the original RGB values assigned during the scanning process with standard RGB value combinations.

7. The DRG metadata file is completed.

8. Prior to archiving, selected DRGs undergo various quality assurance procedures for color, coordinates, and so on.

The white area around the paper topographic map sheets outside the map's neatline is called the collar area. This area has information about the map title, scale, revision date, and so on. When the paper maps are scanned, the collar area is scanned as black "no data" pixels. The collar area must be manually cropped if the user wants to make a seamless mosaic of several DRGs. When cropping the maps, the voided areas of the map should be filled with the necessary adjacent map information to make the map and TIFF file rectangular. Some commercial data vendors and GIS data clearinghouses (e.g., PASDA) provide cropped-collar DRGs.

Raster image file size increases with image precision and resolution. For example, the 17.5 in. × 22.5 in. USGS topographic maps scanned at 250 dpi will have 17.5 × 250 × 22.5 × 250 = 25 million pixels. This will create a gray (8 bits/pixel) file size of 200 million bits or 25 million bytes or 24 MB (1 byte = 8 bits; 1MB = 1,048,576 bytes). Similarly, the color (16 bits/pixel) file size will be 400 million bits or 48 MB. The compressed DRG file size is 3–4 MB in zipped format and 7–10 MB in unzipped TIFF format. Modem download times vary from 15 minutes to several hours.

Digital Raster Graphics Applications

DRG applications include the following:

- GIS mapping
- Raster basemaps
- 3D draping

Aerial Photographs

USGS National High Altitude Photography (NHAP) provides 1:58,000 to 1:80,000 scale aerial photos taken at an aircraft altitude of approximately

40,000 feet above mean terrain. The more recent National Aerial Photography Program (NAPP) products were intended to replace the NHAP products (Garland et al., 1990). NAPP provides quarter-quadrangle centered aerial photos (3.75-min of latitude × 3.75-min of longitude in geographic extent) taken at an aircraft altitude of approximately 20,000 feet above mean terrain using a 152-mm focal-length camera. The scale of the NAPP photography is approximately 1:40,000.

DIGITAL ORTHOPHOTOS

Like DRGs, USGS Digital Orthophoto Quads (DOQs) can also be used as raster GIS basemaps. A USGS DOQ is an orthorectified raster image of a low-altitude aerial photograph in UTM projection in North American Datum of 1983 (NAD83). Based on the NHAP aerial photos, DOQs cover an area equal to 7.5-min (1:24,000) USGS quadrangles, hence their name.

The NAPP imagery and NAPP-like photography are the primary sources of aerial photography used in the production of 1-m Digital Orthophoto Quarter Quadrangles (DOQ_Q or DOQQ) for the National Digital Orthophoto Program (NDOP). DOQQs have a UTM map projection system and NAD83 datum. Their resolution is 3.2 ft (1-m) and their accuracy is 33.3 ft (10.15 m). Although this resolution is not high enough for road or bridge design, a civil engineer can use DOQQs for basic resource planning, for preliminary research, as an overview of the area surrounding a project, for modeling, or as a background for finer resolution data. Draping over, merging with, or laying other digital data with DOQs are common practices in the GIS industry (Murphy, 2000). DOQQs are available as 40 to 50 MB uncompressed files that can be purchased for $50–$70 per file on CD-ROM. Some special DOQQ editions are available as highly compressed JPG files. For example, USGS has created a DOQQ CD-ROM for Washington DC, which has each 40 to 50 MB TIFF file compressed to 4 to 5 MB JPG files. Although DOQQ images are usually of good quality, their availability is spotty and the imagery is often several years old (Gilbrook, 1999).

Figure 5-5 shows portions of a GIS basemap created by making a seamless mosaic of two DOQQs for a sewer system mapping project located in southwestern Pennsylvania. This figure also shows overlayed sanitary sewers and manholes. Figure 5-6 shows a snapshot from a 3D fly-through movie of Pittsburgh's downtown created by the University of Pittsburgh, Department of Geology and Planetary Science. The movie was created by draping a DOQ over a DEM using ERDAS Imagine Version 8.4 Virtual GIS Animation. The 3D animations and movies are very effective public presentation tools.

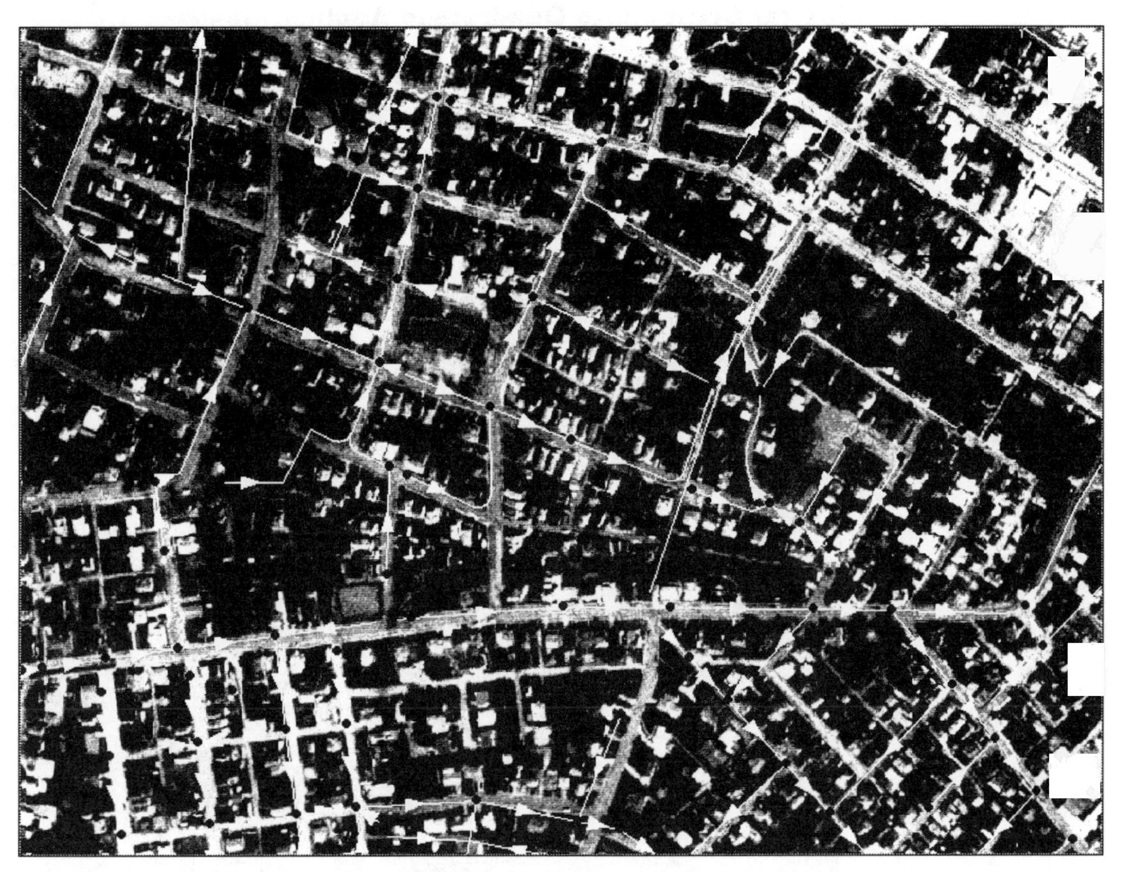

Figure 5-5. USGS DOQQ Image as a Basemap Overlayed by Sanitary Sewers and Manholes

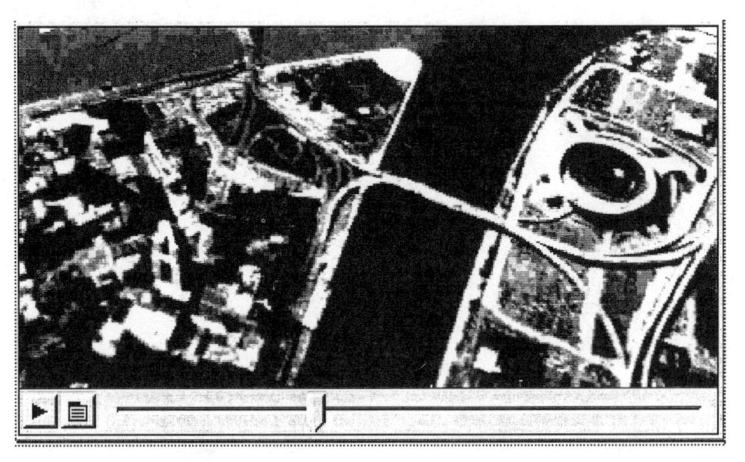

Figure 5-6. A DOQ Draped Over a DEM Showing Downtown Pittsburgh, Pennsylvania (Photo Courtesy of University of Pittsburgh)

Digital Orthophoto Quadrangle Applications

DOQ and DOQQ applications include the following:

- GIS mapping. For more information see Chapter 9.
- Raster basemaps
- 3D draping

DIGITAL LINE GRAPHS

USGS digital line graph (DLG) files are digital vector representations of cartographic information. Figure 5-7 shows a 1:24,000 DLG for West Rapid City (South Dakota, USA).

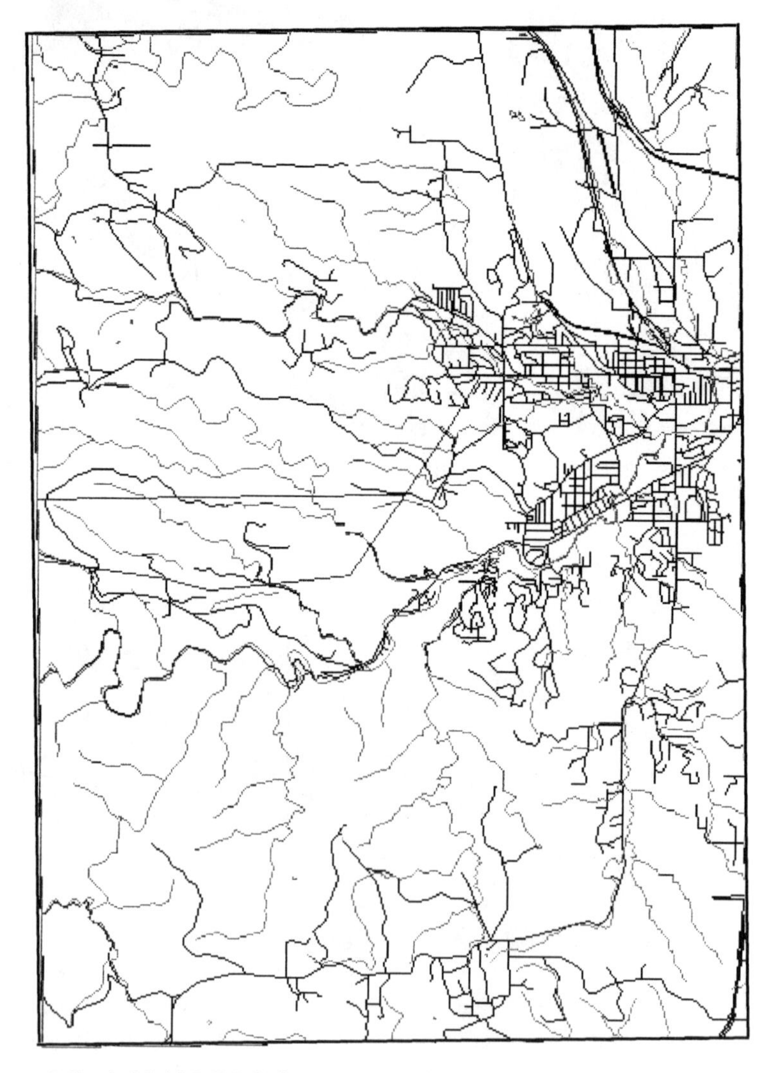

Figure 5-7. 1:24,000 DLG for West Rapid City (Map Courtesy of USGS)

DLG is a vector format developed by the USGS and the DLG files are not images (GIFs, JPEGs, etc.). Data files of topographic and planimetric map features are derived from either aerial photographs or cartographic source materials using manual and automated digitizing methods. DLGs are available in three scales: large, intermediate, and small.

- Large-scale DLG data are derived primarily from USGS 7.5-min topographic quadrangle maps at 1:24,000 and 1:25,000 scales (1:25,000 and 1:63,360 scales for Alaska).
- Intermediate-scale (100,000-scale) DLG data are derived from USGS 1:100,000-scale, 30 × 60 minute quadrangle maps. If these maps are not available, the DLG data are derived from Bureau of Land Management planimetric maps at 1:100,000 scale.
- Small-scale (1:2,000,000-scale) DLG data are organized by section or by state and contain information on planimetric base categories, including transportation, hydrography, and boundaries for all the 50 states.

Large-, intermediate-, and small-scale DLG data are useful for the production of cartographic products, such as basemaps. Also, the data are structured to support GIS applications. A typical use of base category digital cartographic data is to combine them with other geographically referenced data to conduct of various automated spatial analyses.

Large-scale DRGs resemble digitized USGS topo map layers. The DLG data files derived from the 1:24,000 scale maps contain selected base categories of cartographic data in digital form. These categories include

- Political boundaries (e.g., state, county, and city boundaries)
- Hydrography, including all flowing and standing water and wetlands
- Property boundaries (e.g., township, range, and park information)
- Transportation data (e.g., roads and trails, railroads, and pipelines)
- Other significant manmade structures (e.g., schools and churches)
- Hypsography (ground elevation contours)
- Land cover (e.g., woods, scrub, orchards, and vineyards)
- Non-vegetative surface features (e.g., lava and sand)

The DLG contour data can be processed to extract topography, slope, and shape of the watersheds and sewersheds. The DLG hydrography data provide information about the hydrologic features of a watershed. DLGs represent water features, such as streams ditches, and drainage facilities in more detail than EPA Reach File (RF) products described later. However,

DLG data may represent larger water bodies as double lines corresponding to the left and right banks of the river (Slawecki et al., 2001). Such representation may not be conducive to GIS mapping. RF data have a better representation of stream connectivity and network topology. DLG and RF data may not show all the streams because their hydrologic features are based on 1:24,000 and 1:100,000 hydrography (Slawecki et al., 2001). The NHD data described later provide more detailed information about hydrologic features.

DLG files have their own native file format, and cannot be read by every GIS software without data transformation. If you want to use the DLG files, you must have GIS software that has the ability to import DLG format. If your GIS cannot import a DLG file, you will have to convert the DLG file to a GIS-compatible file using one of the data conversion programs described in Chapter 4 (GIS Data).

DLG data are available in the following three formats:

1. Standard format, which was designed to minimize data storage. The topological linkages of the DLG standard format are only contained in the line elements. The files are composed of standard 8-bit ASCII characters organized into fixed-length records of 144 bytes. This format is no longer distributed.

2. Optional or DLG-3 format, which was designed for data interchange, allows for the creation of a vector polygon data structure. The topological linkages are explicitly encoded for node, area, and line elements.

3. In 1982, the National Committee for Digital Cartographic Data Standards (a precursor to FGDC) began work on a data conversion standard called Spatial Data Transfer Standard (SDTS). The standard was adopted in 1993 and federal agencies were mandated to provide data in this format starting in January 1994. SDTS is the latest USGS spatial data file format. SDTS is a robust way of transferring geo-referenced spatial data between dissimilar computer systems and has the potential for transfer with no information loss. Currently, FGDC is promoting the SDTS format because it provides the users with a higher level of detail than the DLG format. SDTS is a transfer standard that embraces the philosophy of self-contained transfers in which spatial data, attribute, georeferencing, data quality report, data dictionary, and other supporting metadata are all included in the transfer. The potential role for one standard that can accommodate all data types is obvious. However, initially SDTS did not find common use in GIS applications. This was partly due to its relatively recent adoption and also because the standard was nearly out of date by the time of its adoption (Huse, 1995). Although inexpensive and user-friendly,

SDTS conversion programs are not readily available to the general public, most GIS software now support the SDTS format (Murphy, 2000). SDTS data are available as tar.gz compressed files. Each compressed file contains 18 ddf files and two readme text files.

DLGs can be downloaded for free or ordered for $40–$60 per CD-ROM. The large-scale DLG data are useful in water, wastewater, and stormwater applications. They are available in the optional DLG-3 format on 8-mm magnetic tape, 3480 cartridge, CD-ROM, and through a semi-anonymous FTP (*edcftp.cr.usgs.gov*) at the EROS Data Center. The large-scale DLG data are also available in the SDTS format on CD-ROM and through anonymous FTP.

Since DLG data are in vector format, the users can attach attributes to DLG features. For example, DLGs can be edited to show property owners or road types, or a DLG of a tax parcel can be linked to a tax assessor's database (Murphy, 2000).

Hydrographic features such as streams or rivers can be defined as centerlines, dual lines, or polygons or as points such as springs or seeps. This leads to potential problems when digital hydrology is developed by different parties depending on their own definition of hydrology. Linking hydrographic data sets across jurisdictional boundaries then becomes a difficult process when each jurisdiction adopts a different hydrology definition. A centerline, dual line, or polygon representation of a stream each has a unique application (PaMAGIC, 2001).

DLG Applications

DLG applications include the following:

- GIS mapping
- Vector basemaps
- Source of cartographic data

DLG Web Sites

USGS DLGs	*edcwww.cr.usgs.gov/glis/hyper/guide/usgs_dlg*
24K DLGs	*edcwww.cr.usgs.gov/glis/hyper/guide/24kdlg*
EROS Data Center FTP	*edcftp.cr.usgs.gov*

LAND USE/LAND COVER

An accurate representation of the land surface is critical to a better characterization of a watershed or sewershed and the hydrologic processes that affect water quantity and quality through surface runoff and drainage. Land use/land cover (LULC) is, therefore, one of the most important data types for hydrologic modeling of watersheds and sewersheds.

A commonly used LULC classification method divides land use into broad categories such as urban, agricultural, forest, and water/wetlands. Such digital representations of land use are available at relatively coarse resolution (about 200 m) from the USGS covering the entire United States (Moglen, 2000). The USGS is the primary federal sources of LULC data in the United States. USGS LULC data are derived from thematic overlays registered to 1:250,000 scale basemaps and a limited number of 1:100,000-scale basemaps. Figure 5-8 shows a 1:250,000 LULC map overlayed with census and political unit boundaries for Spokane (Washington, USA).

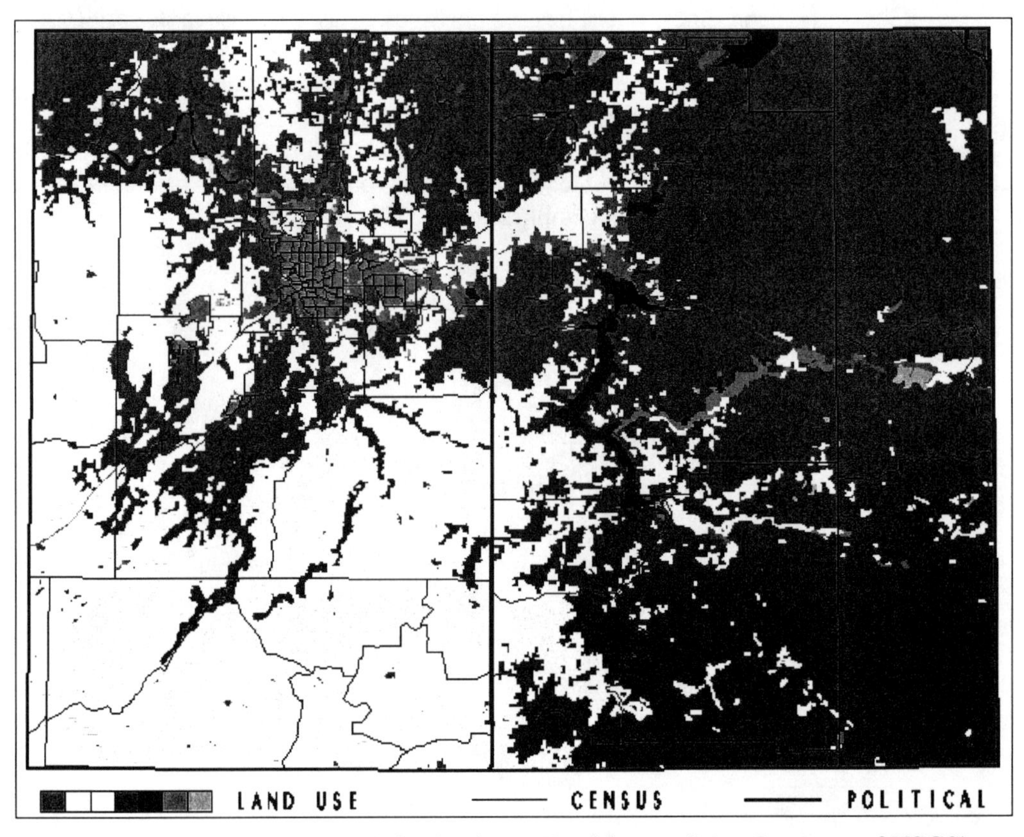

Figure 5-8. 1:250,000 LULC for Spokane, Washington (Map Courtesy of USGS)

LULC data provide information on urban or built up land, agricultural land, rangeland, forest land, water, wetlands, barren land, tundra, and perennial snow or ice. Associated maps display information in five data categories: (1) political units, (2) hydrologic units, (3) census county subdivisions, (4) Federal land ownership, and (5) state land ownership. LULC files are available in Modified UTM projection. LULC files are also available from EPA in the Albers projection system, as described later in this chapter.

The LULC data are available in two formats: GIRAS (vector) and CTG (raster). Geographic information retrieval and analysis system (GIRAS) format involves a standard character fixed-length, usually ASCII-coded, 80-character, logical record. The GIRAS file structure is composed of a map header, section header, arc records subfile, coordinate subfile, polygon records subfile, file of arcs by polygon (FAP) subfile, text subfile, and an associated data subfile. The Composite Theme Grid (CTG) format involves representing data in raster or grid cell form for a given quadrangle. The CTG files are sequential and consist of fixed length records (except for header files) with one grid cell for each logical record. The grid cells are actually regular point samples of the quad where the center point of each cell is 200 meters apart from other center points in adjacent cells. The cells are mapped to the UTM projection and oriented as north-south and east-west. The properties of USGS LULC data are summarized in Table 5-4.

Table 5-4. Properties of USGS Land Use/Land Cover Data

Item	*Property*
Coverage	Available for most of the contiguous United States and Hawaii
Source	National LULC database known GIRAS
Minimum mapping unit	10 acres (4 ha) for urban categories and 40 acres (16 ha) for non-urban categories
Image date	mostly 1970s
Image type	Aerial photographs
Map scale	1:250,000 and 1:100,000
Photo scale	1:24,000 to 1:100,000
Coordinate system	UTM
Projection	Geographic
Classification system	Anderson Level II
GIS format	GIRAS, CTG
Data type	Vector and raster

Land Use/Land Cover Applications

LULC applications include the following:

- Land use change detection: For more information see Chapter 8 (Water System Applications) and Chapter 10 (Stormwater System Applications).
- Planning: For more information see Chapter 8.
- Estimation of stormwater quantity and quality: For more information see Chapter 10.
- Estimation of water consumption (demand) rates.

LULC Web Site

LULC Data	*edcwww.cr.usgs.gov/glis/hyper/guide/1_250_lulc*

HYDROLOGIC UNIT CODE WATERSHEDS

Hydrologic Unit Code (HUC) watershed data provide the boundaries of the major watersheds of the United States. The United States is divided and sub-divided into successively smaller hydrologic units that are classified into four levels: regions, sub-regions, accounting units, and cataloging units. The hydrologic units are arranged within each other, from the smallest (cataloging units) to the largest (regions). Each hydrologic unit is identified by a unique HUC consisting of two to eight digits based on the four levels of classification in the hydrologic unit system. There are 21 regions, 222 sub-regions, 352 hydrologic accounting units, and 2,150 cataloging units (watersheds) in the United States.

HUCs are available in two scales: 1:2M (1:2,000,000) and 1:250K (1:250,000).

- *1:2,000,000-scale hydrologic units (huc2m)*: This is a sophisticated coverage with extensive documentation. Attribute tables show hydrologic unit names and flow direction among cataloging units. The coastline is included and land/water polygons are distinguished. Unofficial offshore extensions of cataloging units are shown. This coverage was developed for work on the National Water Summary.

- *1:250,000 scale hydrologic units (huc250k)*: The data for this large coverage were originally collected for the GIRAS at a scale of 1:250K. Some areas, notably major cities in the west, were recompiled at a scale of 1:100K. The coverage was compiled to provide the National Water Quality Assessment (NAWQA) study units with an intermediate scale river basin boundary for extracting other GIS data layers. These data files include only HUC codes; watershed names are not included.

Figure 5-9 shows the HUC boundaries for two regions HUC02 (Mid Atlantic) and HUC05 (Ohio) that cover the state of Pennsylvania. The HUC coverages are available in the ArcInfo E00 format at no charge from *water.usgs.gov/lookup/getspatial?huc250k*. HUC data may be retrieved as a single file for the entire United States, or by water resources region. More HUC information is available at the HUC home page.

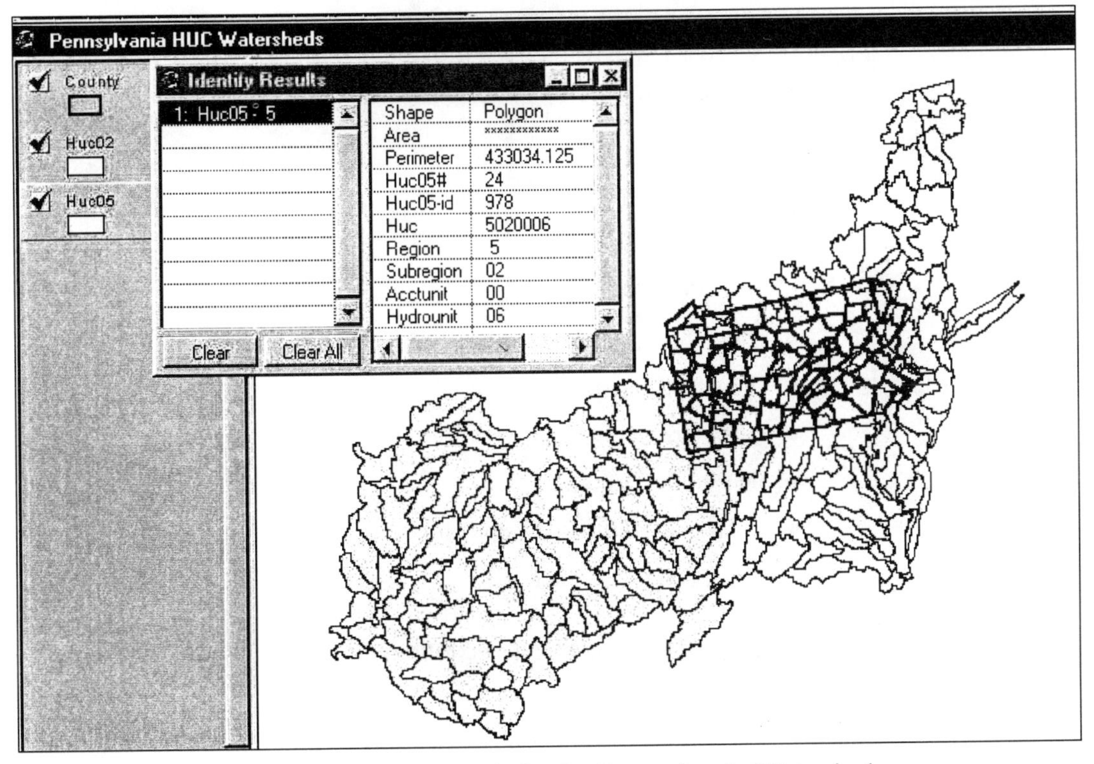

Figure 5-9. HUC Boundaries for Pennsylvania Watersheds

Hydrologic Unit Code Applications

HUC applications include the following:

- Delineation of watershed boundaries
- Stormwater management
- Watershed studies
- Nonpoint source (NPS) pollution studies
- Water quality data management

HUC Web Sites

HUC home page	*Water.usgs.gov/GIS/huc.html*
HUC download	*Water.usgs.gov/lookup/getspatial?huc250k*

Environmental Protection Agency Data

The U.S. EPA provides two types of spatial data services

1. GIS data
   - Better assessment science integrating point and nonpoint sources (BASINS) data
   - LULC
   - Reach files
2. On-line mapping
   - Envirofacts Warehouse
   - EnviroMapper
   - EPA Spatial Data Library System
   - Maps On Demand
   - National Shape File Repository
   - Locational Information

EPA GIS data are available in the following scales:

- County: 1:100,000
- State: 1:250,000
- State/National: 1,200,000

BASINS Data

EPA's BASINS software supports the development of total maximum daily loads (TMDLs) using a watershed-based approach that integrates both point and nonpoint sources. BASINS integrates ArcView GIS with national watershed data, and state-of-the-art environmental assessment and modeling tools into a convenient package. It is a free public domain software. Additional BASINS information is provided in Chapter 7 (Modeling Integration).

BASINS 3.0 is the current (2001) release. It comes with a Web data extractor that allows users to automatically download many national databases on water quality, reach files, NHD, HUC, and so on. The software also has a CD-ROM data extractor for selecting 8-digit HUCs within any of the five EPA regions of the United States.

BASINS Web site (*www.epa.gov/ost/basins/*) provides access to a variety of conveniently packaged meteorological data (e.g., hourly precipitation, temperature, and evapotranspiration) and GIS data (e.g., DEM, LULC, STATSGO, and Reach File Version 3 or RF3). BASINS data also include locations of National Pollution Discharge Elimination System (NPDES) permits, National Oceanic and Atmospheric Administration (NOAA) rain gauges, and USGS stream gauges. This site allows data download for a user-specified watershed.

BASINS Web Sites

BASINS home page	*www.epa.gov/ost/basins/*
Web download	*www.epa.gov/waterscience/basins/b3webdwn.htm*
FTP data download	*www.epa.gov/ost/ftp/basins/*

EPA's Land Use/Land Cover Data

EPA's LULC data are different from the USGS LULC data. Properties of EPA LULC data are summarized in Table 5-5. Many states in the United States have more detailed LULC data based on interpolation at 1:24,000 scale or better. The EPA maintains a digital data directory of the States' LULC data in the form of a clickable map available from their Web site. Data cost and availability for instant download vary from state to state (Slawecki, 2001).

Land Use/Land Cover Applications

HUC applications include the following:

- GIS mapping
- Land use change detection
- Planning
- Estimation of stormwater quantity and quality
- Estimation of water consumption (demand) rates

EPA's LULC Web Site

LULC data	*www.epa.gov/OWOW/watershed/landcover/lulcmap.html*

Table 5-5. Properties of EPA's Land Use/Land Cover Data

Item	*Property*
Coverage	U.S. 48 contiguous states, verified to 110 Longitude line
Source	Derived from GIRAS data without edge matching
Minimum mapping unit	10 acres for urban categories, 40 acres for others
Map date	1994
Image date	mostly 1970s
Image type	aerial photographs, B/W
Map scale	1:250,000
Photo scale	1:250,000
Coordinate system	UTM
Projection	Albers Equal Area
Classification system	Anderson Level II
GIS format	ArcInfo
Data Type	Vector
Availability	Free of charge off the Internet at *www.epa.gov/nsdi*

REACH FILES

A reach is defined as confluence to confluence extent of surface water. Reach Files are a series of national hydrologic databases that uniquely identify and interconnect the stream segments or "reaches" that comprise the United States' surface water drainage system. The three versions of the Reach File that currently exist are known as RF1, RF2, and RF3. They were created from increasingly detailed sets of digital hydrography data produced by the USGS. The EPA enhanced these hydrography datasets by assigning a unique reach code to each stream segment, determining the upstream/downstream relationships of each reach, and, when possible, identifying the stream name for each reach. A variety of other reach-related attributes that support mapping and spatial analysis applications, such as the stream level attribute, are also available. USGS also provides an enhanced version of RF1 data.

EPA's Reach File is a national database of surface water features each identified by a unique reach code. As the Reach File versions have progressed from RF1 to RF3, the scale has improved from 1:500,000 to 1:100,000 and the reach number has increased from 68,000 to 3.2 million. Reach File data format is not very user-friendly and may be too complex for the average GIS user. NHD and ESRI have plans to develop ArcInfo coverages of the Reach File.

Reach Files are provided with EPA's BASINS program. BASINS 2.0 provides the option of modeling stream networks based on RF1 or RF3 Reach File. Basins 3.0 also allows using NHD data. The user defines the Reach File to be modeled when creating (delineating) or importing watershed boundaries. The higher-resolution RF3 allows users to delineate smaller subwatersheds as well as model a more detailed stream network, enabling the user to better assess the distribution and impact of NPS pollutants within a watershed. BASINS reach data allows estimation of stream characteristics, such as channel geometry, width, depth, velocity, and Manning's "n" (roughness coefficient) at bank-full flow and at any other user-specified lower flow.

GIS applications of RF3 files are possible because

- They have coordinates that are placed in a downstream-to-upstream order (tracing the drainage areas)
- They contain names for major rivers and many stream and lakes
- They can help determine where connected water bodies exist when original user data sources show topological discontinuity

Reach File Applications

Reach File applications include the following:

- GIS mapping
- Hydrologic modeling
- Stormwater management
- Watershed studies
- NPS pollution studies
- TMDL modeling

Reach File Web Sites

EPA Reach Files	*www.epa.gov/owow/monitoring/rf/rfindex.html*
USGS enhanced RF1 data	*water.usgs.gov/lookup/getspatial?erf1-2*

ON-LINE MAPPING

On-line mapping services are interactive Internet mapping Web sites and data warehouses that allow creation of a hard copy or image of a thematic map based on user-specified data layers. These maps can be used in decision making, planning, reports, and presentations. Although the maps are created from the Internet GIS applications, no further GIS applications are possible from the map images or hard copies. In the future, EPA's on-line mapping services plan to generate GIS compatible vector files, such as ArcView shapefiles, that can be used as data layers in a GIS project. EPA's online mapping services are listed below.

Envirofacts Warehouse

The EPA created the Envirofacts Warehouse to provide the public with direct access to the wealth of information contained in its databases. The Envirofacts Warehouse allows retrieval of environmental information from EPA databases on air, chemicals, facility information, grants/funding, hazardous waste, risk management plans, Superfund, toxic releases, water permits, drinking water, drinking water contaminant occurrence, and drinking water microbial and disinfection byproduct information (information collection rule or ICR). Information may be retrieved from several databases at once, or from one database at a time. On-line queries allow retrieval of data from these sources and create reports and generation of maps of environmental information by selecting from several mapping applications available through EPA's Maps On Demand described below. Envirofacts Permit Compliance System query page can be used to retrieve NPDES discharger information, including monitoring data and facility location (Slawecki, 2001).

EnviroMapper

Before EnviroMapper EPA's legacy environmental systems were not integrated. EnviroMapper presents a geographic interface for integrating regulatory and monitoring data collected in the Federal government's national systems. EnviroMapper will also benefit other U.S. national databases, such as those from USGS and U.S. Fish and Wildlife Service (specifically the National Wetlands Inventory) that can be accessed through EnviroMapper. For example, the USGS Water Resources Division can use EnviroMapper to identify latitude/longitude coordinates for the intake of water supply facilities (ESRI, 1999). EnviroMapper provides a Web-based interface to environmental information in the Envirofacts database. It places desktop GIS capabilities in the hands of anyone who has access to the Internet, which means it is available to those who do not have desktop GIS software. EnviroMapper is available through the Envirofacts warehouse.

Maps on Demand

The Maps On Demand suite of applications (EnviroMapper, Query Mapper, SiteInfo, BasinInfo, CountyInfo, and ZipInfo) enables users to generate maps displaying environmental information for the entire United States at the national, state, and county levels. These applications integrate data available through the Envirofacts Warehouse, including EPA Spatial Data Library System (ESDLS, described below), the Envirofacts database, EPA-processed U.S. Bureau of Census (USBC) demographic data, and the NGDC. The ESDLS and Envirofacts Warehouse provide user-friendly and free downloads from *www.epa.gov/enviro/index_java.html.* Maps on Demand requires submitting a map request, getting an E-mail notification within 4 hours, and retrieve the map from the Web site. Figure 5-10 shows a map printed using Maps on Demand's SiteInfo service. This map shows watershed boundaries and public water supply locations for a study area in southwestern Pennsylvania.

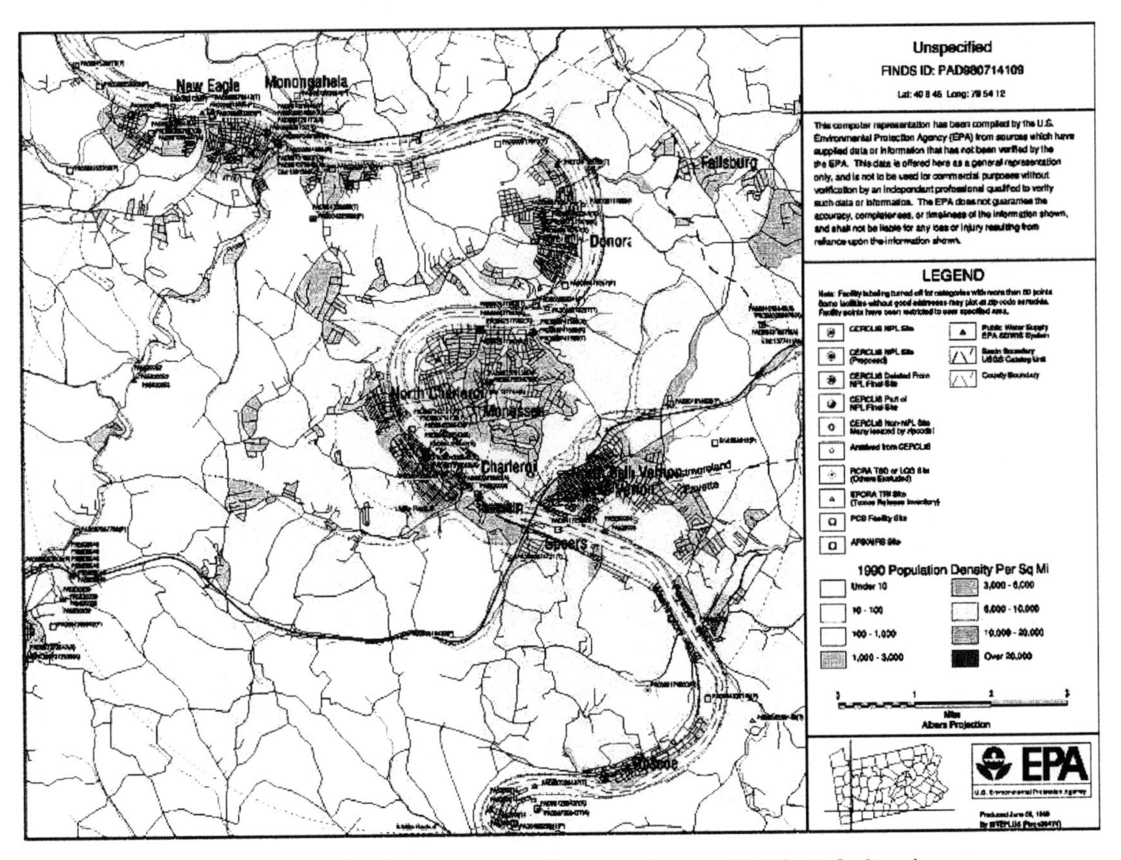

Figure 5-10. Map Printed Using Maps on Demand's SiteInfo Service

Spatial Data Library System

EPA Spatial Data Library System (ESDLS) is a repository for the EPA's new and legacy geospatial data holdings. Data sets are contained at the county, state, and national levels covering the conterminous United States, Alaska, Hawaii, Puerto Rico, and the Virgin Islands. The data are used in the Maps On Demand applications to generate user-specified maps of environmental information. This system allows EPA to access its geospatial data in ArcInfo format and integrates these data in a standardized, consistent, and sustainable framework. In addition, the system also applies standards and guidelines for management and use while managing the permit discovery and retrieval of EPA's geospatial data and associated GIS applications.

On-Line Mapping Web Sites

Envirofacts Warehouse	*www.epa.gov/enviro/index_java.html*
Maps on Demand	*www.epa.gov/enviro/html/mod/index.html*
EnviroMapper	• *maps.epa.gov/enviromapper* • *www.epa.gov/enviro/html/em/index.html*
Query Mapper	*www.epa.gov/enviro/html/multisystem_query_java.html*

NATIONAL SHAPEFILE REPOSITORY

The National Shapefile Repository contains data from USGS, U.S. Department of Transportation, ESDLS, Wessex, and Geographic Data Technologies (GDT). Assorted spatial data layers from ESDLS were converted into ArcView shapefile format, as required by EPA's Office of Information Resources Management. These spatial data sets were then loaded into a national compilation called the National Shapefile Repository. One of the Maps On Demand applications, EnviroMapper, uses this information to provide users with interactive GIS functionality and view spatial data at the national, state, and county levels, as well as display multiple spatial layers, zooming, panning, identifying features, and querying single Envirofacts points. EnviroMapper enables users to visualize environmental data in Envirofacts, view detailed reports for EPA-regulated facilities, eliminate batch-processing mapping, and generate maps dynamically.

National Shape File Repository Web Site

National Shapefile Repository	*www.epa.gov/enviro/html/nsf_index.html*

LOCATIONAL INFORMATION

The Locational Reference Tables are a repository for locational (i.e., latitude and longitude) information that has been collected and documented through EPA's Locational Data Improvement Project. The primary objective of this effort is to identify, collect, verify, store, and maintain an accurate, consistently documented set of locational data for entities of environmental concern. A secondary objective is to support the necessary infrastructure for integration across national, regional, tribal, and state systems. The project intends to support EPA's movement toward data integration based on location, thereby promoting the use of EPA's data resources for a wide array of cross-media analysis, such as community-based ecosystem management and environmental justice. Envirofacts Query can be used to access these latitude/longitude coordinates. The Facility and Geographic Locational Reference Information EZ Query can be used to focus on particular latitude/longitude elements to build a query.

Locational Information Web Site

Locational Reference Tables	*www.epa.gov/enviro/html/locational/lrt/index.html*

USGS AND EPA JOINT DATA

The following data represent joint efforts between USGS and EPA.

NATIONAL LAND COVER DATABASE

Formed in 1992, a multi-agency consortium called Multi-Resolution Land Characteristics (MRLC) Consortium is currently (2002) updating the nationwide LULC coverage for the United States. In this effort, USGS and EPA have teamed up to compile the first seamless National Land Cover Dataset (NLCD) using the 30-m Landsat Thematic Mapper satellite imagery for the conterminous United States. At 30-m resolution, the NLCD is the most detailed land cover information ever compiled at a national level. The CD-ROMs released in 2000 include data for the states east of the Mississippi and Ohio Rivers. Subsequent versions will provide data for the remaining states which are currently being assessed for data accuracy. Once completed, NLCD will be the first national land-cover dataset produced since the early 1970s, effectively replacing the land-cover data known to many as LUDA (land use data) or GIRAS. Available data have been used for many environmental applications, including land use planning, hydrological analyses, and habitat assessments.

NLCD data can be used in a GIS for assessing wildlife habitat, water quality, pesticide runoff, land use change, and so on. However, the user must have a firm understanding of how the datasets were compiled and the resulting limitations of these data. NLCD raster images represent a new type of digital format. They can be downloaded from the USGS FTP site or purchased on a CD-ROM.

The Seamless Data Distribution System (SDDS) provides custom-generated digital products for NLCD and NED data based on user specified geographic extents and user specified datasets. Seamless digital data are available to users in several optional formats for delivery via FTP downloads or CD media. Figure 5-11 shows the user-friendly interface of the SDDS Web site.

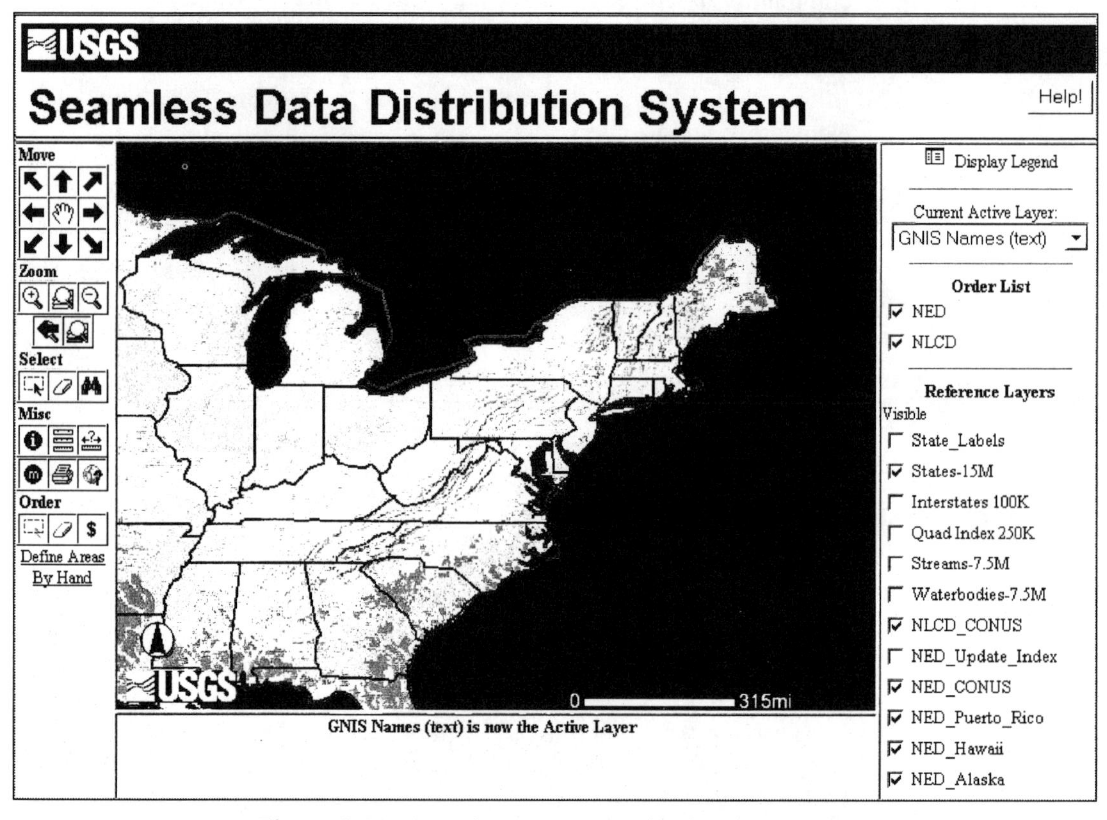

Figure 5-11. Seamless Data Download Using SDDS

The North American Landscape Characterization (NALC) project is another EPA-USGS collaborative effort to provide complete coverage of the conterminous United States and Mexico for mapping LULC changes.

National Land Cover Database Applications

NLCD applications include the following:

- GIS mapping
- Land use change detection
- Planning
- Estimation of stormwater quantity and quality
- Estimation of water consumption (demand) rates

NLCD Web Sites

MRLC	*www.epa.gov/mrlc/data.html*
FTP download	*edcwww.cr.usgs.gov/pub/edcuser/vogel/states/*
CD-ROM orders	*edcwww.cr.usgs.gov/programs/lccp/mrlcreg.html*
SDDS	*edcnts14.cr.usgs.gov/Website/seamless.htm*

National Hydrography Dataset

In partnership with the EPA, the USGS has completed the National Hydrography Dataset (NHD), a geospatial database of all surface water in the conterminous United States. Based on 1:100,000-scale maps, the database is a notable advance from previous databases in that all water features are networked together to establish the flow of water throughout the Nation. NHD data are of vital importance to resource managers, environmental researchers, and regulatory agencies because the flow of water can be traced from origin to destination at any point along the Nation's waterways, making it easier to understand upstream sources and downstream impacts of water quality. EPA is using the new dataset for its Watershed Assessment Tracking and Environmental Results system (*www.epa.gov/waters*).

NHD is a comprehensive set of digital spatial data that contains information about surface water features such as lakes, ponds, streams, rivers, springs, and wells. NHD provides a more comprehensive model of streams than those described earlier in this chapter, such as DLG and RF data. It represents the stream network as a collection of surface water features connected by junctions. Within the NHD, surface water features are combined to form "reaches" that have a permanent identification number (reach ID). The reach ID is the key for unambiguously relating features of other data sets, such as sampling locations and wastewater outfalls, to the river network (Lanfear, 2000). Reaches provide the framework for linking water-related data to the NHD surface water drainage network. These linkages enable the analysis and display of water-related data in upstream and downstream direction.

NHD is based upon the content of USGS DLG hydrography data integrated with reach-related information from the EPA RF3. The NHD supersedes DLG and RF3 by incorporating them, not by replacing them. Users of DLG or RF3 will find the NHD both familiar and greatly expanded and refined. While initially based on 1:100,000-scale data, the NHD is designed to incorporate and encourage the development of higher resolution data required by many users. NHD data are considered to be highly attributed and very accurate.

NHD data can be downloaded at no charge from the NHD home page along with demonstrations, tutorials, technical references, and data tools.

National Hydrography Dataset Applications

NHD applications include the following:

- GIS mapping
- Hydrologic modeling
- Stormwater management
- Watershed studies
- NPS pollution studies
- TMDL modeling

NHD Web Site

NHD home page	*nhd.usgs.gov/data.html*

Census Bureau TIGER Data

Every 10 years in the United States, U.S. Bureau of Census (USBC) creates a census geography that defines the areas it uses to tabulate its counts. The census provides a 100% count of population and housing subjects in the United States. Population items tabulated include age, race, sex, marital status, origin, household type, and household relationship. These data are cross-tabulated by age, race, sex, or Hispanic origin. Housing items tabulated include occupancy and vacancy status, tenure, units in structure, value, contract rent, meals included in rent, and number of rooms in housing unit. Other population characteristics available are ancestry, language, housing, and education. USBC's Population Estimates

Program produces annual, quarterly, and monthly estimates for the United States, areas within the United States (states and counties), and Puerto Rico (Thrall, 2001a).

The digital socio-demographic data are available in USBC Topographically Integrated Geographic Encoding Referencing System (TIGER) files. TIGER digital map database was developed to support USBC's mapping needs for the Decennial Census and other Bureau programs. TIGER automates the mapping and related geographic activities required to support the census and survey programs of USBC.

USBC tabulates decennial census information at three levels: tracts, blocks, and block groups

1. Census tract is a small, relatively permanent statistical subdivision of a county in a metropolitan area (MA) or a selected non-metropolitan county. It is delineated by a local committee of census data users called Census Statistical Area Committee (CSAC) for the purpose of presenting decennial census data. Census tract boundaries typically follow visible features, but may follow governmental unit boundaries and other nonvisible features in some instances. They always nest within counties. Designed to be relatively homogeneous units with respect to population characteristics, economic status, and living conditions at the time the CSAC established them, census tracts usually contain between 2,500 and 8,000 inhabitants.

2. Census block is the smallest entity for which USBC collects and tabulates decennial census information. In dense urban areas, census blocks correspond with individual city blocks. In rural areas, they typically represent larger areas. A census block is bound on all sides by visible and nonvisible features shown on the USBC maps. Census blocks provide a significantly higher resolution than block group for basic demographic and housing data.

3. Census block group consists of several census blocks. A census group is larger than a census block but smaller than a census tract. Figure 5-12 shows the difference between census blocks and block groups.

The data for blocks and block groups are extensive. Block attributes are the more basic, collected for 100% of the people and/or housing units in the smaller unit blocks. The block level demographic parameters are extracted from the public law demographic database. Block group attributes, on the other hand, are more complex demographic data that are sampled within the larger composite block groups to maintain a higher degree of privacy.

Figure 5-12. Difference Between Census Blocks and Census Block Groups

The scale of TIGER data is 1:100,000, which may not be suitable for high-precision measurement applications such as engineering problems, property transfers, or other uses that might require highly accurate measurements of the Earth surface.

TIGER/LINE FILE

A TIGER file is an extract of selected geographic and cartographic information from the TIGER database. In order for others to use the information in the TIGER database in a GIS, the USBS releases periodic extracts of this database to the public known as TIGER/Line files. The most recent version is TIGER/Line 2000. From the GIS applications perspective, TIGER/Line is one of the most important products produced by the USBC. The TIGER/Line files contain the following features for the entire United States:

- Geographic features, such as roads, railroads, rivers, and lakes
- Political boundaries, such as cities and townships
- Statistical boundaries (collection geography), such as census tracts and blocks

The TIGER/Line database contains information about these features, such as location in latitude and longitude, name, type of feature, address ranges and ZIP codes for most streets, and geographic relationship to other features, and other related information. TIGER file also includes the outlines of the census blocks.

A TIGER/Line file contains data describing three major types of features:

- Line features
  1. Roads
  2. Railroads
  3. Hydrography
  4. Transportation and utility lines

- Boundary features
  1. Statistical boundaries, such as census tracts and blocks
  2. Local government boundaries, such as places and counties
  3. Administrative boundaries, such as congressional and school districts

- Landmark features
  1. Point landmarks, such as schools and churches
  2. Area landmarks, such as parks and cemeteries
  3. Key geographic locations, such as apartment buildings and factories

TIGER/Line files are not graphic images of maps, but rather digital data in vector format. To make use of these data, you must have mapping or GIS software that can read TIGER/Line data. USBC does not provide these data in any vendor-specific format. The TIGER/Line product does not include demographic statistics.

The TIGER/Line format has been used by USBC, with some modifications, for more than a decade to distribute the information in its TIGER database to the public. This format is not the most efficient for distributing geospatial data because it does not allow for easy data enhancement and rapid application development (Dougherty, 2000). Original TIGER/Line data are unprojected, i.e., they are stored in a geographic (decimal-degree) coordinate system. GIS applications of TIGER data, therefore, may require projection of TIGER data into other projection systems.

Most TIGER files start with “tgr” and have a predefined format (e.g., tgr21017.f61). The five digits after tgr refer to a specific geographic area, such as Federal Information Processing Standards (FIPS) code + county

FIPS code. The last character in the file extension refers to the type of records in the file. TIGER 95 files have a *.bw* extension (e.g., tgr21017.bw1).

Before the 2000 census, census geography was released in hard copy maps and TIGER/Line files. These TIGER/Line files had to be translated into other formats to be used in a GIS, which can be a cumbersome process. Because TIGER/Line data had its own native file format, it could not be read directly in a GIS without a prior data conversion step using one of the data transformation programs described in Chapter 4 (GIS Data). For example, ArcView GIS cannot read TIGER files directly, but other options are available. ArcInfo and Data Automation Kit (DAK) can process TIGER files and create datasets that will work with ArcView GIS (Hansen, 1998). Alternatively, one may use a data conversion program, such as the TIGER to ArcView shapefile (TGR2SHP) utility from GIS Tools (*www.gistools.com*).

USBC created approximately 8.5 million blocks for Census 2000 (Dougherty, 2000). Census 2000 is the first census in which people can access maps down to the block level without having to order paper maps from USBC. The new Census 2000 TIGER/Line data released in 2001 were greatly enhanced. TIGER/Line 2000 benefits from the massive filed work carried out in the past few years in preparation for the 2000 Census. Users can expect a significant improvement in address-range information, completeness, and feature coverage (Kavaliunas and Dougherty, 2000). Free download of TIGER/Line 2000 data in shapefile format is available from ESRI's Geography Network Web site *www.geographynetwork.com/data/tiger2000/*.

TIGER DATA AVAILABILITY

TIGER data can be downloaded from the USBC Web sites given below. The USBC Web site currently offers all 3,200 TIGER/Line 1998 and 1999 files for free downloading. The USBC also sells the TIGER data on CD-ROMs at $1,500 for 6 CD-ROMs covering all the states or $250 for one CD-ROM containing approximately 8 states.

TIGER Map Service (TMS), an on-line mapping service, provides a public resource for generating high-quality, detailed maps of anywhere in the United States, using public geographic data. This is an on-line interactive mapping service that allows creating and printing census tract and block group level thematic maps as an image (GIF) file. Although these image files can be used in decision making, reports, and presentations, no significant GIS applications are possible. The American FactFinder Web site is a similar service from USBC.

A substantial amount of TIGER data are also distributed with ESRI GIS software. For example, ArcView Version 3.2 comes with five CD-ROMs called "1999 ESRI Data and Maps" that include boundaries and attributes for census tracts and census block groups. A selection of 1990 Census attributes from Summary Tape File 1 and 1997 population estimates from Claritas are also included as basic demographic information for census tracts and block groups. For the most detailed assessment of population, census block centroids and their 1990 population are also included for more than 7 million blocks. These data can be used to create detailed population density maps. Intergraph's GeoMedia GIS also provides TIGER data in Microsoft Access MDB database format along with their sample GIS data.

ESRI's ArcData Online provides both free and commercial data that may be licensed and downloaded. Although ArcData Online is part of ESRI's commercial site, it can be considered a contribution to the GIS community. ArcData provides free download of the TIGER/Line 1995 data in compressed shapefile format for a selected county or data layer. Figure 5-13 shows census block data for Washington County, Pennsylvania, downloaded from ArcData Online and imported in ArcView GIS.

Washington County (PA) Census Blocks

Blk42125.shp

Identify Results

1: Blk42125.shp - 42.125.0

Shape	Polygon
Record_id	1
Key	42.125.007140.101
State90	42
County90	125
Tract90	007140
Block90	101
Sumlev	100
Mcd90	65376
Place90	
Vtd90	1530

Clear | Clear All

Attributes of Blk42125.shp

Shape	Record_id	Key	State90	County90	Tract90	Block90	Sumlev	Mcd90	Place90	Vtd90
Polygon	1	42.125.007140.101	42	125	007140	101	100	65376		1530
Polygon	2	42.125.007140.102	42	125	007140	102	100	65376		1530
Polygon	3	42.125.007140.103	42	125	007140	103	100	65376		1530
Polygon	4	42.125.007140.104	42	125	007140	104	100	65376		1530
Polygon	5	42.125.007140.105	42	125	007140	105	100	65376		1530
Polygon	6	42.125.007140.106	42	125	007140	106	100	65376		1530
Polygon	7	42.125.007140.107	42	125	007140	107	100	65376		1530
Polygon	8	42.125.007140.108	42	125	007140	108	100	65376		1530

b9042125.dbf

Housing	Hu_dens	One_unit	Units2_9	Ten_unit	Rooms	Occupied	Owneroc	Rentocc	Meanvalu	Aggvalu	Meanrent	Singlehh	Oneparent	Key
1	10.28	1	0	0	8.00	1	0	1	0	0	0	0	0	42.125.007140.101
4	412.37	3	1	0	7.80	4	4	0	71300	285200	0	0	1	42.125.007140.102
18	2727.27	17	1	0	5.60	17	13	4	63500	1143000	363	3	1	42.125.007140.103
23	901.96	23	0	0	5.50	22	19	3	44800	1030400	246	6	2	42.125.007140.104
7	1129.03	7	0	0	6.00	7	7	0	64200	449400	0	0	0	42.125.007140.105
15	1500.00	15	0	0	4.80	14	9	5	46700	700500	243	5	1	42.125.007140.106
17	1465.52	16	1	0	5.30	17	15	2	49200	836400	175	2	3	42.125.007140.107

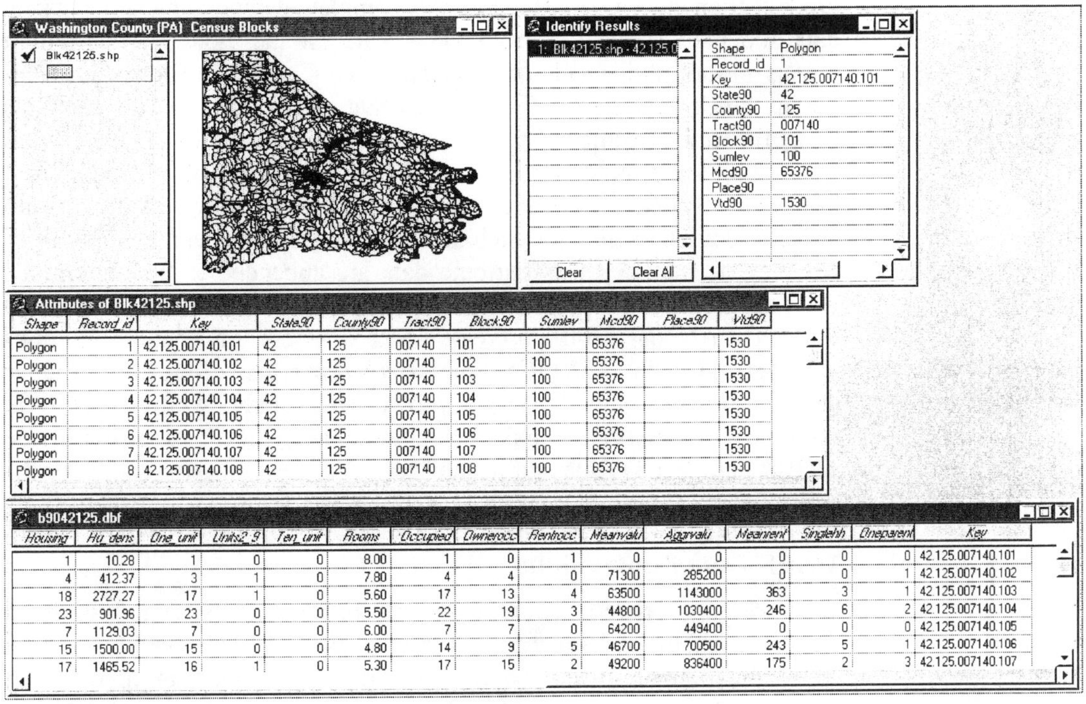

Figure 5-13. TIGER Census Block Data for Washington County, Pennsylvania, Imported in ArcView GIS

Note that the bottom table in Figure 5-13 is an external table that contains extensive census attributes for demographic parameters. This external table has been linked to the Census Block theme table shown in the middle using the common attribute "Key" (third column of the middle table and the last column of the bottom table).

ArcData Online also offers an enhanced version of the TIGER data, the GDT Dynamap/2000 database. GDT has performed extensive updates to many of the TIGER data layers to add new geographic features, improve accuracy and representation of existing features, and update feature attributes such as address ranges. The cost of the GDT's TIGER data is $15 for a single ZIP code or $10 each for multiple ZIP codes.

Projected and enhanced TIGER data in GIS-compatible files can be purchased from commercial data vendors as described in Chapter 4 (GIS Data). Commercial TIGER data are more expensive than the free USBC data, but the data enhancements (e.g., double-line roads), data projections, and GIS-compatible file formats may justify the additional cost. Commercial data vendors provide more accurate and current TIGER roadway network data for address-matching applications. Commercially repackaged data may be cost effective because it saves the time and cost associated with data conversion and re-projection. CensusCD Blocks product from GeoLytics, Inc. provides complete demographic and housing data and map boundaries from the USBC for the more than 7 million census blocks nationwide on a single CD. CensusCD Blocks provides census block GIS data with attributes that can be exported to ArcView Shapefile and MapInfo. The cost of CensusCD Blocks is $1,000 for all states and $500 for a single state CD-ROM. CACI One from CACI Marketing Systems, allows extraction of Census 2000 data along with polygon boundary files for each scale of geography in either MapInfo or ArcView format. Census attribute data are appended to the appropriate polygons (Thrall, 2002). CACI Marketing Systems Group was acquired by ESRI in January 2002 to create ESRI Business Information Solutions (ESRI BIS) Group. ESRI BIS (*www.esribis.com)* also offers 200 variables for current-year and five-year projections, along with current-year estimates for daytime population. Some local agencies also provide TIGER data on CD-ROM for a small fee for a user-specified county or the entire state.

TIGER Data Applications

With the appropriate software a user can produce maps ranging in detail from a neighborhood street map to a map of the United States. To date, many local governments have used the TIGER data in applications requiring digital street maps. Software companies have created software products that allow consumers to produce their own detailed maps. There are many other possibilities, such as

- GIS mapping
- Source of cartographic data
- Vector basemaps
- Water and sewer system planning
- Estimation of quantity and quality of domestic sanitary sewage
- Estimation of water consumption (demand) rates
- Estimation of present and potential future development
- Creation of auxiliary watershed themes for visualization purpose

Estimation of sewage quantity and quality is required when modeling combined and sanitary sewer areas in urban sewersheds. For example, Storm Water Management Model's (SWMM) TRANSPORT Block may need input for some or all of the following demographic parameters to estimate quantity and quality of domestic sanitary sewage:

- Dwelling units
- Persons per dwelling unit
- Market value of average dwelling unit
- Average family income

All of these input parameters can be estimated from the TIGER data.

TIGER Web Sites

U.S. Decennial Census	*www.census.gov/main/www/cen2000.html*
TIGER/Line data download	*www.census.gov/geo/www/tiger/index.html*
	www.census.gov/ftp/pub/geo/www/tiger/
ESRI Census Watch	*www.esri.com/censuswatch*
ESRI ArcData Online	*www.esri.com/data/online/tiger/index.html*
Census 2000 shapefiles	*www.geographynetwork.com/data/tiger2000/*
Population Estimates	*eire.census.gov/popest/data/counties.php*
TIGER Map Service	*tiger.census.gov*
American FactFinder	*factfinder.census.gov*
GeoLytics	*www.geolytics.com* **or** *www.censuscd.com*
Wessex	*www.wessex.com/wessex/index.cfm*
Intergraph GeoMedia	*www.intergraph.com/gis/demos/*

HOUSING AND URBAN DEVELOPMENT E-MAPS

U.S. Department of Housing and Urban Development (HUD) Healthy Communities Environmental Mapping E-Maps provide on-line information about HUD and EPA projects in communities throughout the United Sates. Data can be viewed from any community by using a variety of HUD and EPA categories. Maps can be scaled from a full region all the way down to the neighborhood level. E-Maps are an on-line interactive mapping service like USBS's TIGER Map Service or EPA's Envirofacts. E-Maps consist of layers, which are discussed below.

HUD PROGRAM LAYERS

- Public and Native American housing
- Multifamily housing
- Community development projects

CENSUS BUREAU LAYERS

- Demographic data for states, counties, and census tracts

EPA LAYERS

- Superfund sites
- Brownfields site assessment pilots
- Brownfields tax incentive zones
- Air Releases/Aerometric Information Retrieval System (AIRS) facilities
- Toxic Releases/Toxic Release Inventory (TRI) facilities
- Hazardous waste handlers/Resource Conservation and Recovery Information System (RCRIS) facilities
- Discharges to water/Permit Compliance System (PCS) facilities
- Hazardous waste generators/Biennial Reporting System (BRS) facilities

E-Maps Web Site

HUD E-Maps	*www.hud.gov/emaps*

NATURAL RESOURCES CONSERVATION SERVICE SOILS DATA

Soils data are used to estimate infiltration and erosion both of which affect the quantity and quality of runoff from watersheds and sewersheds.

Based on their resolution, there are three types of U.S. Natural Resources Conservation Service (NRCS) (formerly the U.S. Soil Conservation Service or SCS) soils data that are useful in GIS applications:

- National Soil Geographic (NATSGO)
- State Soil Geographic (STATSGO)
- Soil Survey Geographic (SSURGO)

SSURGO provides the highest resolution soils data at scales ranging from 1:12,000 to 1:63,360. This resolution is appropriate for watersheds a few squares miles in area. STATSGO data are digitized at 1:250,000 scale, which is useful when analyzing large regional watersheds. NATSGO data describe variations in soil type at the multi-state to regional scale, which is not suitable for wastewater and stormwater modeling applications (Moglen, 2000). Currently, the NRCS clearinghouse has approximately 1,100 digital soil datasets on-line with many more being processed.

STATSGO

STATSGO is a soil maps database designed for use in a GIS. This data set consists of georeferenced digital map data and attribute data. The map data are collected in 1- by 2-degree topographic quadrangle units and merged and distributed as statewide coverages. STATSGO represents a digital general soil association map developed by the National Cooperative Soil Survey.

STATSGO consists of a broad-based inventory of soils and non-soil areas that occur in a repeatable pattern on the landscape and that can be cartographically shown at the scale mapped. The soil maps for STATSGO are compiled by generalizing more detailed soil survey maps. Where more detailed soil survey maps are not available, data on geology, topography, vegetation, and climate are assembled, together with LANDSAT satellite images. Soils of like areas are studied, and the probable classification and extent of the soils are determined. Map unit composition for a STATSGO map is determined by transecting or sampling areas on the more detailed maps and expanding the data statistically to characterize the whole map unit.

The STATSGO attributes are contained in the 16 relational tables shown in Table 5-6. File No. 2 (comp.dbf) contains data for hydrologic soils group (A, B, C, D). File No. 6 (layer.dbf) contains data for soil texture (sand, silt, and so on). The hydrologic soil group and texture data can be used to estimate watershed subbasin hydrologic parameters, such as runoff curve number and Green-Ampt infiltration parameters for input to hydrologic models.

Table 5-6. STATSGO Database Relational Tables

No.	*Name*	*Description*
1	codes.dbf	Database code: stores information on all codes used in the database
2	comp.dbf	Map unit component: stores information that will apply to a specific component of a soil map unit
3	compyld.dbf	Component crop yield: stores crop yield information for soil map unit components
4	forest.dbf	Forest understory: stores information for plant cover as forest understory for soil map unit components
5	interp.dbf	Interpretation: stores soil interpretation ratings (both limitation ratings and suitability ratings) to soil map unit
6	layer.dbf	Components layer: stores characteristics that apply to soil layers for soil map unit components
7	mapunit.dbf	Map unit: stores information that applies to all components of a soil map unit
8	plantcom.dbf	Plant composition: stores plant symbols and percentage of plant composition associated with components of soil map units
9	plantnm.dbf	Plant name: stores the common and scientific names for plants used in the database
10	rsprod.dbf	Range site production: stores range site production information for soil map unit components
11	taxclass.dbf	Taxonomic classification: stores the taxonomic classification for soils in the database
12	windbrk.dbf	Windbreak: stores information on recommended windbreak plants for soil map unit components
13	wlhabit.dbf	Wildlife habitat: stores wildlife habitat information for soil map unit components
14	woodland.dbf	Woodland: stores information on common indicator trees for soil map unit components
15	woodmgt.dbf	Woodland management: stores woodland management information for soil map unit components
16	yldunits.dbf	Yield units: stores crop names and the units used to measure yield

Each STATSGO map is linked to the Soil Interpretations Record (SIR) attribute database. The attribute database gives the proportionate extent of the component soils and their properties for each map unit. The STATSGO map units consist of 1 to 21 components each. The SIR database includes more than 25 physical and chemical soil properties, interpretations, and productivity. Examples of information that can be queried from the database are available water capacity, soil reaction, salinity, flooding, water table, bedrock, and interpretations for engineering uses, cropland, woodland, rangeland, pastureland, wildlife, and recreation development.

Figure 5-14 shows a STATSGO soils map and database imported in ArcView GIS for a study area in southwestern Pennsylvania.

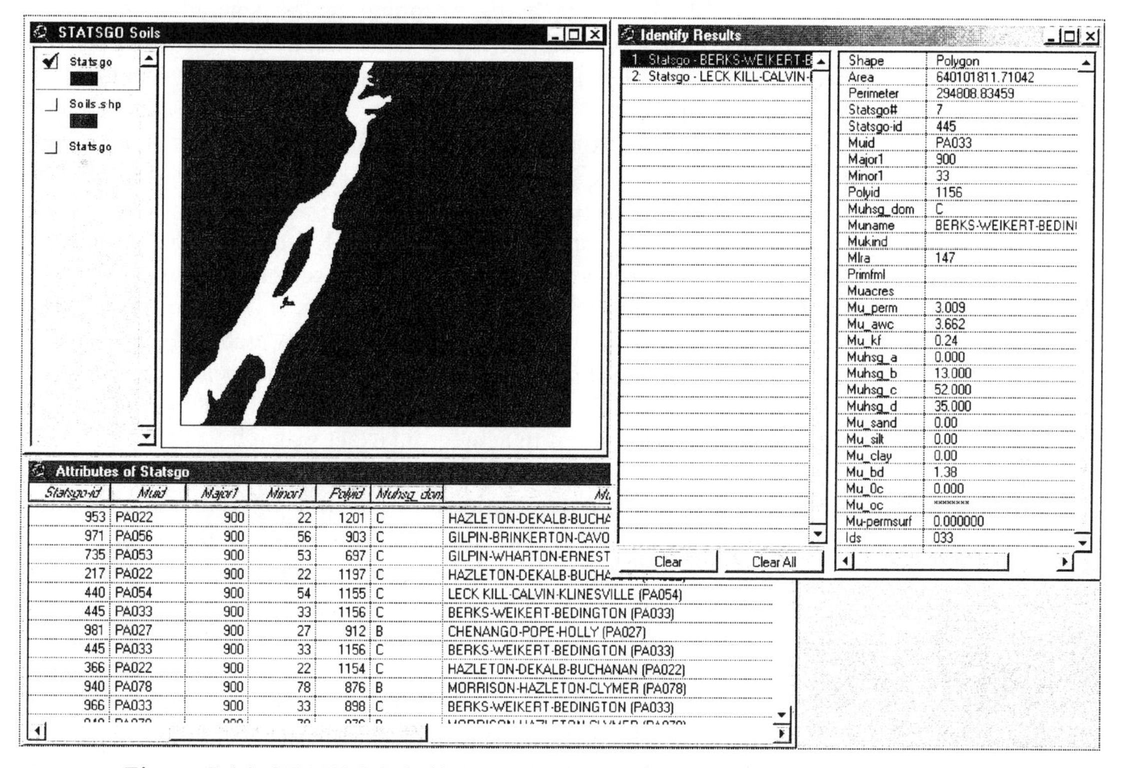

Figure 5-14. STATSGO Soils Map And Database Imported Into ArcView GIS for a Study Area In Pennsylvania

STATSGO data are available for the conterminous U.S., Hawaii, and Puerto Rico. STATSGO uses 1:250,000 USGS topographic quadrangles as basemap, so the mapping scale for the STATSGO geographic data is also 1:250,000. This scale is more suitable for regional planning applications covering state and multi-state areas. Each quadrangle area contains 100 to 400 soil polygons. The smallest mapped area is about

1,500 acres. STATSGO data are available in USGS DLG-3 optional distribution format, ArcInfo coverage, and GRASS vector formats (ASCE, 1999).

STATSGO data can be downloaded free from the local GIS data clearinghouse sites. For example, PASDA provides free STATSGO data for Pennsylvania in the ArcInfo exchange (E00) format. NRCS provides STATSGO data for 49 states and Puerto Rico on a CD-ROM for $50. Data for Alaska is available on a separate CD-ROM.

SSURGO

SSURGO is the most detailed level of soil mapping done by NRCS. Field mapping methods using national standards are used to construct the soil maps in the SSURGO database. Mapping scales generally range from 1:12,000 to 1:63,360; SSURGO digitizing duplicates the original soil survey maps. This level of mapping is designed for use by landowners, townships, and county natural resource planning and management. SSURGO users are expected to be familiar with soils data and their characteristics.

Digitizing is done by line segment (vector) format in accordance with NRCS digitizing standards. The basemaps meet national map accuracy standards and are either orthophoto quadrangles or 7.5-min topographic quadrangles. SSURGO data are collected and archived in 7.5-min quadrangle units, and distributed as complete coverage for a soil survey area. Soil boundaries ending at quadrangle neatlines are joined by computer to adjoining maps to achieve an exact match.

SSURGO is linked to the National Map Unit Interpretation Records (MUIR) attribute database. The attribute database gives the proportionate extent of the component soils and their properties for each map unit. The SSURGO map units consist of one to three components each. MUIR data contain about 88 estimated soil physical and chemical properties, interpretations, and performance data. These include available water capacity; soil reaction; soil erodibility factors (K, Kf, and T); hydric soil ratings; ponding, flooding, water table depth and duration; bedrock; interpretations for sanitary facilities, building site development, engineering, cropland, woodland, and recreational development; and yields for common crops, site indices of common trees, and potential production of rangeland plants.

The map extent for a SSURGO data set is a soil survey area, which may consist of a county, multiple counties, or parts of multiple counties. A SSURGO data set consists of map data, attribute data, and metadata. SSURGO map data are available in modified DLG-3 optional and Arc interchange file formats. Attribute data are distributed in ASCII format with DLG-3 map files and in ArcInfo interchange format with Arc interchange map files. Metadata are in ASCII format. SSURGO data can be downloaded from the NRCS Web site given below or ordered on a CD-ROM as described above for the STATSGO data.

STATSGO and SSURGO Data Applications

STATSGO and SSURGO data applications include the following:

- Soil mapping
- Estimation of soil infiltration parameters
- Estimation of runoff curve numbers
- Runoff estimation
- Hydrologic modeling

NRCS Web Sites

STATSGO	*www.ftw.nrcs.usda.gov/stat_data.html*
SSURGO Download	*www.ftw.nrcs.usda.gov/ssurgo_ftp3.html*
SSURGO Data Access	*www.ftw.nrcs.usda.gov/ssur_data.html*
National Soil Survey	*www.statlab.iastate.edu/soils/nsdaf/*

FEDERAL EMERGENCY MANAGEMENT AGENCY FLOOD DATA

The Federal Emergency Management Agency's (FEMA) Map Service Center (MSC) provides on-line distribution of their products. MSC products include: Digital Flood Insurance Rate Maps (DFIRM), Flood Insurance Rate Maps (FIRM), Flood Insurance Study (FIS) reports, Digital Q3 flood data, Community Status Book, Flood Map Status Information Service (FMSIS), Letters of Map Change (LOMC), and National Flood Insurance Program (NFIP) insurance Manuals. DFIRM and Q3 data have GIS applications and are described below.

DIGITAL FLOOD INSURANCE RATE MAPS DATA

The DFIRM is composed of all digital data required to create the hardcopy FIRM. This includes basemap information, graphics, text, shading, and other geographic and graphic data required to create the final hardcopy FIRM product to FEMA FIA-21 standards and specifications. This product serves the purpose of map design and provides the database from which the flood risk thematic data are extracted to create the DLG file for DFIRM. This product is generally produced in a county-wide format.

Specifications for digitizing DFIRMs are consistent with those required for mapping at a scale of 1:24,000 or larger.

DIGITAL Q3 FLOOD DATA

Digital Q3 Flood Data are developed by scanning the existing FIRM hardcopy and vectorizing a thematic overlay of flood risks. Vector Q3 Flood Data files contain only certain features from the existing FIRM hardcopy. These features include

- Annual chance floodplain areas of 1 and 0.2% (100- and 500-year return periods), including Zone V areas, certain floodway areas, and zone designations
- Political areas, including community identification numbers
- FIRM panel areas, including panel number and suffix
- USGS 7.5-min corner points and neatline
- Mapable LOMCs

Q3 vector data are contained in one single county-wide file, including all incorporated and unincorporated areas of a county. Q3 flood data are also available from ESRI's ArcData Online for approximately 1,200 counties across the United States. The data may be downloaded in compressed shapefile format for a selected county. In 2002, the price of the data was $35 per county.

HAZUS

FEMA, under a cooperative agreement with the National Institute of Building Sciences (NIBS), has developed a standardized, nationally applicable earthquake-loss estimation methodology implemented through GIS software called HAZUS. HAZUS is an add-in for MapInfo or ArcView and also includes data about point-loss estimates for specific facilities (hospitals, fire and police stations, schools, and bridges) from winds (hurricanes, thunderstorms, tornadoes, extratropical cyclones, and hailstorms) floods (riverine and coastal), and earthquakes. Geographic scale of HAZUS data is at the census-tract level for the entire United States (Thrall, 2001a).

FEMA Data Web Sites

FEMA Map Service Center	*web1.msc.fema.gov/MSC/*
HAZUS	*www.fema.gov/hazus/*
ESRI ArcData Online	*www.esri.com/data/online/fema/femadata.html*

National Wetlands Inventory Data

The National Wetlands Inventory (NWI) of the U.S. Fish and Wildlife Service produces information on the characteristics, extent, and status of the Nation's wetlands and deepwater habitats. NWI Center provides free downloading of wetlands data in a variety of formats, including DLG and Arc Export.

The data are available as 7.5-min quadrangle maps that have been digitized and converted to DLG format. The scale of the data is 1:24,000. As of this writing (March 2002) about 44% of the lower 48 states and 13% of Alaska have been digitized. The data is organized by USGS 1:250,000 map name, so the user should refer to the USGS index book for his/her state. The NWI Web site also provides an interactive on-line mapping tool for finding wetlands. Wetlands data can be merged with a LULC layer to increase the accuracy of land cover classes.

NWI Web Sites

National Wetlands Inventory Center	*wetlands.fws.gov*
	enterprise.nwi.fws.gov

National Imagery and Mapping Agency Data

The U.S. National Imagery and Mapping Agency (NIMA) produces and distributes Mapping, Charting and Geodesy (MC&G) products to the Armed Forces and to all other national security operations. Among these products are paper maps and charts of many scales, printed aeronautical and nautical information, and geodetic and gravity data. NIMA products also include MC&G digital data on computer-readable media such as magnetic tape and CD-ROM.

The raster products supported by NIMA include

- Arc digitized raster graphics (ADRG)
- Arc digitized raster imagery (ADRI)
- Controlled image base (CIB)
- Compressed aeronautical chart (CAC)
- Compressed arc digitized raster graphics (CADRG)
- Digital bathymetric data base 5 minute (DBDB5)
- Digital bathymetric data base variable spaced (DBDBV)
- Digital terrain elevation data (DTED)

The International raster products supported by NIMA include

- Arc standard raster product (ASRP) edition 1.2 (United Kingdom)
- UTM standard raster product (USRP) (France)
- Standard raster graphics (SRG) (Italy)
- Compressed raster product (CRP) (United Kingdom)
- IT Format (IT) (Denmark)

The NIMA vector data products called Vector Product Format (VPF) include

- Digital chart of the world (DCW)
- Digital nautical chart (DNC)
- World vector shoreline (WVS)
- Vector smart map series (VMAP0 through 2)
- Urban vector map (UVMAP)
- Tactical terrain data (TTD)
- Digital flight information publication (DFLIP)

NIMA MC&G Utility Software Environment (NIMAMUSE) Version 2.1 is a self-contained set of free computer programs and computer utilities designed to work with MC&G data and information. It provides new and experienced users with three distinct activities that can be performed with NIMA data: (1) build your own map, (2) access and prepare NIMA digital data, and (3) run specialty applications.

NIMAMUSE Web Site

NIMAMUSE Download Page	*www.nima.mil/geospatial/SW_TOOLS/NIMAMUSE/*

SUMMARY

This chapter provided information about public domain GIS data that are available on the Internet and are generally useful in developing GIS applications for water, wastewater, and stormwater systems. Major sources of free GIS data included USGS, EPA, USBC, and NRCS Web sites. Local and regional GIS data clearinghouse Web sites are also useful. At a minimum, GIS users should search and bookmark at least one clearinghouse Web site for their own state, county, and city.

SELF EVALUATION

1. Using your favorite search engine, search the Internet for Web sites that provide free public domain data suitable for use in your GIS software. Use key words such as, GIS, data, shapefile, free, download, and so on. Then document your findings in a comparison matrix (table) with Web sites as rows and their features as columns. Based on your matrix, identify the best Internet site. Some suggested site features are:

   - Data type (e.g., DRG, counties, and watersheds)
   - File format (e.g., shapefile, DXF, TIFF, and JPEG)
   - Projection
   - Scale
   - Resolution
   - Free download? (Yes or No)
   - Map printing? (Yes or No)
   - Interface (e.g., user-friendly or confusing?)
   - Data extent (e.g., all states or only your state?)

2. Download a DOQQ and corresponding metadata for one of your project sites from the Internet. Is it accurate enough for: (1) a construction project (2) a sewer mapping project, and (3) a watershed hydrologic modeling project?

3. Create a list showing the names and addresses (URLs) of the GIS data Web sites for your state, county, and city.

Chapter 6

GIS DATABASE DESIGN

A GIS database, though not as visible as a GIS map, is an equally critical GIS component and must be designed carefully. The design of a database, like that of a GIS map, is driven by the intended GIS application.

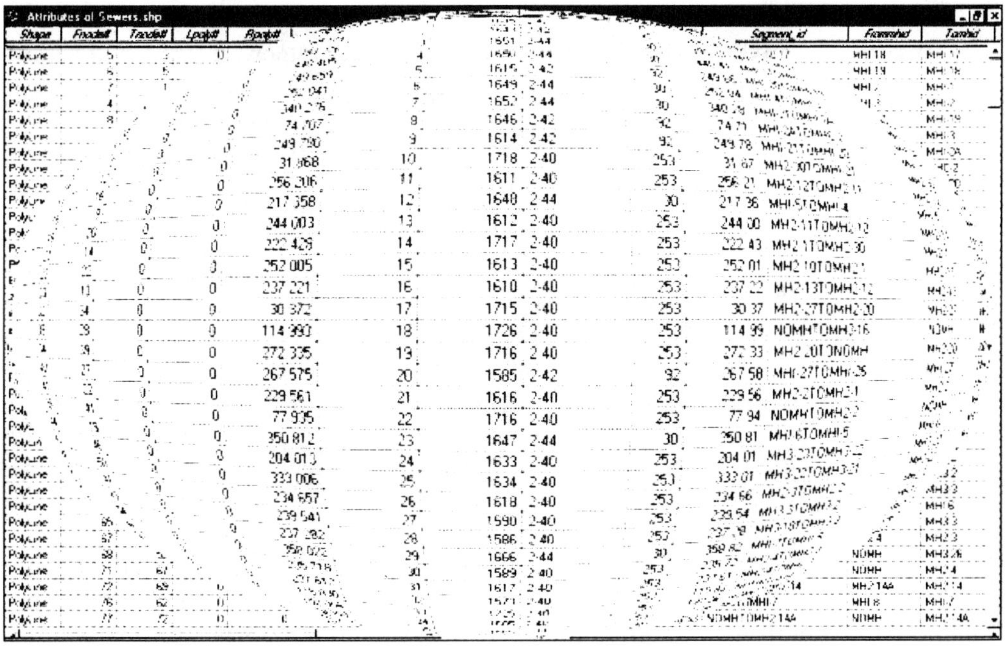

Attributes of Sewers.shp

Shape	Fnode#	Tnode#	Lpoly#	Rpoly#							Segment_id	Fromnhd	Tomhd
Polyline	5	[illegible]	0	[illegible]		4	1690	2.44	[illegible]	[illegible]	[illegible]	MH1-18	MH1-17
Polyline	6	5	[illegible]	[illegible]		5	1615	2.42	92	[illegible]	[illegible]	MH1-19	MH1-18
Polyline	7	1	[illegible]	[illegible]	252.041	6	1649	2.44	[illegible]	252.04	[illegible]	[illegible]	[illegible]
Polyline	4	[illegible]	[illegible]	[illegible]	340.275	7	1652	2.44	30	340.28	[illegible]	[illegible]	[illegible]
Polyline	8	[illegible]	[illegible]	0	74.707	8	1646	2.42	92	74.71	[illegible]	[illegible]	[illegible]
Polyline	[illegible]	[illegible]	[illegible]	0	249.790	9	1614	2.42	92	249.78	[illegible]	[illegible]	MH1-3
Polyline	[illegible]	[illegible]	3	0	31.868	10	1718	2.40	253	31.67	[illegible]	[illegible]	[illegible]
Polyline	[illegible]	[illegible]	[illegible]	0	256.206	11	1611	2.40	253	256.21	MH2-12TOMH2-11	[illegible]	[illegible]
Polyline	[illegible]	[illegible]	0	0	217.358	12	1648	2.44	30	217.36	MH1-5TOMH1-4	[illegible]	[illegible]
Polyline	[illegible]	[illegible]	0	0	244.003	13	1612	2.40	253	244.00	MH2-11TOMH2-10	[illegible]	[illegible]
Polyline	[illegible]	14	0	0	222.429	14	1717	2.40	253	222.43	MH2-1TOMH2-30	[illegible]	[illegible]
Polyline	[illegible]	[illegible]	0	0	252.005	15	1613	2.40	253	252.01	MH2-10TOMH2-9	[illegible]	[illegible]
Polyline	[illegible]	11	0	0	237.221	16	1610	2.40	253	237.22	MH2-13TOMH2-12	[illegible]	[illegible]
Polyline	[illegible]	34	0	0	30.372	17	1715	2.40	253	30.37	MH2-27TOMH2-20	[illegible]	[illegible]
Polyline	[illegible]	28	0	0	114.990	18	1726	2.40	253	114.99	NOMHTOMH2-16	NOMH	[illegible]
Polyline	[illegible]	39	0	0	272.335	19	1716	2.40	253	272.33	MH2-20TONOMH	MH2-20	[illegible]
Polyline	[illegible]	27	0	0	267.575	20	1585	2.42	92	267.58	MH1-27TOMH1-25	MH1-27	[illegible]
Polyline	[illegible]	[illegible]	0	0	229.561	21	1616	2.40	253	229.56	MH2-2TOMH2-1	[illegible]	[illegible]
Polyline	[illegible]	[illegible]	0	0	77.935	22	1716	2.40	253	77.94	NOMHTOMH2-2	NOMH	[illegible]
Polyline	[illegible]	[illegible]	0	0	350.812	23	1647	2.44	30	350.81	MH1-6TOMH1-5	[illegible]	[illegible]
Polyline	[illegible]	[illegible]	0	0	204.013	24	1633	2.40	253	204.01	MH3-23TOMH3-22	[illegible]	[illegible]
Polyline	[illegible]	[illegible]	0	0	333.006	25	1634	2.40	253	333.01	MH3-22TOMH3-21	[illegible]	MH3-2
Polyline	[illegible]	[illegible]	0	0	234.657	26	1618	2.40	253	234.66	[illegible]	[illegible]	MH3-3
Polyline	[illegible]	[illegible]	[illegible]	0	239.541	27	1590	2.40	253	239.54	[illegible]	[illegible]	MH1-6
Polyline	66	[illegible]	[illegible]	0	237.282	28	1586	2.40	253	237.28	[illegible]	[illegible]	MH3-3
Polyline	67	[illegible]	[illegible]	[illegible]	358.822	29	1666	2.44	30	358.82	[illegible]	[illegible]	MH2-3
Polyline	68	[illegible]	[illegible]	[illegible]	[illegible]	30	1589	2.40	253	[illegible]	[illegible]	NOMH	MH3-26
Polyline	71	67	[illegible]	[illegible]	[illegible]	31	1617	2.40	253	[illegible]	[illegible]	NOMH	MH2-4
Polyline	72	69	0	[illegible]	[illegible]	[illegible]	1571	2.40	[illegible]	[illegible]	[illegible]14	MH2-14A	MH2-14
Polyline	76	62	0	[illegible]	[illegible]	[illegible]	[illegible]	2.40	[illegible]	[illegible]	[illegible]TOMH1-7	MH1-8	MH1-7
Polyline	77	72	0	0	[illegible]	[illegible]	[illegible]	2.40	[illegible]	[illegible]	NOMHTOMH2-14A	NOMH	MH2-14A

LEARNING OBJECTIVE

The learning objective of this chapter is to design a GIS database with a special emphasis on listing water, wastewater, and stormwater devices, features, and attributes. Major topics discussed in this chapter include

- Database types
- Assets databases
- Applications databases
- Database design standards, steps, and examples
- Data dictionaries
- Object-oriented databases
- Geodatabases

HONOLULU'S GIS DATABASE

Application	Wastewater Information Management System (WIMS)
Year Published	1998
Reference	Ono et al., 1998
GIS software	ArcInfo 7.0.1 Application Development Framework
Hardware	IBM RS/6000, Hewlett-Packard DesignJet 650C
Source of data	Honolulu Land Information System (HoLIS)
Study area	Honolulu (Hawaii, USA)
Organization	City and County of Honolulu, Hawaii

The City and County of Honolulu developed one of the most comprehensive GIS databases for any municipality of its size. The Honolulu Land Information System (HoLIS) is an enterprise-wide system serving more than 14 city departments with land use, permit, tax, infrastructure, and environmental data. Geographically referenced information links existing city records to precise locations on the island of Oahu for spatial query and analysis.

HoLIS provides a rich, full source of geographically referenced information. The GIS serves the public by collecting, maintaining, and distributing georeferenced information necessary to support city operations. Now this information is also being made available to the public via the Internet (*www.co.honolulu.hi.us/planning/gis/index.html*) to promote its use for developing local business and marketing opportunities. HoLIS water and wastewater applications include:

- Automated Permit Management and Tracking System (POSSE)
- Wastewater Information Management System (WIMS)
- Water Pipe Replacement Analysis

These applications can print maps showing all sewer pipes based on pipe diameter for each mainline segment. Such GIS maps and their associated databases are being used to plan improvements to existing wastewater infrastructure that will protect beaches and bays from sewer discharge and promote more efficient sewer permitting programs.

GIS Database Definition

The creation of an appropriate GIS database is the most difficult and expensive part of developing GIS applications. A GIS database represents how things are on the surface of the Earth by using binary digits to approximate the real world (Goodchild, 1998). It stores descriptive information about the map features as attributes (ESRI, 1992). For example, a utility database contains the objects that compose the utility network. A water system database includes attributes for pipes, valves, meters, hydrants, and so on. A sewer system database contains attributes for pipes, manholes, catch basins, outfalls, and so on.

The database concept is central to a GIS and is the main difference between a GIS and a CADD system. Successful GIS management requires knowledge of more than just mapping. In addition to networking, programming, and GIS analysis, a GIS manager must also understand GIS database design, data management, and data conversion techniques (Zimmer, 2001). The design of a database is crucial to the consistency, integrity, and accuracy of its data. An improperly designed database may not retrieve certain types of information or may retrieve inaccurate information. Inaccurate information is probably the most detrimental result of improper database design (Hernandez, 1997).

Databases store information about features and their relationships with each other. The database design specifies the file structure for each graphic and attribute data file and defines how the files will be linked together (Wells, 1991). The GIS database design is the basis for GIS database development and implementation. A GIS database, though not as visible as the GIS features (map), is an equally critical GIS component. The most difficult job in creating a GIS is the enormous effort required to enter the large amount of data and to ensure both its accuracy and proper maintenance (Walski and Male, 2000). Moreover, construction of an appropriate GIS database is the most expensive part of developing GIS applications. Database creation costs usually account for two-thirds or more of overall costs (Gilbrook, 1999). However, because the GIS costs are dropping steadily, most utilities may be able to afford a GIS (Walski and Male, 2000). Successful GIS applications require a database that provides appropriate information in a useful and accessible form. The design of the database is, therefore, driven by application needs.

Conventional GIS databases consist of graphic features with links or pointers (usually facility identification numbers, or IDs) to related attribute or tabular data. Modern object-oriented databases described later do not require links between features and attributes because they can store both the features and attributes in the same file. The features are usually organized into a series of vector layers consisting of points, lines, polygons, or objects that are registered to a geographic control framework and a set of base information called a basemap. The common reference base provides registration between geographic features, which enables combinations of layers to be overlayed, viewed, analyzed, and plotted together (Cannistra, 1999).

DATABASE TYPES

The main objective of a water or sewer system database is to create a decision support system for efficient infrastructure asset management. Two types of databases are required to accomplish this goal: assets databases and applications databases.

ASSETS DATABASES

An assets database contains data for the utility assets or infrastructure inventory data that define the physical characteristics of the system. Table 6-1 shows a list of typical water, wastewater, and stormwater devices.

Table 6-1. Typical Water, Wastewater, and Stormwater Devices

Water	*Wastewater*	*Stormwater*
Flowmeter	Emergency equipment	Flowmeter
Hydrant	Flowmeter	Motor
Motor	Generator	Pump
Pressure gauge	Grease separator	Vault
Pump	Motor	Weir
Service meter	Pump	
Valve	Regulator	
Vault	Reuse meter	
	Tide gate	
	Valve	
	Weir	

Assets data make a GIS and are, therefore, necessary to develop a GIS. Because assets data should be present in a GIS before any applications are developed, they can be considered "necessary" data. In addition to utility network components like pipes and manholes, assets also include devices like valves and pumps.

Assets may also be defined by structure types. For example, a stormwater drainage structure may be defined as a structure in which surface water enters or exits through one or more connected drainage pipes. Manholes, catch basins, drop inlets, junction boxes, and headwalls are examples of stormwater drainage structures. A sanitary sewer structure may be defined as a structure in which sanitary sewage enters or exits through one or more connected sewer pipes. Manholes, pumps, and junction boxes are examples of sanitary sewer structures.

Before creating a database, we must decide on the basic size of the asset for which data will be stored. For example, we should decide how a pipe segment will be defined (Walski and Male, 2000). Will it be defined by a city block or by the distance between valves or intersections? Should we draw the pump and the motor as one feature or as separate objects? Or should we simply represent the entire pump station as a single feature? Should the pump station be drawn as a point feature, as a polygon feature, or as an object?

For water systems, pipe segments are typically defined by end points, crosses, tees, reducers, or major changes in alignment, age, or material. Valves are discouraged as end points. Sanitary sewer lines are typically segmented by manholes, clean outs, pump stations, and junctions. Manholes, inlets, and outfalls typically segment storm sewer lines. For combined sewer and storm sewer systems, the sewer and storm components are normally placed on the same layer and inlets and outfalls are also used to segment the lines. The majority of segments are from manhole to manhole. Sewer lines should be digitized to reflect the direction of flow (i.e., upstream and downstream node).

All system components should be uniquely identified so that they can be linked to other data sources, such as maintenance activities, capital planning, and condition assessment (PaMAGIC, 2001). All features should be linked together in a topological structure that supports other applications, such as hydraulic modeling, water main isolation (capability to isolate sections of main for repairing pipe breaks by closing down valves), and network tracing (capability to identify upstream or downstream pipes).

APPLICATIONS DATABASES

Applications data are required for creating GIS applications, such as work order management, planning, and hydraulic modeling. Applications data may be considered "optional" because they are not "necessary" to build a utility GIS and are needed at the time of developing applications. Deciding which data belong to the assets or applications categories is often subjective. For example, some people may consider a digital orthophoto basemap necessary to build accurate GIS layers of the physical system and therefore put it in the assets category. Others may regard orthophotos as optional luxuries to enhance data visualization and include them in the applications category.

The next four sections list typical features and attributes for both assets and applications data types. They provide a detailed inventory of water, wastewater, and stormwater devices, features, and attributes. This inventory includes a large number of features and attributes, not all of which are necessary for a database. These lists are general and can be customized by adding and deleting features or attributes according to project requirements. You can use this inventory as a checklist to design your own database by selecting the features and attributes that are appropriate to your particular application.

WATER SYSTEM ASSETS DATA

The first step in designing an assets database is to make a list of system features. Table 6-2 shows typical water system features and how they are modeled in a GIS. Note that some point features that are spatially related to a line feature are modeled as nodes.

Table 6-2. Typical Water System Features

Point	*Node*	*Line*
Backflow preventor	Analysis/monitoring point	Casing
Flowmeter	Fire hydrant	Fire line
Leaks	Pipe fitting	Hydrant line
Manhole	Pump node	Raw water intake
Pump station	Service connection	Service
Reservoir	Surge relief	Service main
Vault	Tank	Water main
	Tower	
	Treatment plant	
	Valve	
	Well	

Typical attributes for some common water system features are listed below.

- Pipes
  - Structure ID
  - Hydraulic model ID (if different from structure ID)
  - Source ID (drawing or map number)
  - Street address or street location
  - Owner
    - Public
    - Private
  - Status
    - Active
    - Inactive
  - Pressure zone
  - Upstream (from) ID
  - Downstream (to) ID
  - Diameter (2 in. to several feet)
  - Diameter units
  - Length
  - Length units
  - Upstream invert elevation
  - Downstream invert elevation
  - Depth
  - Date installed or installation year
  - Pipe age in years
  - Friction factor (C-factor)
  - Corrosion factor
  - Pipe type
    - Transmission main
    - Supply main
    - Distribution main
    - Service main
    - Hydrant leg
    - Casing
    - Reservoir piping
    - Pump station suction/discharge piping
  - Material
    - Cast iron
    - Ductile iron
    - Cement mortar-lined iron

        - Iron with organic coatings
        - Asbestos-cement (A-C)
        - Plastic
    - Joint
    - Frost depth
    - Valve shutoff
    - Critical service and hydrant cross-referencing
- Fittings
    - Device ID
    - Type
        - Cross
        - Tee
            - Hydrant connection tee
            - Service connection tee
        - Coupling
        - Reducer
        - Plug
        - Pipe cap
        - Offset
        - Bend
        - Sleeve
        - Wet tap
        - Flange
        - Joint
- Valves
    - Device ID
    - Type
        - Check
        - Ball
        - Butterfly
        - Pressure-reducing valve
        - Pressure-sustaining valve
        - Stop valve
        - Vent valve
        - Air release valve
        - Nonreturn valves

    - Status
        - Open
        - Closed
        - Open (but) inoperable
    - Direction to open
    - Number of turns
    - Manufacturer
    - Size
    - High/low pressures
    - Operator depth
    - Water main and valve cross-referencing
- Hydrants
    - Device ID
    - Type
    - Owner
        - Public
        - Private
    - Status
        - Active
        - Inactive
    - Size
        - Barrel size
        - Height
    - Flow rate
    - Color
    - Packing
    - Feeder information
    - Number of ports or outlets
    - Make (manufacturer)
    - Serial number
    - Model number
    - Parcel
    - Water main and hydrant cross-referencing
- Supply source
    - Facility ID
    - Type
        - Groundwater (wells)

    - Surface water
        * River
        * Reservoirs
        * Lake
    - Treatment plant
    - Intake
  - Size
- Storage facilities
  - Facility ID
  - Type
    - Standpipe
    - Water tank
    - Basin
    - Clear well
  - Size
  - Number
  - Pumps
  - Motors
- Services
  - Customer name
  - Customer address
  - Account number
  - Telephone
  - Fax
  - E-mail
  - Tap location
  - Curb stop
  - Backflow
  - Water main cross-reference
  - Service line (connection between the property and distribution main)
    - Diameter (0.75 to 3 in.)
    - Material
        * Lead (mostly prior to early 1950s)
        * Copper
        * Plastic
        * Galvanized steel
        * Ductile iron
        * Cement mortar-lined iron
        * Iron with organic coatings

        * Asbestos-cement (A-C)
    - Meter
        - Meter type
        - Size
        - Number of dials
        - Serial number
- Pumping stations
    - Facility ID
    - Number of pumps
    - Capacity
- Monitoring point
    - Facility ID
    - Type
    - Date installed
- Motors

WASTEWATER SYSTEM ASSETS DATA

Table 6-3 shows typical wastewater system features and how they are modeled in a GIS. Note that some point features that are spatially related to a line feature are modeled as nodes.

Table 6-3. Typical Wastewater System Features

Point	*Node*	*Line*
Cleanout	Cleanout	Casing
Monitoring location	Combined sewer overflow (CSO) regulator	Force main
Sampling location	Diversion chamber	Gravity sewer
Tunnel door	Lift station	Reuse main
Tunnel shaft	Manhole	Reuse service
Tunnel vent	On-lot disposal system	Service lateral
Valve	Outfall	Siphon
Vault	Pipe fittings	Tunnel
	Pumping station	
	Treatment plant	
	Valve	
	Wet well	

Typical attributes for some common wastewater system features are listed below.

- Sewers
  - Structure ID
  - Hydraulic model ID (if different from structure ID)
  - Source ID (drawing or map number)
  - Street address or street location
  - Owner
    - Public
    - Private
  - Status
    - Active
    - Inactive
  - Collection system (watershed, sewershed, subbasin, subarea, or municipality)
  - Upstream (from) ID
  - Downstream (to) ID
  - Pipe type
    - Combined
    - Sanitary
    - Force main
    - Gravity pipe
    - Collector sewer
    - Interceptor sewer
    - Relief sewer
    - Outfall pipe
    - Siphon
    - Casing
    - Tunnel
    - Stub
    - Wye (service)
    - Pump station suction/discharge piping
  - Cross-sectional shape
    - Basket-handle
    - Catenary
    - Circular
    - Egg
    - Gothic
    - Horseshoe

    - Irregular (natural)
    - Parabolic
    - Rectangular
    - Rectangular with round bottom
    - Rectangular with triangular bottom
    - Semicircular
    - Semielliptical
    - Trapezoidal
  - Diameter, if circular
  - Dimensions, if not circular
  - Dimension units
  - Length
  - Length units
  - Upstream invert elevation (from elevation)
  - Downstream invert elevation (to elevation)
  - Slope
  - Depth
  - Depth units
  - Roughness coefficient
  - Pipe material
    - Asbestos cement pipe (ACP)
    - Cast iron pipe (CIP)
    - Corrugated metal pipe (CMP)
    - Ductile iron pipe(DIP)
    - Plastic pipe
    - Polyvinyl chloride (PVC)
    - Reinforced concrete pipe (RCP)
    - Terra cotta pipe (TCP)
    - Vitrified clay pipe (VCP)
    - Unknown
  - Joint length
  - Date installed or installation year
  - Pipe age
  - Groundwater level
- Manholes
  - Structure ID
  - Hydraulic model ID (if different from structure ID)
  - Source ID (drawing or map number)

- Street address or street location
- Manhole type
  - Standard
  - Drop
  - Interceptor
  - Overflow
  - Air release
  - Summit
  - Well-hole
  - Other
- Number of inlets
- Number of outlets
- Pipe(s) in and out
  - Invert elevation
  - Diameter
  - Direction
    - In
    - Out
  - Type
  - Optional: sketch showing all pipes leaving and entering the structure
- Cover material
- Cover type
- Top (rim) elevation
- Bottom (invert) elevation
- Depth
- Shape
- Diameter, if circular
- Dimensions, if not circular
- Ring
- Wall
- Frame
- Steps
- Bench
- Channel
- Meter
- Distance to hydrant
- Road traffic conditions
- Date installed or installation year

    - Status
        - Operational
        - Planned
        - Abandoned
        - Buried
- Customer data
    - Account number or customer ID
    - Street address
    - Telephone
    - Type of customer
        - Residential
        - Commercial
        - Industrial
        - Institutional
    - EDUs
- Services
    - Service line
        - Diameter
        - Length
        - Material
        - Pipe type
        - Point of connection (sub-unit and stationing)
        - Digital photo ID
    - Tap location
    - Cleanout
    - Mainline cross-referencing
    - Chemical abstract ID
- Diversion chambers
    - Facility ID
    - Type
    - Permit No.
- Outfalls
    - Facility ID
    - Type
        - Combined sewer overflow (CSO)
        - Sanitary sewer overflow (SSO)
    - Permit No.

- Pumping stations
  - Facility ID
  - Number of pumps
  - Capacity
- Treatment plants
  - Facility ID
  - Owner
  - Type
    - Municipal sewage
    - Industrial waste
    - CSO
    - SSO
  - Capacity
  - Permit No.
  - Compliance (yes/no)
- On-lot sewage disposal systems
  - Facility ID
  - Owner
  - Type
  - Capacity
  - Age
- Monitoring locations
  - Facility ID
  - Type
    - Flowmeter
    - Sampler
    - Rain gauge
    - Permanent
    - Temporary
    - Digital (electronic)
    - Analogue (strip chart)
  - Flow type
    - Sanitary
    - Combined
    - Overflow
      - CSO
      - SSO
  - Installation Date

    - Device
        - Manufacturer
        - Model No.
- CSO locations
- SSO locations
- CSO treatment facilities
- SSO Treatment facilities
- Wet wells
- Motors
- Cleanouts
- Grit chambers
- Detention basins
- Retention treatment basins (RTBs)
- Wet weather equalization facilities

STORMWATER SYSTEM ASSETS DATA

Table 6-4 shows typical stormwater system features and how they are modeled in a GIS. Note that some point features that are spatially related to a line feature are modeled as nodes.

Table 6-4. Typical Stormwater System Features

Point	*Node*	*Line*
Basin	Cleanout	Casing
Cleanout	Infiltration basin	Channel
Monitoring location	Inlet	Culvert
Sampling location	Manhole	Pipe
Vault	Outfall	Siphon
	Pipe fittings	Stream
	Pumping station	Swale
	Retention basin	

Typical information required to manage a stormwater system includes:

- Physical characteristics of the stormwater system
- Receiving waters
- Outfall locations
- Service area land use

- Service area polluters
- Concentrations and annual loads of pollutants
- Pollution control measures

Many stormwater system attributes are the same as the wastewater system attributes described above, especially for pipes and manholes. Typical attributes for some common stormwater system features are listed below.

- Storm sewers
  - Structure ID
  - Hydraulic model ID (if different from structure ID)
  - Source ID (drawing or map number)
  - Street address or street location
  - Owner
    - Public
    - Private
  - Status
    - Active
    - Inactive
  - Collection system (watershed, sewershed, subbasin, subarea, or municipality)
  - Upstream (from) ID
  - Downstream (to) ID
  - Pipe type
    - Force main
    - Gravity pipe
    - Collector sewer
    - Relief sewer
    - Outfall pipe
    - Siphon
    - Casing
    - Tunnel
    - Stub
    - Wye (service)
    - Culvert
    - Pump station suction/discharge piping
  - Cross-sectional shape
    - Basket-handle
    - Catenary

  - Circular
  - Egg
  - Gothic
  - Horseshoe
  - Irregular (natural)
  - Parabolic
  - Rectangular
  - Rectangular with round bottom
  - Rectangular with triangular bottom
  - Semicircular
  - Semielliptical
  - Trapezoidal
- Diameter, if circular
- Dimensions, if not circular
- Dimension units
- Length
- Length units
- Upstream invert elevation (from elevation)
- Downstream invert elevation (to elevation)
- Slope
- Depth
- Depth units
- Roughness coefficient
- Pipe material
  - Asbestos cement pipe (ACP)
  - Cast iron pipe (CIP)
  - Corrugated metal pipe (CMP)
  - Ductile iron pipe(DIP)
  - Plastic pipe
  - Polyvinyl chloride (PVC)
  - Reinforced concrete pipe (RCP)
  - Terra cotta pipe (TCP)
  - Vitrified clay pipe (VCP)
  - Unknown
- Joint length
- Date installed or installation year
- Pipe age
- Groundwater level

- Open channels
  - Source ID (drawing or map number)
  - Street address or street location
  - Owner
    - Public
    - Private
  - Type
    - Channel
    - Ditch
    - Swale
- Outfalls
  - Facility ID
  - Permit No.
- Manhole
  - Attributes are the same as for the wastewater system listed above
- Pumping station
  - Attributes are the same as for the wastewater system listed above
- Basins
  - Facility ID
  - Type
    - Detention pond
    - Retention pond
- Treatment plants
  - Facility ID
  - Owner
  - Type
    - Municipal
    - Industrial
- Monitoring locations
  - Facility ID
  - Type
    - Flowmeter
    - Sampler
    - Rain gauge
    - Permanent
    - Temporary
    - Digital (electronic)

        - Analogue (strip chart)
    - Flow type
        - Pipe flow
        - Overflow
    - Installation date
    - Device
        - Manufacturer
        - Model No.
- Inlet
- Catchbasin
- Tide gate
- Best management practice location
- Dam
- Culvert

APPLICATIONS DATABASES

The design of applications databases cannot be generalized because it really depends on the application, be it hydraulic modeling, work order management, or document management. For example, hydraulic modeling—one of the most demanding applications of GIS data—requires strict topology and nearly perfect data accuracy (Graybill, 1998). The general data that may be required in all the three areas of water, wastewater, and stormwater systems are listed first. Data that are specific to applications for one of these three systems are then listed.

GENERAL APPLICATIONS DATA

- Basemap
    - Digital orthophoto basemap
    - Aerial photos
    - USGS digital raster graphics (DRG) or topographic map images
    - USGS digital line graphs (DLG)
    - USGS digital orthophoto quarter quads (DOQQ)
    - TIGER/Line data
    - Satellite imagery
- USGS Digital Elevation Model (DEM)
- Streets (geocoded)
- Land use/land cover

- Industries
- Soils
- Hydrography
- Geology
- Parcels
- Census blocks and block groups
- Watershed boundaries
- Subbasin / sewershed boundaries
- Building footprints
- Right of way
- Wetlands
- Floodplains
- Flow monitors
- Water quality samplers
- Rain gages
- Field inspections
- Attributes required for hydraulic and hydrologic modeling using EPANET; Storm Water Management Model (SWMM); Hydrologic Engineering Center's (HEC) HEC-HMS and HEC-RAS; and so on

Water System Applications Data

- Attribute data required for hydraulic modeling (e.g., node demands and elevations)
- Pressure zone (district) boundaries
- Field inspections
    - Water quality tests
    - Hydrant tests
    - Pressure tests
    - Flow tests
- Preventive maintenance
    - Valve exercising
    - Flushing
    - Pipe breaks
    - Leak survey
    - Service interruptions

  - Damaged pipes
  - Repairs
- Monitored constituents
  - Hardness (calcium and magnesium)
  - pH
  - Iron
  - Bicarbonate and carbonate ions
  - Chlorine residuals
  - Microbiological parameters
    - Total plate counts
    - Heterotrophic plate counts
    - Assimilable organic carbon
- Disinfection by-products (DBPs)
- Trace metals (lead and copper)

WASTEWATER SYSTEM APPLICATIONS DATA

- Attribute data required for hydraulic modeling (e.g., sewage flow rates and inflow/infiltration estimates)
- Condition ratings to prioritize critical pipes:
  - Structural
  - Root
  - Hydraulic
- Preventive maintenance
  - Periodic cleaning
  - Flushing
  - Inspections
- Dye test
  - Status
  - Findings
- Sources of illegal flow
  - Roof drains
  - Floor drains
  - Foundation drains
- TV inspections
  - Date of last TV inspection
  - Findings
  - Root intrusion data

    - Noted structural problems
    - Number of taps by type (fitting or break-in)

Stormwater System Applications Data

- Attribute data required for hydraulic modeling (e.g., inlet capacities)
- Watersheds
- Drainage areas
- Landfills
- Parks
- Recreation areas
- Open lands
- National Pollution and Discharge Elimination System compliance data
    - Roadways
        - Sweeping schedule
        - Debris/litter removal program designations and responsibilities
        - Drainage structures
    - Construction sites
        - Standard Industrial Classification code
        - Owner/occupant information
        - Inspection schedule
        - Inspection dates
        - Comments
    - Field data
    - Dry weather testing data
    - Wet weather testing data
    - Land use
- Identify potential sources of stormwater pollution and drainage defects
    - TV inspections
        - Attributes same as for wastewater system listed above
    - Manhole inspections
    - Smoke testing
    - Dye testing

Stormwater Management Applications Data

- Attribute data required for hydraulic modeling (e.g., runoff curve numbers and stream cross sections)

- Land use
- Stream centerline
- Cross-section location
- Field surveys
- Benchmarks
- Watershed boundaries
- Subwatershed / subbasin boundaries
- Photograph locations
- Outfall locations
- Hydraulic obstructions (culverts and bridges, etc.) and their capacities
- Floodplain boundaries
  - Existing
  - Future
- Improvement locations

DATABASE DESIGN STANDARDS

Certain design standards should be followed in creating a GIS database. For example, the design should adhere to computer industry standards. The standards set forth guidelines on system interoperability and integration, which are critical for the success of a GIS application project. There are four important standards for modern GIS software (ESRI, 2000):

1. Microsoft Windows for interface
2. Structured Query Language (SQL) for data access
3. Component Object Model (COM) for tools
4. Transmission control protocol/Internet protocol (TCP/IP) and hyper text transfer protocol (HTTP) for network data transfer

Interoperability is not a reality today because different standards are advocated by various groups. The major players are Microsoft, with its ActiveX standards, and the open standards bodies, which are promoting the use of Common Object Request Broker Architecture (CORBA) standards. The software industry has recognized the need to provide seamless interoperability between products developed by different vendors. Mechanisms are required to allow components developed by different vendors to communicate, wherever those components reside, be it on the same device or distributed across a local or wide area network.

There are significant benefits in this "component-based" approach, but it requires the adoption of a common standard for the interoperability of component objects manufactured by different vendors. The computer industry standards used in GIS software procurement and integration should also be used for the database design. Additional database design standards and requirements are described in the next section.

GIS DATA MODELS

For data storage and manipulation, a database management system (DBMS) uses a data model, such as a hierarchical, network, or relational data model. In the late 1990s, a new object-oriented (OO) data model was introduced. This model, which can store spatial data inside a relational DBMS (RDBMS), is described at the end of this chapter.

The relational model is the most widely used data model. An RDBMS is a software program that is used to create, maintain, modify, and manipulate a relational database. An RDBMS is also used to create the applications that will enable users to interact with the data stored in the database (Hernandez, 1997). It allows for easy data entry and manipulation, provides fast query and display, and maintains data integrity and security (EPA, 2000). Relational database systems have become the commercial de facto standard because of ease of use and implementation, ability to be modified, and flexibility.

Some applications use information from several different databases or tables. For example, a smoke testing application might require information from manhole and customer account databases. It is inefficient and cumbersome to enter the data from different tables in one table. Relational database tables allow information to be accessed from different tables without joining them together physically.

A relational database stores information in records (rows) and fields (columns). An RDBMS conducts searches by using data in specified fields of one table to find additional data in another table. To accomplish this, there should be at least one "key" or "common" field in each table that uniquely identifies the records. The common field can be used to link the GIS database tables to virtually any external database table. This linkage capability allows the GIS to make effective use of existing databases without requiring new data entries in the GIS database. Once the GIS and external database tables have been linked, the external data can be queried or mapped from within the GIS. Figure 6-1 shows how a GIS table for valves can be linked to an external table containing valve maintenance history. In this example, the "Valve ID" field is the key field common to both the GIS and the external database tables. Figure 6-2 shows the linkage of a manhole theme table in ArcView with an external manhole inspection table.

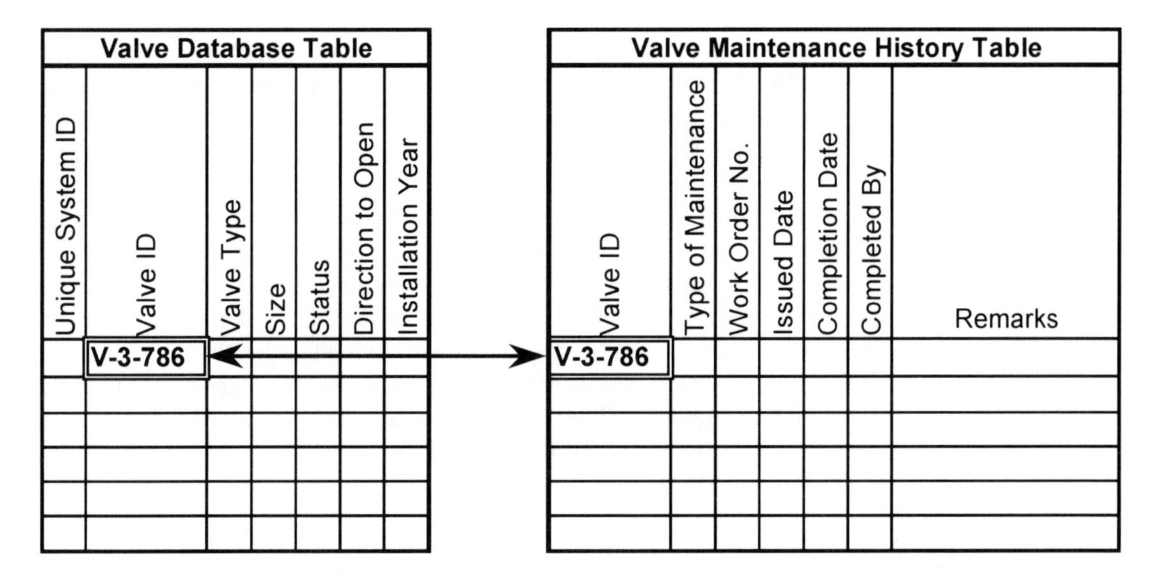

Valve Database Table
Unique System ID
Valve ID
Valve Type
Size
Status
Direction to Open
Installation Year
V-3-786
Valve Maintenance History Table
Valve ID
Type of Maintenance
Work Order No.
Issued Date
Completion Date
Completed By
Remarks
V-3-786

Figure 6-1. Linking GIS and External Tables in Relational Databases

COMMON FIELD

Enhanced Sewer System

MANHOLE THEME TABLE

Attributes of Mhsp

Shape	Area	Perimeter	Mhsp#	Mhsp-id	Mhid
Point	0.000	0.000	220	238	MH7-8
Point	0.000	0.000	221	239	MH6-77
Point	0.000	0.000	222	240	MH9-2
Point	0.000	0.000	223	241	MHI-48

mh_insp2.dbf

Mh_id	Num	Mapgrid	Strunum	Street	Date	Strutype	Diameter
MH9-2	155	S-67	MH-2	Along median strip on Rte. 88	05/23/96	Manhole	26-3/4"
MH9-3	156	S-67	MH-3	Middle intersect. E. Main near I	05/30/96	Manhole	24-1/2"
MH9-3A	157	S-67	MH-3	E. Main near junction of Route	05/23/96	Manhole	24"
MH9-4	158	S-67	MH-4	Intersection of Jones St. & E. M	05/30/96	Manhole	25-1/2"
MH9-4A	159	S-67	MH-4	E. Main St. & Jones St.	05/23/96	Manhole	25"
MH9-5	160	S-67	MH-5	Intersection of Orange St. & Jo	05/30/96	Manhole	22-1/2"
MH9-8	161	S-67	MH-8	Intersection of Gregg St. & Sta	05/30/96	Manhole	22-1/2"
MH9-9	162	S-67	MH-9	Gregg Street & Staib Street	05/23/96	Manhole	24"
MH9-10	163	S-67	MH-10	Slow lane of Route 88	05/24/96	Manhole	23-3/4"
MH9-13	164	S-67	MH-13	Intersection Country Club Rd.	02/05/99	Manhole	23-1/2"

Text
Rivers
Subb.
Railroads
Study Area Boundary

MANHOLE INSPECTION TABLE (EXTERNAL)

Figure 6-2. Linking Manhole Theme Table and External Manhole Inspection Table in ArcView

The linkage shown in Figure 6-2 is accomplished using the fields "Mhid" of the manhole theme table and "Mh_id" of the manhole inspection table. It shows that when manhole "MH9-2" is selected in the theme table, the corresponding records in the manhole inspection table are also selected.

The traditional file system (hybrid) database design approach saves the results of spatial queries, such as an overlay operation, in a new file. This approach requires a GIS manager to keep all derivative files (such as a GIS overlay) synchronized with their source data. Any change in the source coverage would also require updating the overlay map—a laborious process! The RDBMS approach saves the query rather than the results of the query. Thus, an update of source data would automatically update the query view and there would be no need to change the overlay map (Lowe, 2000a).

DATABASE DESIGN STEPS

Database design involves three steps: (1) conceptual design, (2) logical design, and (3) physical design. Conceptual design does not depend on hardware or software. It defines graphic features, attributes, labels, data format, reports, feature placement rules and guidelines, and database output. Logical design depends on software only and defines the logical structure of the database elements according to the software requirements. The logical database design describes the size, shape, and necessary systems for a database; it addresses the informational and operational needs of a business or utility. Logical design reduces the time required to implement the database structure. This frees up time for developing applications that will be used to interact with the data in the database (Hernandez, 1997). Physical database design depends on the hardware. It specifies system-specific file structure, database structure, data layers, feature symbology, file names, table names, attribute names, and key IDs. Physical design also provides specifications on memory, disk space, access, and speed. You can think of the logical database design as the architectural blueprints and the physical database implementation as the completed home (Hernandez, 1997).

Conventional database design methods involve three phases: needs analysis (or assessment), data modeling, and normalization. Needs analysis clarifies the project's specific needs; identifies and quantifies the GIS needs of an organization and its stakeholders; and defines how a GIS will benefit an organization by relating specific organizational resources and needs to specific GIS capabilities (Wells, 1991). Needs analysis is analogous to strategic planning. Careful needs assessment is critical to successful GIS implementation. Before a systemwide implementation, the initial database design and data conversion methodology should be evaluated for a small pilot area and fine-tuned if necessary. Additional needs analysis and pilot project testing is provided in Chapter 9 (Wastewater System Applications).

Depending on the design goals, basic database design may be divided into three categories: standalone, giant, and joint (most complex). Today, the trend is to use standard database management systems, such as Oracle, Informix, DB2, or SQL Server, to store both attribute and geometrical data. The dependence between the basic data model and applications can be reduced by using simple storage structures in a standard DBMS (Bernhardsen, 1999). A detailed discussion of database design phases and categories is outside the scope of this book. Please read one of the database books (e.g., Hernandez, 1997) listed in Chapter 12 (GIS Resources) for more information.

Each feature of a database must have a unique identifier or "structure ID." Each feature should also have a "source ID" attribute in the database. This attribute represents an identification number for the source material from which the feature was digitized. The source IDs should be explained in a separate table called a "source table" that accompanies the database design report. This table should have information about the source title, number, creator, date, system (water, wastewater, or stormwater), and comments. The columns and sample entries for a typical source table are shown in Table 6-5.

Table 6-5. Typical Source Table for a Database Design Report

Column	*Sample Entry*
Source ID	1001
Source title	River Park Water Service Extension
Creator	ABC Engineers
Date created	January 15, 1959
Date modified	October 21, 1960
Drawing No.	2252-24-59-13
Sheet No.	13
System	Water
Comments	River Park is now called the Reagan Park

Once you have created tables, set up table relationships, and established the appropriate levels of data integrity, your database is complete. Now you are ready to develop applications that will allow you to interact easily with the data stored in your database, and you can be confident that these applications will provide you with timely and—most important—accurate information.

The database design should also support the intended applications, such as work order management. The assets and applications database design should be fully compliant with the data requirements of the facilities management software, such as Cityworks® (more information in Chapter 3). Each feature type should contain certain mandatory attribute fields that facilitate some function within the application software. In addition to these, user-defined attribute fields should be allowed, which can be added to or subtracted from the design. For example, database design for a layer of combined sewer overflow (CSO) events is presented in Table 6-6.

Table 6-6. Database Design for Combined Sewer Overflow (CSO) Events

This layer describes CSO events and supports links to the CSO database.

Layer Name: CSOEVENT.PAT

Layer Type: POINT

Attribute Name	*Definition*	*Description*
Arc#	10 10 I	System-generated ID
Csoevent#	10 10 I	System-generated ID
Csoevent_id	10 10 I	User-specified ID
Structure code	3 3 I	Code representing structure type
Source ID	9 12 F	Source drawing
X_coord	8,20,F,5	State plane *x* coordinate
Y_coord	8,20,F,5	State plane *y* coordinate
Address	32 32 C	Address nearest event
City	32 32 C	City where event occurred
Zip	10 10 C	Zip code where event occurred
Parcel_ID	48 48 I	Tax parcel ID number of event location
County	32 32 C	County where event occurred
Type	32 32 C	Type of event
CSO_Num	16 16 C	CSO event number
Comments	64 64 C	Data converter comments

Typical structure codes referenced in Table 6-6 are shown in Table 6-7. The second column (definition) of the above table consists of three elements:

- Item width: the number of bytes to store the item
- Output width: the number of columns to display the item value
- Item type: the type of data stored in the item

  B = Binary
  C = Character
  F,n = Floating with n decimal places
  I = Integer

Table 6-7. Typical Sewer System Structure Codes

Structure Code	*Structure Description*
101	Manhole
102	Buried manhole
103	Lamphole
104	Cleanout
105	Force main valve
106	Air release valve
107	Pump
108	Diversion chamber
109	Manhole overflow
110	Combined sewer overflow
111	Sanitary sewer overflow
112	Pump station overflow
113	Grease trap
114	Grinder pump
115	Catch basin
116	Head wall
117	End wall
118	Meter pit

DATA DICTIONARY

The first step in designing a database is the development of a data dictionary, which is defined as a catalog that explains the data. It can be a paper or computer document. A data dictionary delineates the specific categories of descriptive information required for each map feature. Data that do not come with a data dictionary may not be usable (ESRI, 1997). Each unique category of descriptive information is called an attribute.

The database design provides a detailed definition of the structure and content of the database. The database design is presented with the help of a data dictionary, which documents the logical and physical structure of the layers of the GIS. It includes descriptions of individual layers, specifications of all tables associated with the layers, physical definitions of the items contained within those tables, listings of the valid codes or codes associated with those items, and diagrams graphically showing the logical relationships between the items and tables.

Table 6-8 shows a portion of a data dictionary for a sewer system documenting pipe material codes. Table 6-9 shows a sample database for a water system (ESRI, 1996a).

Table 6-8. Sample Data Dictionary for a Sewer System

Data Layer	*Sanitary Sewer Pipes*
Name	SANSEWER
Source	Public Works Department
Date	1985
Projection	State Plane
Item	PIPEMAT
Code = ACP	Asbestos cement pipe
Code = CIP	Cast iron pipe
Code = CMP	Corrugated metal pipe
Code = DIP	Ductile iron pipe
Code = PVC	Polyvinyl chloride
Code = RCP	Reinforced concrete pipe
Code = TCP	Terra cotta pipe
Code = VCP	Vitrified clay pipe

Table 6-9. Sample Water System Database

Department	*Layer*	*Abbreviation*	*Theme*	*Features*
Public Service (PS)	WATERPOT	WP	Potable water system	Lines Nodes Points
PS	WATERFACI	WF	Water facilities	Polygons
PS	WATERSERV	WS	Water service areas	Polygons
PS	WATERSV	WV	Water services	Lines Nodes
PS	WATERLEAK	WL	Water system leaks	Points

The data dictionary provides the physical description of each layer using three types of documents (ESRI, 1996a):

1. A layer fact sheet that provides basic information about the layer. Figure 6-3 provides a sample fact sheet for the WATERLEAK layer of the sample database shown in Table 6-9.

2. A data diagram that illustrates the logical relationships among the layer's feature attribute tables (FATs) and related tables. Figure 6-4 provides a sample data diagram for the WATERLEAK layer.

3. A data dictionary template(s) that define(s) the FATs and related tables associated with the layer. Table 6-10 gives a sample FAT template for the WATERLEAK layer.

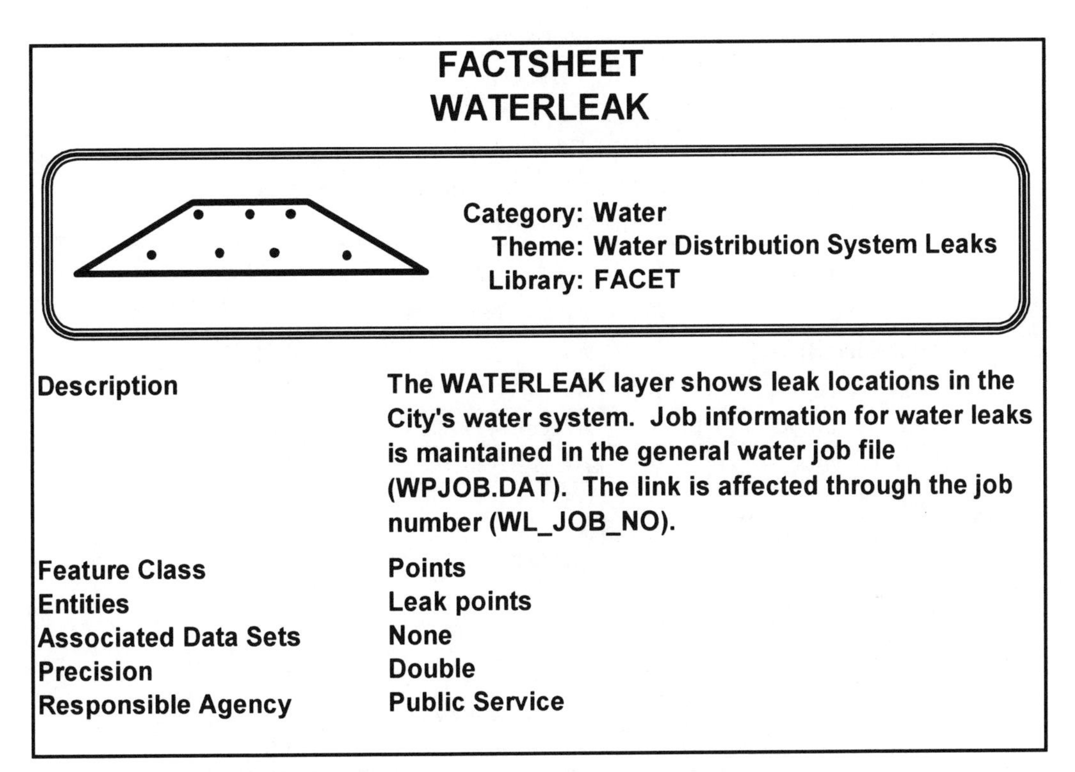

Figure 6-3. Data Dictionary Fact Sheet for Water Leak Layer

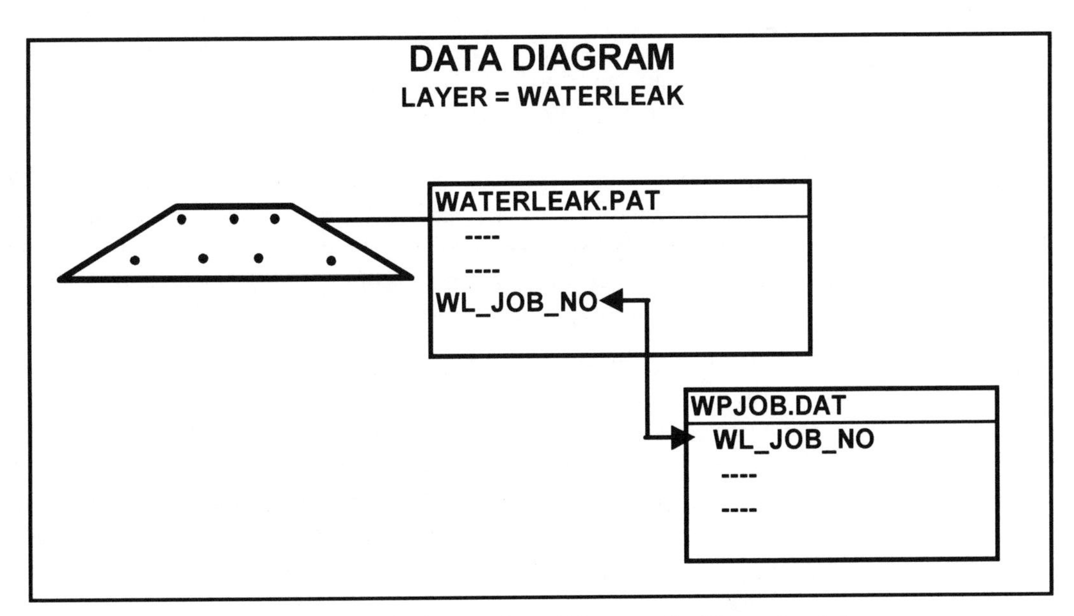

Figure 6-4. Data Dictionary Data Diagram for Water Leak Layer

Table 6-10. Sample Water Leak Feature Attribute Table Template

Layer	WATERLEAK (WL)		
Feature class	Points		
Table name	WATERLEAK.PAT		
Data source	Public Service Department Leak Survey		
Description	Water leak attributes		
Description	*Begin Column*	*Defined Item Name*	*Item Definition*
Leak identifier	1	WL_ID	4,10,B
Leak type	5	WL_TYPE	2,2,I
Feature number	7	WL_FEATURE_NO	17,17,C
Feature type	24	WL_FEATURE_TYPE	3,3,C
Feature flag	27	WL_FEATURE_FLAG	2,2,C
Feature condition	29	WL_COND	1,1,I
Infrastructure system	30	WL_SYSTEM	2,2,C
Location address	32	WL_LOCADD	30,30,C
Repair type	63	WL_RPR_TYPE	2,2,I
Repair cost	65	WL_RPR_COST	4,12,F,2
Job number	69	WL_JOB_NO	9,9,C

Object-Oriented Database Design

During the late 1990s, many GIS software packages started switching to a new object-oriented (OO) data model, which offers common storage of spatial data inside an RDBMS. The main features of OO design are given below (DeMartino and Hrnicek, 2001):

1. A traditional GIS data model uses points, nodes, lines, and polygons to represent real-world objects, such as hydrants, sewers, and watersheds. An OO data model stores these objects as hydrants, sewers, and watersheds rather than as generic points, lines, or polygons. An OO model is, therefore, more user friendly.

2. Database design of conventional data models requires setting up and joining a series of related tables, columns, and numeric codes. The OO model stores "intelligent" objects in logical groups called "classes." Each feature class has predefined properties, behavior, and relationships, which are embedded in the initial database design. The OO model database design can be automated using computer-aided software engineering (CASE) and Unified Modeling Language (UML) tools.

3. In traditional GIS data models, the network connectivity (e.g., topological relationship among the pipes) should be created by manual post-processing of GIS data, often by running external network connectivity routines. This approach is time consuming because it requires network connectivity to be rebuilt every time a change is made in the database (e.g., a pipe is changed). The OO model can generate the network connectivity "on-the-fly," which eliminates manual post-processing and saves time.

4. In the traditional data models, applications are developed by writing custom computer code and scripts. It is easier to develop applications in the OO data model because the rules and relationships required for applications are part of the database. For example, the valves needed to isolate and repair a broken water main can be identified without resorting to GIS software extensions.

5. The traditional data models generally use proprietary database management systems for storing data, which are strange to the information technology (IT) personnel of an organization. The OO model can use a standard corporate relational database that can be operated and maintained by the IT department, thereby saving money for the GIS department.

The OO data model concepts and jargon are new to most water industry professionals and GIS technicians. Implementing or migrating to an OO data model therefore will require extensive training in OO database design. Converting a traditional data model to an OO model also requires extensive quality control measures to prevent loss of data during the translation process. The data creation or migration problems can be reduced by undertaking a pilot project for a small study area before full implementation. A pilot project provides an excellent opportunity to experiment with the data and learn from the mistakes.

GEODATABASE DESIGN

Chapter 1 (GIS Basics) and Chapter 2 (GIS Development Software) provided information about ESRI's new OO GIS data model, called a geodatabase. Geodatabase is used in ESRI's new software suite, called ArcGIS. Although, geodatabase design is very new, water and wastewater applications are among the first groups taking advantage of the geodatabase technology. For example, the City of Fort Lauderdale, (Florida, USA) has recently completed a geodatabase pilot project to integrate its water, wastewater, and stormwater systems with the city's asset management and utility billing systems (DeMartino and Hrnicek, 2001).

ArcFM Water and ArcGIS Hydro are examples of geodatabase models. ArcFM Water is a data model for water and sewer systems, whereas ArcGIS Hydro is applicable to surface water hydrology and hydrography. Additional information on geodatabase design is available from the ESRI website (*www.esri.com*) and in several white papers and books published by ESRI, such as "Building a Geodatabase" (ESRI, 2000a; ESRI, 2000b).

ArcFM Water Geodatabase Model

ArcFM Water was used by the City of Germantown (Tennessee, USA) for asset inventory, operation and maintenance (O&M), hydraulic modeling, and capital improvement planning of its water distribution system. Germantown used ArcFM because it wanted an efficient database design but did not have the time or money to develop a proprietary design from scratch. ArcFM required minimal customization, and the database design took weeks instead of months. (DeMartino and Hrnicek, 2001).

ArcGIS Hydro Geodatabase Model

ArcGIS Hydro model was developed through collaboration between ESRI and the Center for Research in Water Resources (CRWR) of the University of Texas at Austin. The ArcGIS Hydro data model was created jointly by a Consortium for GIS in Water Resources consisting of partners from industry, government, and academia.

The ArcGIS Hydro data model provides a framework for integrating geospatial and temporal data describing surface water hydrography and hydrology. The objectives of this model are: (1) mapping of water features, (2) linear referencing on the river network, and (3) dynamic modeling of water resources. The backbone of the ArcGIS Hydro model is the Hydro Network, which is a combination of the Flow Network through streams and water bodies, and the Shorelines of the water bodies. The Flow Network is represented by ArcGIS network features, edges, and junctions, which are special forms of line and point features. An ArcGIS network has three associated models:

1. A geometric model, which defines the (x, y, z) spatial locations of its junctions and edges;
2. A logical model, which defines the connectivity of the edges and junctions; and
3. An addressing model, which defines how location and distance are measured on the network using the methods of linear referencing.

Any location on a network can thus have four dimensions associated with it: (x, y, z, m), where m is the measure of distance established by linear referencing, such as the flow distance in feet or miles to a downstream reference point on the network. Network analyses, including directional tracing and material tracking, can be performed with the logical network. Certain relationships and connectivity rules ensure that all data connections are made smoothly and that the Hydro Network functions as an integrated whole (Maidment, 2000).

SUMMARY

This chapter provided information on how to design a database for water, wastewater, and stormwater system applications. Generally, two types of databases are required to develop applications: assets and applications. Assets data are considered necessary because they define the physical characteristics of utility networks and because applications cannot be developed without them. Applications data are considered optional because they really depend on your applications and are not needed until you are ready to develop your applications. Traditional database design involves three steps: (1) conceptual design, (2) logical design, and (3) physical design. Database design is presented with the help of a data dictionary. The recent object-oriented data model offers several advantages over traditional relational or coverage-based data models.

SELF-EVALUATION

1. What are the different types of databases for developing GIS applications?
2. What type of database is more important: assets or applications? Why?
3. What are the typical steps in designing a traditional GIS database?
4. What is the difference between conceptual and physical design?
5. What is a data dictionary? Why is it an important part of GIS database design?
6. Create a data dictionary for your water or wastewater system applications database.
7. What are the benefits of using an object-oriented data model?
8. What is a geodatabase and how does it differ from the ESRI coverage model?

Chapter 7

MODELING INTEGRATION

The integration of a GIS with computer models allows users to be more productive. Integrated models enable users to devote more time to understanding problems and less time to the mechanical tasks of preparing input data and interpreting the output.

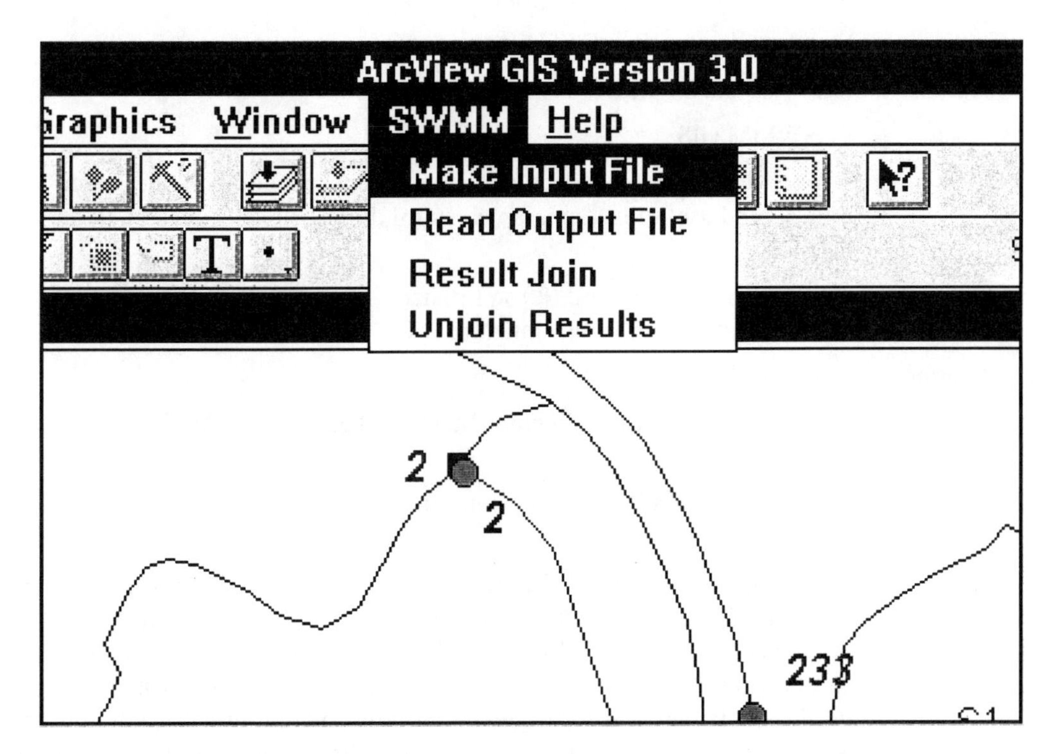

Integration of U.S. EPA's Storm Water Management Model (SWMM) With ArcView® GIS

LEARNING OBJECTIVES

The learning objectives of this chapter are to classify the methods of integrating hydrologic and hydraulic (H&H) computer models with a GIS and to understand the differences among various methods. Major topics discussed in this chapter include

- Interchange method
- Interface method
- Integration method
- Examples and case studies of above methods

VIRGINIA'S GIS-BASED URBAN HYDROLOGY MODEL

Application	H&H modeling of urban watersheds with storm drains and open channels
Reference	Waye, 2001
Project Status	Ongoing (as of May 2001)
Web site	*www.novaregion.org/4MileRun/4mr-swmm.htm*
Hardware	Standard desktop PC (Dell Dimension P3)
GIS software	ArcView and SWMMTools ArcView extension
Other software	PCSWMM GIS
GIS data	1-m resolution DOQs, subcatchment polygons, conduit polylines, node points, land use, soils, topography, storm drains, streams, manholes, and catchbasins
Study area	Four Mile Run watershed (20 sq mi) across four Northern Virginia localities, just south of Washington, D.C.
Project duration	Ongoing since 1977
Project budget	About $60,000 a year (GIS component has been about $10,000 a year since 1999)
Organization	Northern Virginia Regional Commission

The commission's Four Mile Run Model is one of the oldest continually used applications of the U.S. EPA's legacy program Storm Water Management Model (SWMM). What began in 1977 as a mainframe application of WREM (a precursor to SWMM) has grown with advances in the modeling field to become a more useful and user-friendly tool for stormwater management. The original model was used to analyze management of a flood control channel built by the U.S. Army Corps of Engineers. The analysis was done on a yearly basis by the four local governments that share the watershed. Today, it is also used to support floodplain analysis and site-level stormwater management for numerous developments, redevelopments, and road and bridge projects throughout the 20-sq mi urban watershed. It has recently undergone tight integration with GIS with a twofold benefit: faster and more accurate model enhancement, and vastly improved visualization of model results. The model has migrated to GIS in two software environments: ArcView and PCSWMM GIS, both of which have provided complementary benefits. The free SWMMTools extension for ArcView has greatly facilitated this migration. PCSWMM GIS supports SWMM model runs and dynamic playback from within a GIS interface. Figure 7-1 shows a screenshot of the SWMTools Extension (Heineman, 2000).

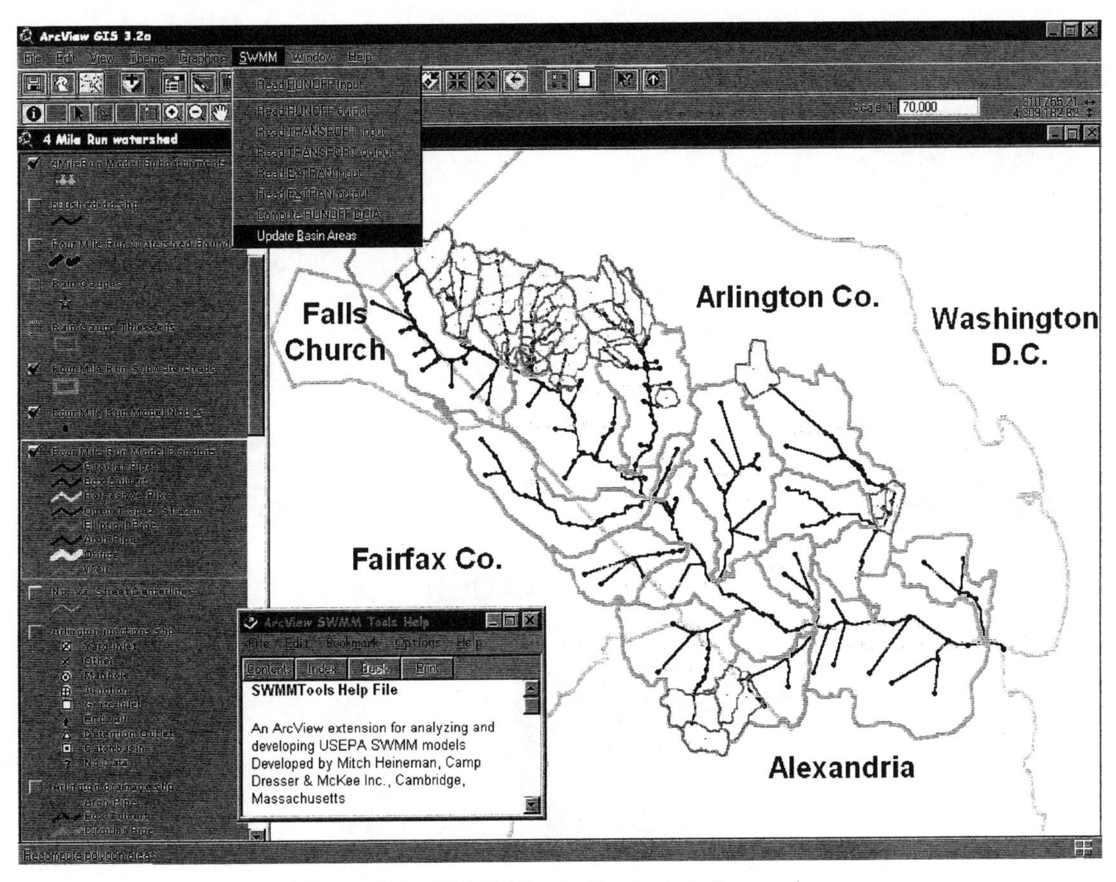

Figure 7-1. SWMMTools Extension Screenshot

Modeling Integration Taxonomy

According to a literature review of GIS applications in computer modeling conducted by the EPA, Shamsi (1998, 1999) offers a useful taxonomy to define the different ways a GIS can be linked to computer models (Heaney et al., 1999). The three methods of GIS application defined by Shamsi are:

1. Interchange method
2. Interface integration method
3. Integration method

Figure 7-2 shows the differences among these methods.

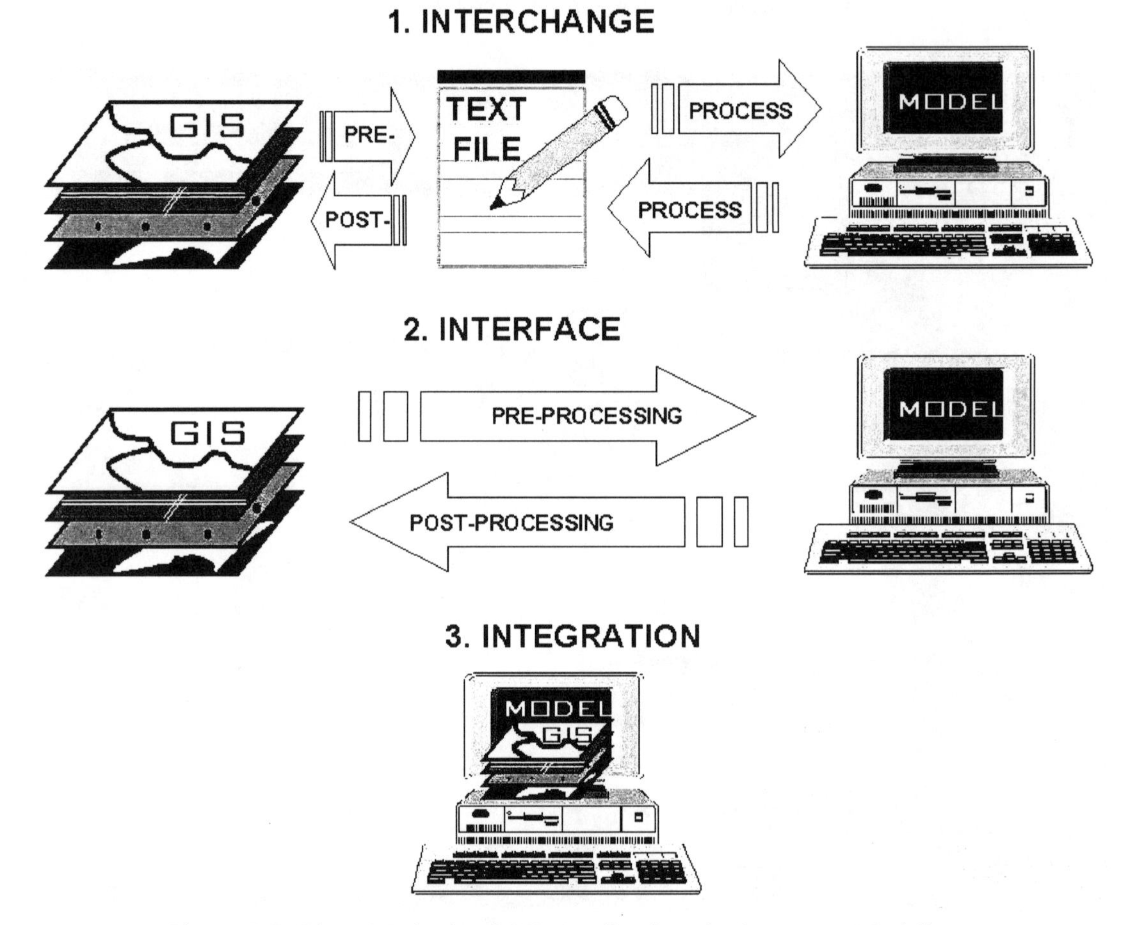

Figure 7-2. Three Methods of GIS Applications in Computer Modeling

Interchange Method

The interchange method employs a batch process approach to interchange (transfer) data between a GIS and a computer model. In this method, there is no direct link between the GIS and the model. Both the GIS and the model are run separately and independently. The GIS database is pre-processed to extract model input parameters, which are manually copied into a model input file. Similarly, model output data are manually copied in the GIS to create a new layer for presentation mapping purposes. Script programming is not necessary for this method, but it may be done to automate some manual operations, such as derivation of runoff curve numbers (described below). This is often the easiest method of using a GIS in computer models, and it is the method used most at the present time.

Any GIS software can be used in the interchange method, but a GIS with both vector and raster capabilities provides more interchange options. Chapter 10 (Stormwater System Applications) presents several applications of the interchange method in stormwater management. Another example of the interchange method is described below.

Runoff Curve Number Estimation

Runoff curve numbers are a series of standard empirical curves used to estimate runoff. GIS estimation of subbasin runoff curve number, a critical input parameter in many rainfall–runoff models, is perhaps the best example of the interchange method. The estimation approach is based on land use, hydrologic soil group (HSG), and runoff curve number relationships developed by the U.S. Natural Resources Conservation Service (NRCS), formerly known as the Soil Conservation Service (SCS). These relationships are available in the form of runoff curve number tables (U.S. Department of Agriculture, 1986). These tables provide runoff curve numbers for a large number of land uses and four hydrologic soil groups: A, B, C, and D. They also list average percent imperviousness values for various land use classes. Table 7-1 gives imperviousness and runoff curve number data for some typical land use classes.

A vector layer for subbasin runoff curve numbers is created by overlaying the layers for subbasins, soils, and land use to delineate the runoff curve number polygons. Each polygon should have at least three attributes: subbasin ID, land use, and HSG. The SCS land use–HSG–curve number matrix can now be used to assign runoff curve numbers to each polygon according to its land use and HSG. Polygon runoff curve number values can be area-weighted to compute the mean runoff curve number for each subbasin. These subbasin runoff curve numbers can then be entered into the model input file. The runoff curve number estimation technique is shown in Figure 7-3. The first map shows the GIS layers for subbasins and land use. The second map shows the GIS layers for subbasins and hydrologic soil groups. The last map shows the runoff curve number polygons created by intersection of subbasins, land use, and hydrologic soil groups.

Table 7-1. Runoff Curve Numbers

Land Use	*Percent Imperviousness*	*Runoff Curve Number for Hydrologic Soil Group*			
		A	B	C	D
Open space (lawns, parks, etc.)		49	69	79	84
Impervious areas (parking lots, roofs, etc.)		98	98	98	98
Paved streets and roads		98	98	98	98
Dirt roads		72	82	87	89
Commercial and business	85	89	92	94	95
Industrial	72	81	88	91	93
Residential: 1/8 acre lots or smaller (townhouses)	65	77	85	90	92
Residential: ¼-acre lots	38	61	75	83	87
Residential: ½-cre lots	25	54	70	80	85
Residential: 1 acre lots	20	51	68	79	84
Residential: 2 acre lots	12	46	65	77	82
Newly graded areas		77	86	91	94
Row crops (straight row)		72	81	88	91
Meadow		30	58	71	78
Brush (good condition)		30	48	65	73
Woods (good condition)		30	55	70	77
Farmsteads		59	74	82	86

Source: U.S. Natural Resources Conservation Service

ESTIMATING IMPERVIOUSNESS

Some rainfall–runoff models also need an input for the subbasin percent imperviousness, which can also be estimated in a GIS using the SCS runoff curve number tables. A layer for the subbasin percent imperviousness can be created by overlaying the layers for subbasins and land use to delineate the polygons for percent imperviousness. Each polygon should have at least two attributes: subbasin ID and land use. The land use–percent imperviousness matrix can then be used to assign percent imperviousness values to the polygons. Polygon percent imperviousness values can be area-weighted to compute the mean percent imperviousness value for each subbasin.

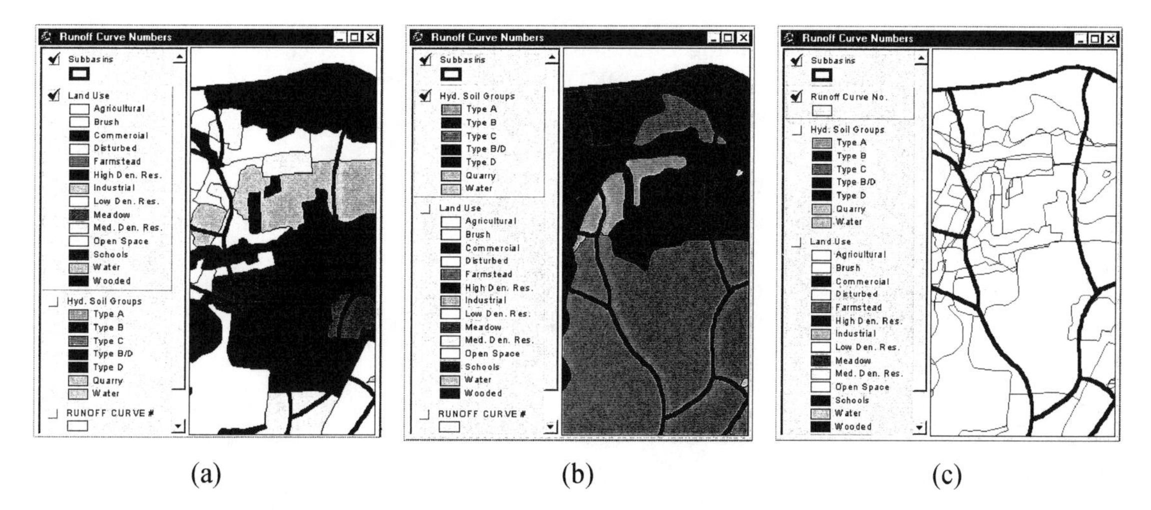

Figure 7-3. Runoff Curve Number Estimation: (a) Subbasins and Land Use; (b) Subbasins and Hydrologic Soil Groups; (c) Runoff Curve Number Polygons

INTERFACE METHOD

The interface method provides a direct link to transfer information between the GIS and the model. The interface method consists of at least the following two components: (1) a pre-processor, which analyzes and exports the GIS data to create model input files; and (2) a post-processor, which imports the model output and displays it as a GIS theme. Enabling the pre- and post-processing capabilities requires computer programming using the GIS software's scripting language, such as Avenue or Visual Basic for Applications (VBA). End users do not need to know how to program—all they have to do is point and click.

The interface method basically automates the data interchange method. The automation is accomplished by adding model-specific menus or buttons to the GIS software. The model is executed independently from the GIS; however, the input file is created, at least partially, from within the GIS. The main difference between the interchange and interface methods is the automatic creation of a model input file. In the data interchange method, the user finds a portion of a file and copies it. An interface automates this process, so that the pre-processor and post-processor find the appropriate portion of the file automatically. Let us look at some interface examples.

SWMM AND ARCVIEW INTERFACE

The EPA's Storm Water Management Model (SWMM) is the most widely used urban hydrologic/hydraulic model in the United States (Huber and Dickinson, 1988). A sample interface developed by Shamsi for SWMM is shown in the figure on the first page of this chapter (Shamsi, 1998). The interface adds a SWMM menu to the main menu of ESRI's ArcView GIS software. The SWMM menu has the following options: Make Input File, Read Output, Join Results, Unjoin Results. These options allow a user to create a model input file, import an output file, join output results to the GIS themes, and remove the results joined to the themes.

SWMMTools, a free and open-source SWMM extension for the ArcView GIS extension described above, is another example of the interface method (Heineman, 2000).

HEC GEO-HMS And GEO-RAS Interfaces

U.S. Army Corps of Engineers' Hydrologic Engineering Center (HEC) has developed HEC Geo-HMS and HEC Geo-RAS as geospatial hydrology toolkits for HEC-HMS and HEC-RAS users, respectively, who have limited GIS experience. These toolkits—developed as ArcView GIS extensions—allow users to expediently create hydrologic input data for HEC-HMS and HEC-RAS models. Additional HEC Geo-HMS and HEC Geo-RAS information is available in Chapter 3 (GIS Applications Software).

Interface For Management of Rainfall Data

All rainfall–runoff models obviously have a critical need for precipitation data to drive the model. Continuous simulation, now becoming more common, requires hourly or sub-hourly rainfall data for many years (1–50) (Shamsi and Scally, 1998). GIS can link the rain gauge locations with the rainfall database to facilitate rain gauge selection and data retrieval. Figure 7-4 shows such a utility called GeoSelect, which consists of the following two parts (Hydrosphere Data Products, 1996):

1. Hydrodata: a standalone windows software consisting of a relational database model and interface to retrieve stations and rainfall time-series data; and

2. ArcData: GIS layers for rain gauges, rivers, lakes, watersheds, and counties; and an ArcView interface for transferring data to and from Hydrodata.

Historical rainfall and GIS data for a entire state are provided on a CD-ROM. The rainfall data correspond to the National Climatic Data Center (NCDC) archives of U.S. National Weather Service (NWS) gaging stations. The hourly data in these files date from as early as 1900, with most stations' digitized records dating from 1948. The 15-minute data are from 1971 on. GeoSelect can export rainfall data in standard NCDC formats, which can be read by many computer models. For example, rainfall events data exported in the NCDC format can be read directly by SWMM as Post-1980 NWS Format. In GeoSelect, Hydrodata and GIS are not integrated. Both ArcView and Hydrodata must be run independently. The main purpose of the ArcView interface is to select stations for data retrieval in Hydrodata. GeoSelect adds two new menu items and two new buttons to the standard interface of ArcView 3.x: Export Selection (E button) and Import Selection (I button). Export Selection will transfer a list of selected rain gauge stations from ArcView to Hydrodata. Import Selection will transfer a list of selected rain gauge stations from Hydrodata to ArcView.

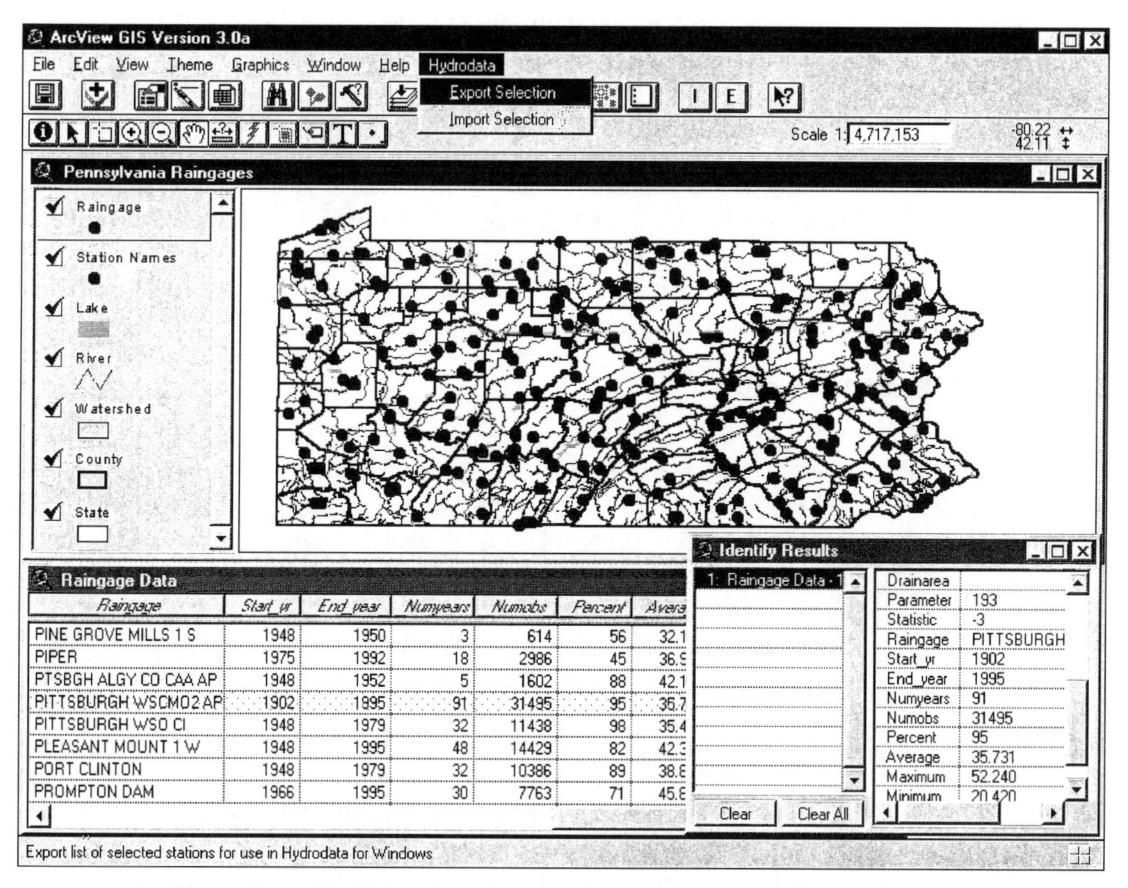

Figure 7-4. GeoSelect Interface for the Management of Rainfall Data

INTEGRATION METHOD

In the interface method options for data editing and launching the model from within the GIS are not available. The interface simply adds new menu options or buttons to a GIS interface to automate the transfer of data between a computer model and a GIS. GIS integration, conversely, is a combination of a model and a GIS such that the combined program offers both the GIS and the modeling functions. This method represents the closest relationship between GIS and model. Two integration approaches are possible:

1. GIS-based integration: In this approach, modeling modules are developed in or are called from a GIS. All the four tasks of creating model input, editing data, running the model, and displaying output results are available in the GIS. There is no need to exit the GIS to edit the data file or run the model.

2. Model-based integration: In this method, GIS modules are developed in or are called from a computer model.

Because development and customization tools within most GIS packages provide relatively simple programming capability, the first approach provides limited modeling power. Because it is difficult to program all the GIS functions in a hydrologic model, the second approach provides limited GIS capability. Furthermore, the availability of the source code for a large number of public-domain computer models makes the first approach more feasible.

EPA's BASINS Program

The best example of GIS integration with multiple models is the EPA's Better Assessment Science Integrating Point and Nonpoint Sources (BASINS) program. BASINS supports the development of total maximum daily loads (TMDLs), which require a watershed-based approach that integrates both point and nonpoint sources. BASINS integrates the ArcView GIS with many national watershed databases and with advanced environmental assessment and modeling tools in one convenient package. The heart of BASINS is its suite of interrelated components essential for performing watershed and water-quality analysis. These components are grouped into five categories: (1) national databases; (2) screening-level assessment tools for evaluating water quality and point source loadings at a variety of scales; (3) utilities such as land use and digital elevation model (DEM) reclassification and watershed delineation; (4) watershed and water-quality models, including HSPF, TOXIROUTE, and QUAL2E; and (5) post-processing output tools for interpreting model results. Figure 7-5 shows the target feature of the BASINS program, which broadly evaluates a watershed's water quality and point source loadings.

BASINS 3.0 is the current release at press time. This release includes additional functional capabilities as well as an updated and expanded set of national data layers. New data and functions include 1-degree DEM grids, an automatic delineation tool that creates watershed boundaries based on DEM grids, and new watershed report function for land use, topography, and hydrologic response units. In addition, a new watershed model, the Soil and Water Assessment Tool (SWAT), has been added. SWAT is a watershed scale model developed to predict the impact of land management practices on water, sediment, and agricultural chemical yields in large complex watersheds with varying soils, land use, and management conditions during long periods of time.

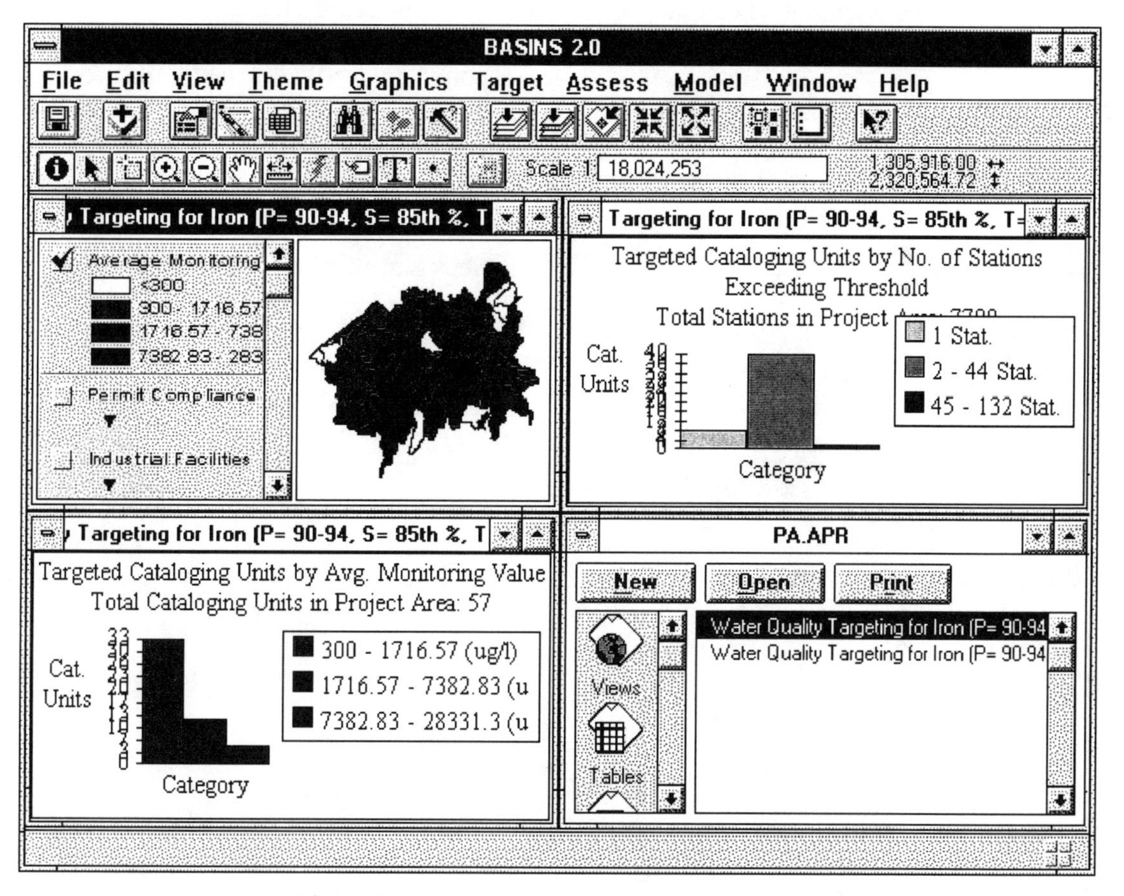

Figure 7-5. ArcView and BASINS Integration

Plans to add the EPA's SWMM program to future releases of BASINS are on again or off again, although EPA Region 4 is working on an upcoming release of their own Watershed Characterization System that is similar to BASINS and may finally integrate SWMM with GIS.

BASINS is distributed on a CD-ROM containing the program and data for an EPA region of interest. It can also be downloaded from the BASINS Web site (*www.epa.gov/OST/BASINS*).

PCSWMM GIS

PCSWMM GIS (Computational Hydraulics International, Guelph, Canada; *www.chi.on.com*) is a good example of the model-based integration method. PCSWMM GIS is a pre-processor for EPA's SWMM, and it also facilitates output visualization through a variety of plug-in tools. Model input parameters (node, conduit, and subcatchment) are extracted from an ODBC-compliant database, such as Microsoft Access, using SQL queries. Model input data can also be imported from an underlying dBASE-compatible GIS database. Extracted data are saved in an intermediate database (MS Access) for pre-processing into a useful model. Processed data are exported to a SWMM input file (Runoff, Transport, or Extran). The cost of PCSWMM GIS 2000 is $400. Figure 7-6 shows the PCSWMM GIS interface that has been developed using Visual Basic and resembles the ArcView GIS interface.

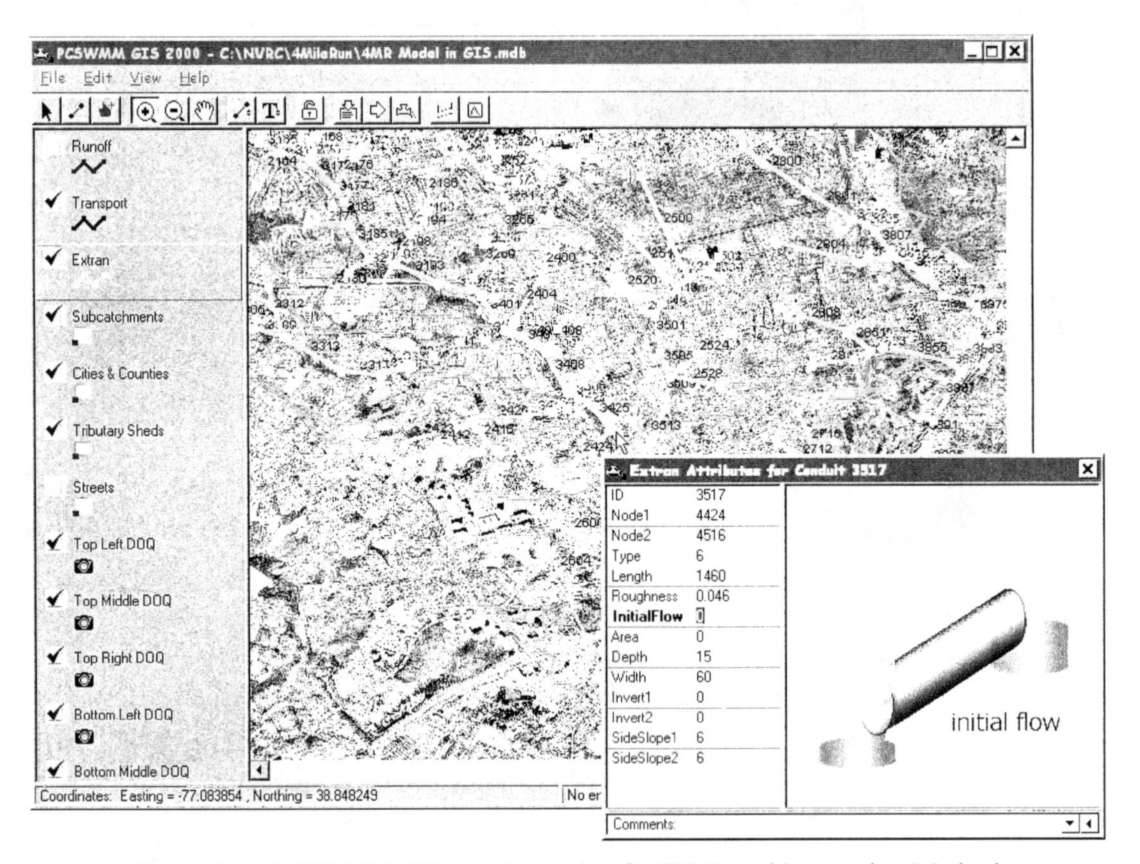

Figure 7-6. PCSWMM GIS: An Example of a GIS Based Integration Method

WHICH METHOD TO USE?

Each method has its pros and cons. Table 7-2 lists the advantages and disadvantages of the three methods of linking hydrologic and hydraulic (H&H) models with a GIS. Integrated systems are easy to learn and use and actually make the excruciating task of H&H modeling a fun activity. Integrated systems save model development and interpretation time and make modeling cost-effective. Seamless model integration with GIS is not without some disadvantages. The simplistic modeling approach and user-friendly tools, provided by integrated systems like BASINS, may encourage inexperienced users to become instant modelers, which could be dangerous (Thuman and Mooney, 2000).

Table 7-2. Pros and Cons of GIS Linkage Methods

Feature	*Interchange Method*	*Interface Method*	*Integration Method*
Automation	None; requires manual batch processing to copy data	Some; must frequently switch between H&H and GIS software	Full; all tasks can be performed from within one software program
Ease of use	Cumbersome	Easy	Very user-friendly
Learning curve	Steep	Average	Short
Data entry error potential	High	Moderate	Low
Data error tracking	Easy	Difficult	Difficult
Misuse potential by inexperienced users	Low	Moderate	High
Development of linkage between H&H models and GIS	Easy	Moderately difficult	Very difficult
Computer programming and scripting requirements to create the linkage	Optional	Moderate	Extensive

The GIS can easily convert reams of computer output into thematic maps. The beautiful GIS maps can both highlight and hide data errors. Inexperienced H&H modelers should exercise caution when using integrated systems, making sure to browse the model input and output files for the obvious errors, and avoiding over-reliance on thematic mapping and graphing of model output. Modification and recompilation of the computer codes of existing legacy programs like SWMM or EPANET should be avoided to create the integrated systems. If recoding is necessary, developers should be extremely cautious to avoid coding errors and associated liability. Users should remember that a model that is difficult to use is much better than an inaccurate model (Shamsi, 2001).

SUMMARY

This chapter demonstrated that GIS and H&H model linkage allows users to be more productive because they can devote more time to understanding the problem and less time to the mechanical tasks of data input and checking, getting the program to run, and interpreting reams of output. There are three methods of model integration: interchange, interface, and total integration. Interchange systems have existed since the 1980s in all domains. There are a few interface systems in the public domain and many in commercial and proprietary domains. In the fully integrated category, BASINS is the only public-domain modeling system. A few commercial systems are available, priced starting at $5,000. There are many in-house and proprietary integrated systems. The number of public-domain and commercial integration packages is expected to grow steadily during the next 5 to 10 years.

SELF-EVALUATION

1. What are the different ways of linking H&H models with GIS?

2. Give an example of each method of GIS linkage using programs other than those described in this chapter.

3. Which GIS linkage method is the easiest to use? Why?

4. Which GIS linkage method is the easiest to implement? Why?

5. Make a list of H&H modeling software programs being used by your organization and the type of GIS linkage method used by each program.

Chapter

8 WATER SYSTEM APPLICATIONS

Satellite imagery data obtained through remote sensing can be integrated with a GIS to assist with the planning and hydraulic modeling tasks of a water distribution system.

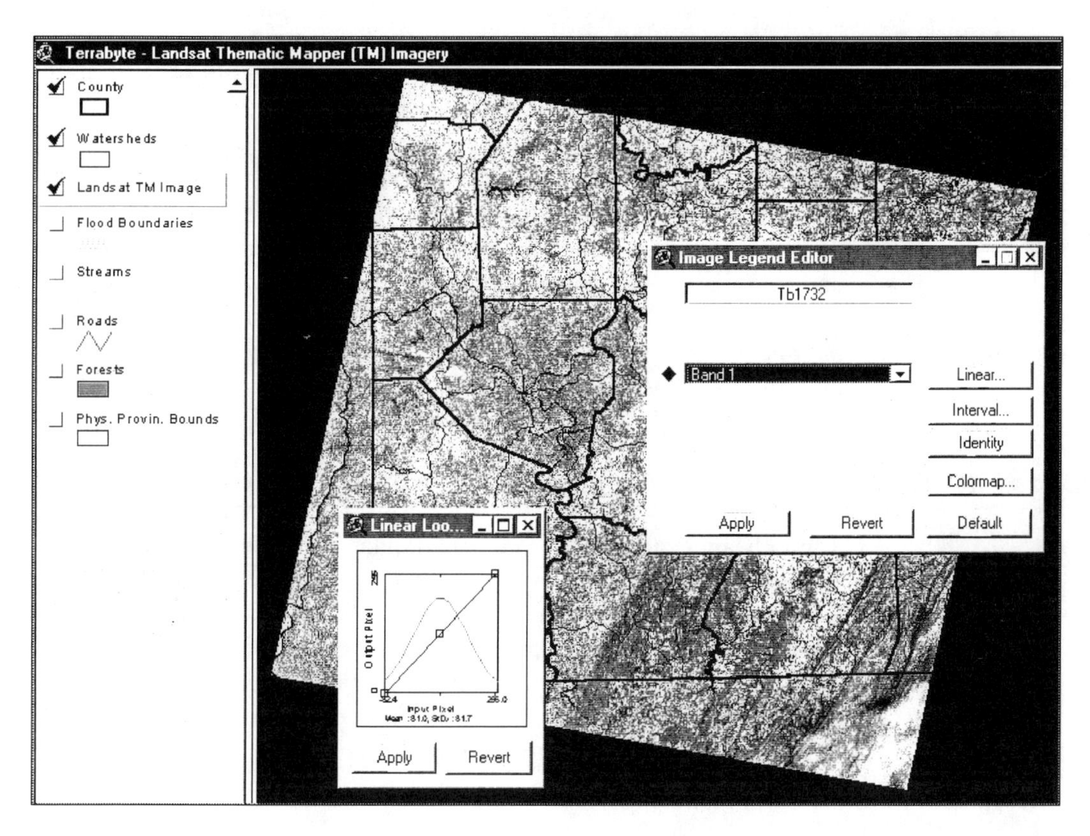

Integration of Remotely Sensed Satellite Imagery Into a GIS

LEARNING OBJECTIVE

The learning objective of this chapter is to illustrate GIS applications for water distribution systems, with a special emphasis on master planning. Major topics discussed in this chapter include

- Master planning
- Remote sensing and satellite imagery
- Land use/land cover classification
- Three-dimensional (3D) applications
- Digital elevation models (DEM)
- Calculating node elevations using a GIS

SAN DIEGO USES SATELLITE IMAGERY

The San Diego Association of Governments (SANDAG) deals with one of the nation's largest county jurisdictions, covering more than 4,200 square miles. Before 1988, SANDAG used costly and time-consuming aerial photography and photo interpretation techniques to create land use/land cover (LULC) maps and updated them only once every five years. To meet the special challenge of keeping track of this rapidly changing area in a cost-effective manner, SANDAG turned to GIS. They used ERDAS, which is a raster GIS and image-processing software program, along with ESRI's ArcInfo vector GIS software, color infrared aerial photographs, and satellite images. Switching to satellite imagery and a GIS as a land inventory tool allowed SANDAG to see the region in a new way and permitted rapid change detection. The GIS-based mapping approach provided SANDAG with current and verified LULC data for modeling transportation, infrastructure, and water needs (Kindleberger, 1992).

WATER SYSTEM APPLICATION EXAMPLES

The GIS applications that are of particular importance for water utilities include mapping, facilities management, work order management, and short- and long-term planning. The planning activities of a water distribution system can be greatly improved through the integration of these applications. By using information obtained with these applications, a water system manager can develop a detailed capital improvement program or operations and maintenance plan (Morgan and Polcari, 1991).

GIS and mapping have wide applicability to drinking water system studies. Some typical examples are listed below.

- Assessing the feasibility and impact of system expansion;
- Hydraulic modeling of water distribution systems;
- Estimating node demands from land use, census data, or billing records;
- Estimating node elevations from digital elevation model (DEM) data;
- Model simplification or skeletonization (i.e., reducing the number of nodes and links to be included in the hydraulic model);
- Determining water main isolation (i.e., identifying the valves that must be closed to isolate a broken water main for repair);
- Identifying dry pipes (i.e., locating customers or buildings that would not have any water due to a broken water main);
- Preparing a work order management using a point-and-click approach;
- Providing the basis for investigating the occurrence of regulated contaminants for estimating the compliance cost or evaluating human health impacts (Schock and Clement, 1995);
- Investigating process changes for a water utility or to determine the effectiveness of some existing treatment such as corrosion control or chlorination; and
- Developing wellhead protection plans.

In this chapter, we will focus on the first two applications involving water system expansion and hydraulic modeling. This is a popular application of GIS in the area of planning. GIS application in LULC classification will also be demonstrated. Although the planning application will be illustrated for water systems, the methodology is quite general and can be applied to wastewater and stormwater systems.

REMOTE SENSING

Remote sensing is the process of observing and mapping from a distance that allows obtaining data for a process from a location far away from the user. Remote sensing is, therefore, defined as the detection, identification, and analysis of objects through the use of sensors located remotely from the object. The most common remote sensing (or remotely sensed) data consist of digital images of the Earth's surface called *satellite imagery* acquired from airplanes and satellites.

Although vector GIS data are still an important and vital tool for many water, wastewater, and stormwater system applications, the newer raster GIS applications of remotely sensed images are beginning to make a major move into the GIS and mapping industry. The remote sensing systems that are useful in water, wastewater, and stormwater applications are aerial photography, satellite imagery, and radar data. Satellite imagery data have several benefits (Schultz, 1988):

1. They enable aerial measurements in place of point measurements.
2. They are collected and stored in one place.
3. They offer high spatial and temporal resolution.
4. They are available in digital form.
5. Their acquisition does not interfere with data observation.
6. They can be gathered for remote areas that are otherwise inaccessible.
7. Once remote sensing networks are installed, their measurement is relatively inexpensive.

Satellite data became available in 1972 when the U.S. government launched the first Landsat satellite, which was specifically designed to provide scientifically valuable high-altitude imagery of the Earth that was previously unavailable (Miotto, 2000). Figure 8-1 shows a Landsat Thematic Mapper (or TM, a sensor onboard Landsat) image called "Terrabyte" overlayed by floodplain boundaries for the City of Pittsburgh (Pennsylvania, USA). This image was created from the data compiled by Pennsylvania State University. Until recently, satellite images had relatively low resolution. In January 2000, high resolution satellite images became available in the commercial marketplace for the first time. Thanks to the September 1999 launch of IKONOS (Space Imaging, Thornton, Colorado; *www.spaceimaging.com*) anyone can now purchase 1-m black and white imagery and 4-m multispectral (color) imagery at a reasonable cost. One-meter imagery represents an accuracy level commensurate with 1:2,400 mapping, which is more than adequate for many planning and hydraulic modeling applications.

LAND USE/LAND COVER MAPPING

LULC information is critical in master planning, watershed analysis, stormwater management, hydrologic modeling, and urban growth assessment. The LULC resolution requirements vary from user to user. For example, a data source acceptable for a regional study may be inadequate for a local study (Garland et al., 1990). USGS and EPA provide public-domain LULC data, as described in Chapter 5 (Internet GIS). However, the small scale of these data (1:100,000 to 1:250,000) may not be appropriate for local planning studies. For more detailed studies, custom LULC maps should be created using aerial photographs and high resolution satellite imagery. Identification and quantification of LULC classes is a labor-intensive and expensive process. Derivation of LULC classes from the low-level aerial photography is referred to as the "conventional" method described in Chapter 10 (Stormwater System Applications). Satellite imagery provides a cost-effective alternative to conventional hands-on techniques (ERDAS, 2001). Studies have shown that the remote sensing techniques for LULC classification are more cost-effective than the conventional method. The cost benefits have been estimated on the order of 6 to 1 in favor of the satellite imagery approach (Engman, 1993).

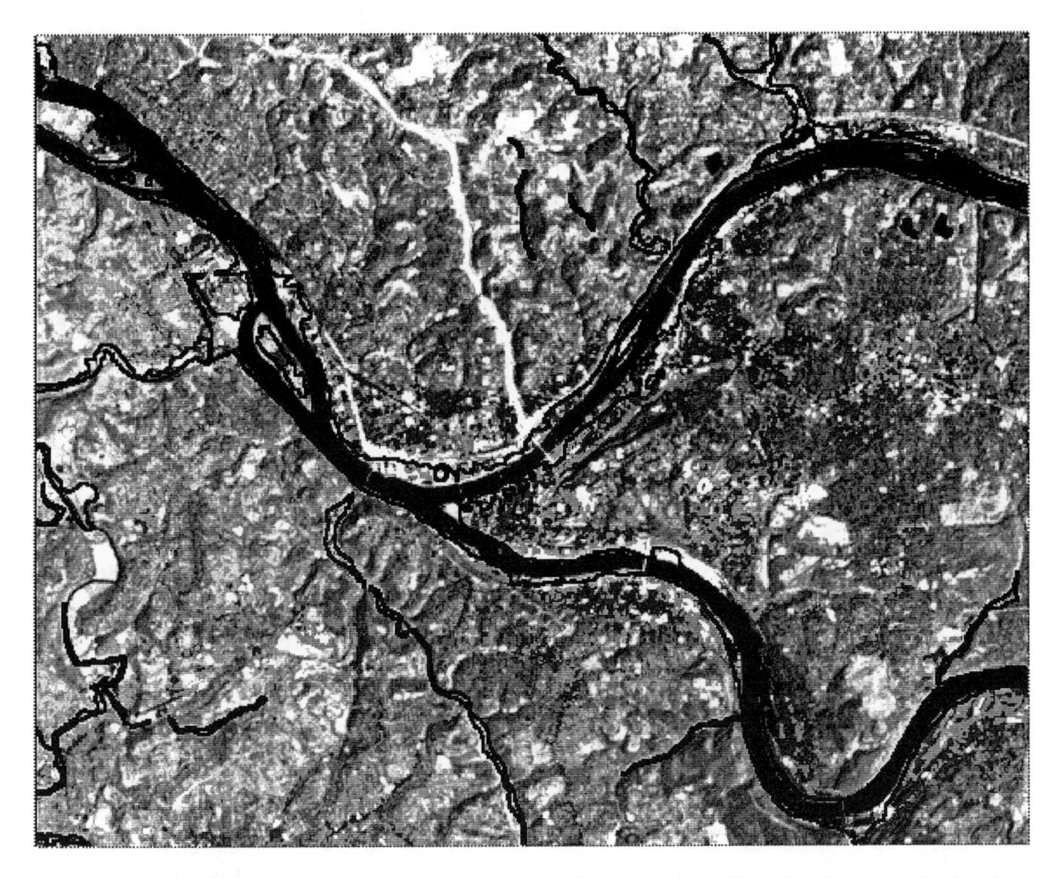

Figure 8-1. Landsat Thematic Mapper Image Overlayed by Floodplain Boundaries for Pittsburgh (Map Courtesy of Pennsylvania State University).

Remote sensing data can be analyzed in a GIS or image processing package to prepare an LULC layer. LULC changes observed in remote sensing imagery are of interest to planners, ecologists, hydrologists, and atmospheric scientists. Remote sensing imagery taken at different times can be compared to detect changes in LULC, using a process called "change detection." Change detection is an important and growing use of image data in which images captured on different dates are compared to calculate difference in their spectral responses (brightness) over time.

CASE STUDY

There is a popular saying among planning circles, "If you will put in a water pipe, they will come." Simply stated, access to a public water supply promotes new development. This case study illustrates a GIS application of planning for identifying new water supply areas.

West Deer Township is located about 20 miles north of Pittsburgh in Allegheny County of southwestern Pennsylvania. It is a developing rural community with a 1980 population of approximately 11,000. In 1990, about 10 percent of the township's population had no public water and depended on water from private wells. Some residents supplied by private wells were facing well water contamination problems and had petitioned the township for public water supply extensions. The township's industrial park was showing signs of substantial growth. The industrial park relied on a water supply from wells, which was not sufficient to satisfy day-to-day demands and was far short of meeting the fire demands. Recently, a $20 million sewer project had been completed in the township, which had attracted new residential and commercial developments. These conditions indicated a need for an active waterline extension program. Such a program should, however, be based on a well-planned implementation strategy rather than one driven solely by a particular development. In 1990, the township retained an engineering consultant (Chester Engineers, Pittsburgh) to develop a Master Water Plan as a tool to prioritize the areas within the township that should receive public water (Shamsi, 1990). The major objective of the master plan was to identify the steps to be taken for a planned waterline extension program.

The water service extensions should be given priority in the areas where a need for a public water supply has already been identified. Subsequently, those areas should be considered where future water supply needs can be most reasonably predicted. Short-term water service extension needs were defined as those extending through 2000. Such needs were based on known water supply problems; public water service requests from residents, industry or business; and building permits. Water service needs most likely to be imposed between 2000 and 2010 were referred to as long-term water service extension needs.

SHORT-TERM WATER SERVICE EXTENSIONS

West Deer Township was experiencing significant residential development. Township staff revealed several areas where an extension of the public water supply should be reviewed. These areas were classified into high- and low-priority groups on the basis of the number of houses expected to benefit from the new service, existing water quality problems, and approved and proposed residential and industrial developments.

LONG-TERM WATER SERVICE EXTENSIONS

Long-term water service extensions were identified from an assessment of potential future development. This assessment was conducted by combining remote sensing and GIS techniques.

Software

Both the vector and raster GIS techniques were used to benefit from the best features of each. Vector GIS analysis was performed using PC ArcInfo. Raster GIS analysis and image processing was conducted using ERDAS (now known as IMAGINE). More information about these packages is provided in Chapter 2 (GIS Development Software) and Chapter 3 (GIS Applications Software).

Data Sets

Four data sets were used in this study:

1. The first data set was a 1988 Landsat TM scene, which was a subset based on the location of the township within the image area. It was registered to a Universal Transverse Mercator (UTM) map projection using 31 control points. The root mean square (RMS) error from this registration was less than 0.5 cells or 16.4 ft (5 m). During registration, the cell size was resampled to 32.8 ft by 32.8 ft (10 m by 10 m) to allow raster processing with the other data sets used in this analysis. This resampling procedure does not provide more or better information at the 32.8-ft resolution, but merely allows processing with other data layers of the same size and dimensions.

2. The second data set was the French satellite SPOT-1 1987 panchromatic image. The SPOT High Resolution Visible (HRV) image sensors capture reflectance information in the spectral range of 0.51 to 0.73 μm wavelengths at a spatial resolution of 32.8 ft (10 m). This image was used to obtain qualitative information about the recent changes in the area and the textural diversity in urban areas. This textural information was used to delineate different levels of

residential densities and to check the results of the automated land cover classification.

3. The third data set was derived from a USGS 7.5-min. DEM. These data were also resampled to a 32.8 ft (10 m) resolution to allow digital processing.

4. The fourth data set consisted of water mains, sewers, and roads initially digitized in a vector format using PC ArcInfo digitizing software and later converted to a raster format to achieve compatibility for GIS processing.

Data Processing

All data sets were registered to the UTM projection using control points that were identifiable in both the original image and USGS 7.5-min. quadrangles. The UTM coordinate locations for a total of 31 control points were determined and used to register all the data sets. The high number of control points used is due to the fact that some points were not identifiable on each map layer. Each map was registered using an acceptable RMS error of not more than 0.0382 map in. or 60.5 ft (18.44 m). All GIS processing was performed in meters rather than in feet because of the UTM map projection used.

The percent slope was calculated from the DEM data set to provide information on the availability of suitable sites for future development. The roads and water and sewer line networks were processed to create a buffer area around each of these linear features to provide information about proximity.

GIS Analysis

Land cover classes were derived using a multistage approach to processing many levels of map data, providing different kinds of spatially oriented information. The initial product used to obtain land cover information was the TM scene, which was classified using a supervised classification procedure in which selected areas of the scene, which are believed to be representative of the major classes of land cover in the scene, are delineated. The image processing software computes statistics (mean, covariance, standard deviation) of the population in the enclosed area. The statistics are displayed as a histogram and compared with the ideal normal distribution. By this process, spectral signatures for the land cover classes are chosen. The signatures are then analyzed, cell by cell, and classified using a maximum likelihood decision rule.

These derived classes of land cover were further refined using rules derived from specific surface reflectance properties in specific spectral bands. For example, to separate the cells, which were statistically similar enough to be classified in the same land cover class but which actually represent cells of vegetation and bare soil, the reflective differences in the photographic infrared wavelengths were used to separate the growing vegetation from the bare soil. The most common and pervasive problem in any spectral classification is the presence of mixed cells. A 32.8 by 32.8 ft (10 by 10 m) location on the ground may be half lawn and shrub and half rooftop. Obviously, because the spectral reflectance value for this cell cannot show a half and half value, the value becomes an average of the two surface types and becomes statistically dissimilar from both the vegetation and urban areas. Taking into consideration the fact that cells are not spectrally divided into only two distinct surface types, but into every possible combination, the difficulties in spectrally based classification procedures become apparent.

A key challenge to using the satellite imagery data is to "unmix" (usually urban) land use areas that contain numerous mixed pixels due to a highly variable landscape. The latest software packages, such as IMAGINE software from ERDAS Inc., provide a "Subpixel Classifier" to handle the mixed cells (ERDAS, 2001). As an alternative, the following rule-based approach can be used to classify the mixed cells.

Because of mixed cell effects, additional rules were derived based on the information in the image in specific spectral bandwidths. These were used to separate land cover types from classes that contain many dissimilar cells. For example, after the supervised classification procedure created 17 classes of land cover types, several classes were not easily identified with any type. To pull wooded areas from these classes, a query of reflectance values in the photographic infrared band was used to find actively growing vegetation. Because of mixed cells and other problems, an additional manually interpreted map was used to separate known urban areas from areas not known to be urbanized. This was used to separate cells with a low probability of being correctly assigned to a land cover class to either an urban or nonurban surface type. This map was derived from several sources of information: the USGS quadrangles, the SPOT imagery, and the township road map. The road map provided somewhat current (July 1985) information about areas of development, but as was stated above, the spatial accuracy was not adequate. The SPOT imagery was more current (August 15, 1987) and showed several subdivisions not present on the USGS quadrangle (photorevised, 1979). The USGS quadrangle contained the oldest cultural information but the best spatial accuracy. For this reason, all current development information from the image was compiled onto the USGS quadrangle and digitized.

GIS Results

Present Development

The result of several iterations of land use classification refinements in land use categories are listed in Table 8-1 (Shamsi, 1992). A vector map of present LULC map is shown in Figure 8-2. Woodland is the largest class, with a little more than 50 percent. The delineation of wooded areas required a number of training sites and additional spectral and GIS queries to separate the effects of terrain from all of the other information present in the image.

Table 8-1. Potential Future Land Use/Land Cover in West Deer Township

Land Use Class	*Present*		*Future*	
	Percent	Acres	Percent	Acres
Woodland	50.34	9,265.61	33.81	6,222.35
Agriculture	12.31	2,265.49	7.27	1,338.70
Mine tailings	00.53	98.40	0.08	14.65
Residential	13.21	2,430.88	40.40	7,435.07
Commercial	1.10	202.16		1.57
Industrial	0.23	42.33	2.94	541.01
Urban vegetation	0.57	105.76	0.30	55.15
Urban bare soil / open field	0.68	125.90	0.06	11.12
Mixed low-density residential / vegetation	21.03	3,868.49	13.57	2,497.77
Total	100.00	18,405.02	100.00	18,405.01

Source: After Shamsi, 1992.

Future Development

The result of several iterations of land use classification refinements in potential future land use categories are listed in Table 8-1 (Shamsi, 1992). A vector map of future LULC is shown in Figure 8-3.

The possible future development in the township was derived on the basis of the following rules for likely development constraints:

1. Development would occur within 2,000 ft of a road, a sewer, or a water main.

2. Development would not take place on slopes greater than 25%.

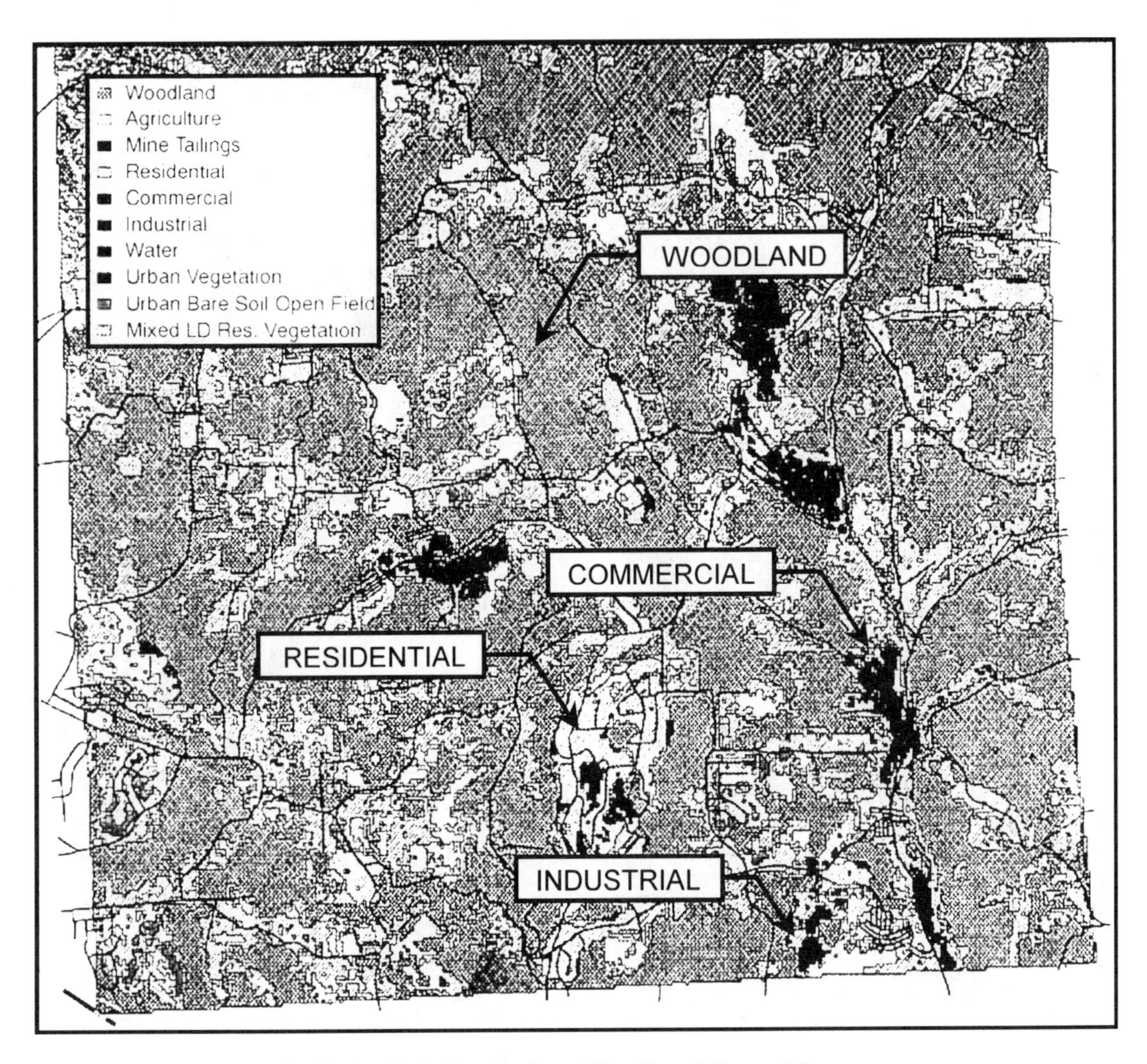

Figure 8-2. Present Land Use/Land Cover Map

3. Development would not take place in floodplain and conservation areas. In these areas, it was assumed that development would not occur in contradiction to the long-term plans of the township as shown in its zoning map.

Figure 8-3 shows that most of the central portion of the township has a potential for residential development, as is indicated by the lightly shaded area. Substantial industrial growth potential exists in the southwestern portion, as is indicated by the dark areas in Figure 8-3. Such distinctions are much easier to see on the color map plots.

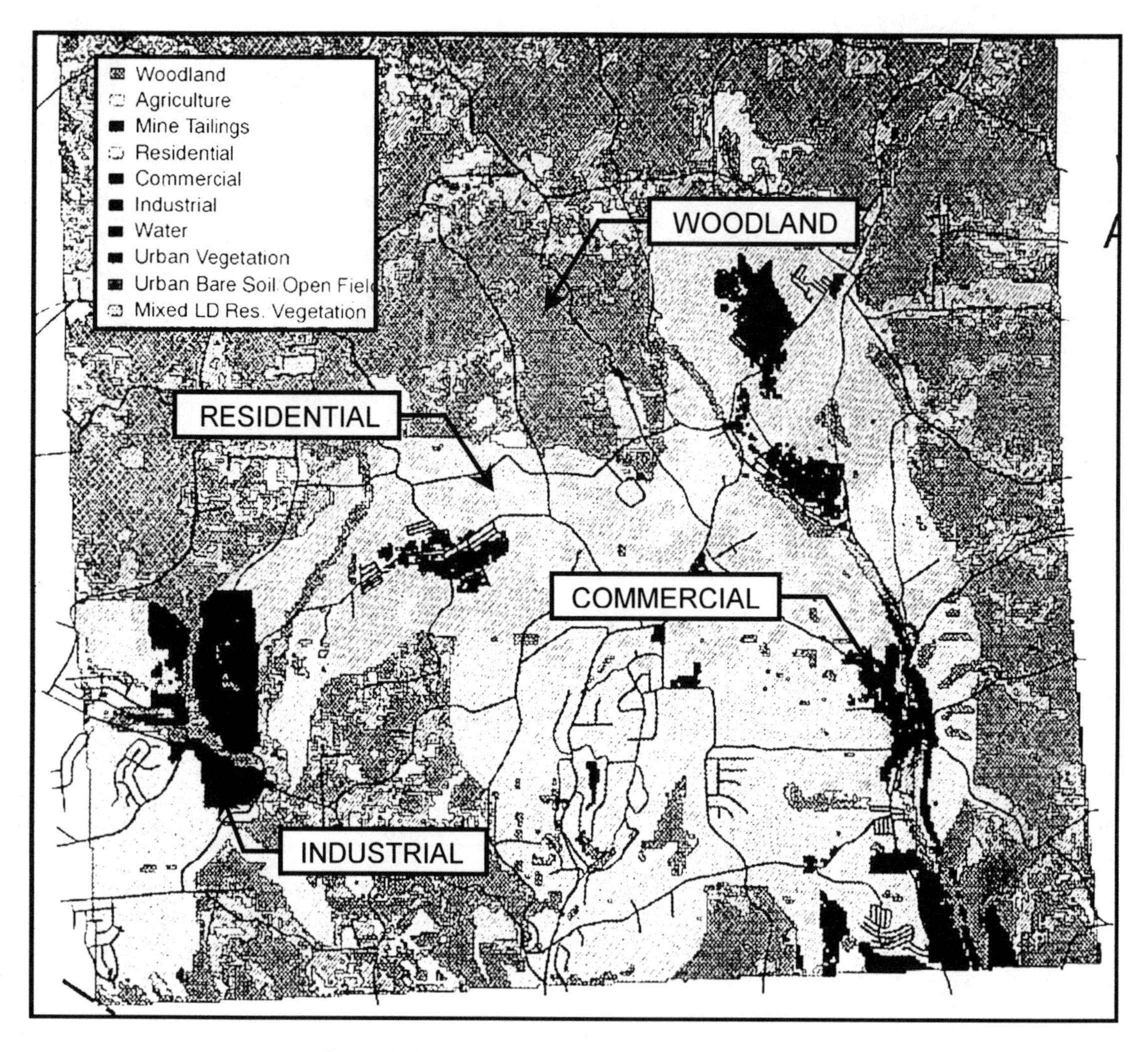

Figure 8-3. Future Land Use/Land Cover Map

A third map (Figure 8-4), showing only the change in development, was constructed next, by subtracting the present developed areas from likely future development areas. This map showed 5,590 acres (residential, 5,004; commercial, 87; industrial, 499) of potential future development areas. The areas of future development can be divided into two categories:

1. Those that will grow in the future but have water service and do not require service extensions.
2. Those that will grow in future, do not have water service, and require new water service extensions.

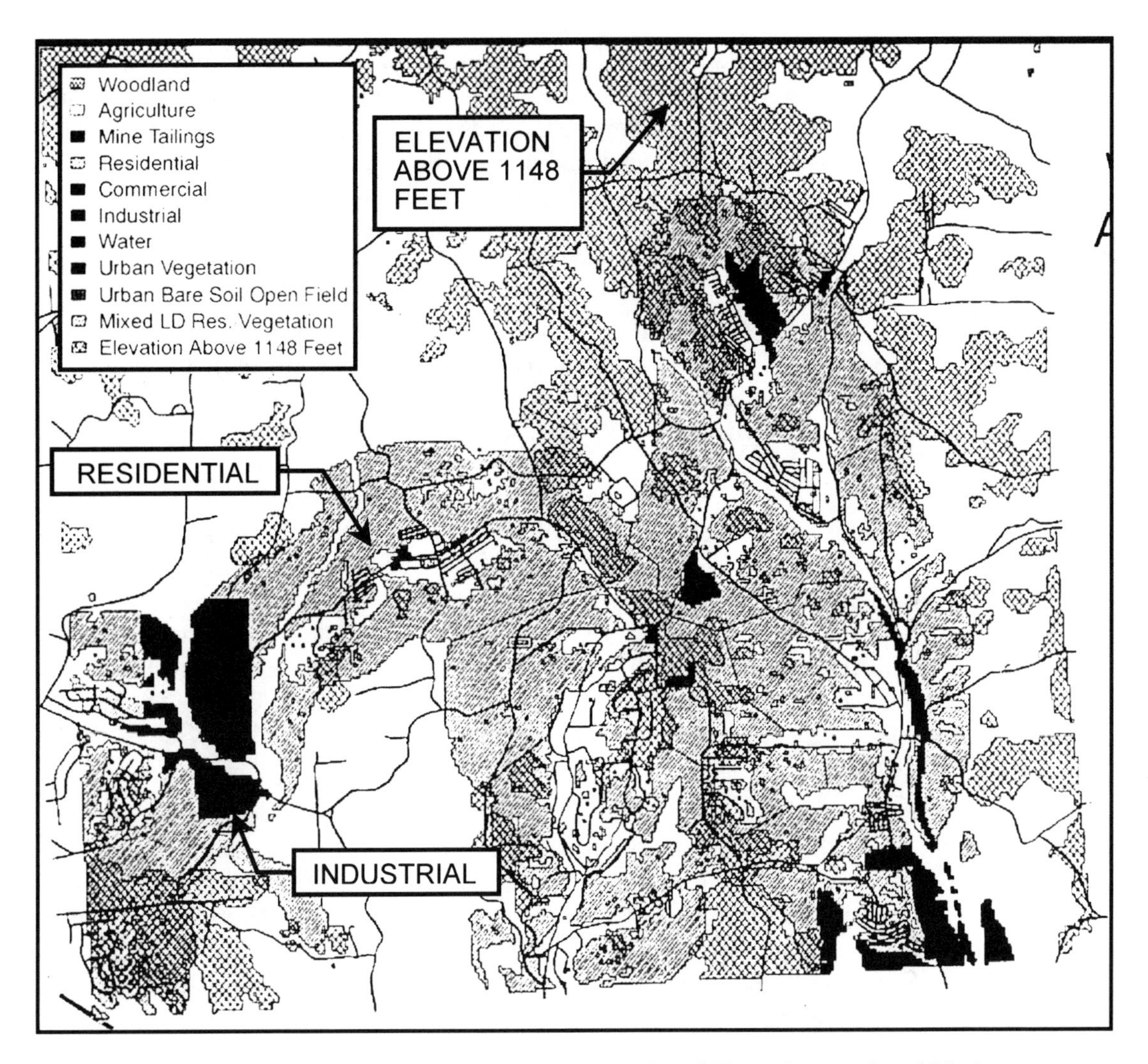

Figure 8-4. Change Detection Map (Future Land Use – Present Land Use)

Only the residential growth areas were found to fall into both of these categories. In future commercial growth areas, water service is already available; therefore, no additional service extensions are required. Such areas with commercial growth potential, but having water service availability were, however, given special consideration in the hydraulic network modeling phase of the project to evaluate the township's fire-fighting capability in anticipation of future development. In all but one future industrial growth area of the township, public water supply was already available. Finally, a list of future development areas was prepared.

The remote sensing and GIS analysis described above shows that, on the basis of assumed criteria, significant future development potential existed in the township. In total, about 5,600 acres of land could be developed in residential, commercial, and industrial classes, which was more than twice the existing development within these three categories (2,675 acres). Most of the projected development will occur in the residential land cover class. Significant industrial growth potential existed in the industrial park area, which confirmed the township's initial assessment.

Next, a hydraulic network model of the water distribution system was developed using the University of Kentucky's legacy KYPIPE program (*www.kypipe.com*). Water service extensions were modeled by introducing hypothetical pipe lines between the existing pipelines and future areas. What-if simulations were performed for each extension to determine a suitable design for providing acceptable flows and pressures. The hydraulic model indicated that areas above an elevation of 1,148 ft could not be served by public water without the installation of new storage and pumping facilities. As shown in Figure 8-4, a layer of areas with elevation above 1,148 ft was developed from the DEM data. These high areas were also excluded from the potential future development areas.

Case Study Conclusions

The LULC classification developed in this study was derived using a multistage approach combining spectral reflectance information from satellite imagery classified using both a supervised classification and spectral rules, and manual interpretations. To derive the potential development map, a zoning map, digital elevation model, and water, sewer, and road layers were used to establish the criteria whereby development may occur in the future.

Development was considered possible in an undeveloped area within 2,000 ft of a water line, sewer line, or a road, and on slopes of less than 25% that did not fall within either the floodplain or conservation areas as outlined on the township's zoning map. The differences between current and future LULC were quite significant in residential, commercial, and industrial sectors of development.

It should be noted that long-term water service extension needs were based on certain assumptions about the likelihood of future development and therefore cannot be considered precise. Priorities cannot be determined without knowing the hydraulic importance of the extensions to the overall hydraulic performance of the system. A prioritized list of the proposed service extensions was developed in the hydraulic network modeling phase of the project.

Three-dimensional GIS Applications

The performance of a water distribution system greatly depends on ground elevation. For example, high elevation areas may experience inadequate pressure if appropriate pumping is not provided. Thus, all distribution system models require input data for the elevation of each model node. DEMs can be used for three-dimensional (3D) representation of water supply networks and estimating elevation at the nodes of a model.

Graphics are employed as a visual communication tool to display a 3D view of an object in two-dimensional media. Until the early 1980s, a large mainframe computer was needed to view, analyze, and print objects in a 3D graphics format. Now, however, hardware and software are available for 3D mapping of water distribution networks on personal computers. Software developers have made dramatic advances in 3D visualization during the past few years, helping thousands of GIS users bring their applications to life. Optional 3D extensions of GIS software, such as ESRI's 3D Analyst, can create 3D maps of terrain, water demand, and modeled pressures. Shamsi (1991) showed applications of PC-based 3D graphics in network and reliability modeling of water distribution systems. GEOWorld (2001) presents several case studies to demonstrate what 3D visualization software can do and how the 3D applications are being used.

How to Calculate Node Elevations?

This example shows how to calculate node elevations for Haestad Method's WaterCAD model using ArcView GIS. If node location are available in an ArcView Shapefile, ArcView 3D Analyst extension (Version 3.x) can be used to calculate node elevations from a DEM grid. This extension uses bilinear interpolation to calculate the node elevation from the grid cell elevation data. The elevation data are pulled from the DEM overlay, and the node file is converted to a 3D file. The steps for this procedure are given below (Walski et al., 2001):

1. Start ArcView and load the 3D Analyst extension.

2. Add both the DEM grid and the node point themes to the same view.

3. Select Theme \ Convert to 3D Shapefile. Select "Surface" for the elevation file type and enter the name of the DEM grid (input). Also provide a name for the 3D shapefile (output).

4. Use the addxyz.ave (3D.ADD_XYZ) script shown in Figure 8-6 to extract elevation data from the 3D shapefile and place them in the node theme table.

```
' Name: 3D.ADD_XYZ
' Title: Add X, Y and Z coord field into a 3D shape file
' Topic: View, 3D
' By:  Yuan Ming Hsu, GIS Data analyst
'      Minnesota Department of Health
'      121 East 7th Place, Suit 210
'      St. Paul, MN 55164
'      Tel:651.215.0737  Fax: 651.215.0979
'      yuanming.hsu@health.state.mn.us
' Date: 06.30.99
' Update: 06.30.99
' Description: This program will add three fields (X,Y and Z) into shape file's attribute table.  The shape file must
' be a 3D point class (pointZ). Program is modified from existing ESRI sample script.  To convert a shape file into
' 3D, see system script: 3D.ConvertTo3D. Requires: 3D shape file
'
' Self:
' Retrun:
' inital setup

theView = av.GetActiveDoc
'must be global to work in Calc exp below
_theProjection = theView.GetProjection
project_flag = _theProjection.IsNull.Not  'true if projected
theTheme = theView.GetActiveThemes.Get(0)

'Check if point or polygon theme
if ((theTheme.GetSrcName.GetSubName = "pointZ").not) then
  MsgBox.Info("Active theme must be a 3D point theme","")
  return nil
end

'get the theme table and current edit state
theFTab = theTheme.GetFTab
theFields = theFTab.GetFields
edit_state = theFTab.IsEditable

'make sure table is editable and that fields can be added
if (theFtab.CanEdit) then
  theFTab.SetEditable(true)
  if ((theFTab.CanAddFields).Not) then
    MsgBox.Info("Can't add fields to the table."+NL+"Check write permission.",
    "Can't add X,Y and Z coordinates")
    return nil
  end
else
  MsgBox.Info("Can't modify the feature table."+NL+
  "Check write permission.","Can't add X,Y and Z coordinates")

  return nil
end
```

Figure 8-6. Sample Avenue Script for Calculating Node Elevations of a Water Distribution System Model from a Digital Elevation Model (DEM)

```
'Check if fields named "X-coord", "Y-coord" and "Z-coord" exist
x_exists = (theFTab.FindField("X-coord") = NIL).Not
y_exists = (theFtab.FindField("Y-coord") = NIL).Not
z_exists = (theFtab.FindField("Z-coord") = NIL).Not

if (x_exists or y_exists or z_exists) then
  if (MsgBox.YesNo("Overwrite existing fields?",
  "X-coord, Y-coord and Z-coord fields already exist", false)) then
    'if ok to overwrite, delete the fields as they may not be defined
    'as required by this script (eg., created from another script).
    if (x_exists) then
      theFTab.RemoveFields({theFTab.FindField("X-coord")})
    end
    if (y_exists) then
      theFTab.RemoveFields({theFTab.FindField("Y-coord")})
    end
    if (z_exists) then
      theFtab.RemoveFields({theFtab.FindField("Z-coord")})
    end
  else
    return nil
  end 'if (MsgBox...)
end 'if

x = Field.Make ("X-coord",#FIELD_DECIMAL,18,5)
y = Field.Make ("Y-coord",#FIELD_DECIMAL,18,5)
z = Field.Make ("Z-coord",#FIELD_DECIMAL,18,5)

theFTab.AddFields({x,y,z})

'Get point coordinates coordinates
if (theTheme.GetSrcName.GetSubName = "pointZ") then
  if (project_flag) then

    'Projection defined
    theFTab.Calculate("[Shape].ReturnProjected(_theProjection).GetX", x)
    theFTab.Calculate("[Shape].ReturnProjected(_theProjection).GetY", y)
    theFTab.Calculate("[Shape].RetrunProjected(_theProjection).GetZ", Z)
  else
    'No projection defined
    theFTab.Calculate("[Shape].GetX", x)
    theFTab.Calculate("[Shape].GetY", y)
    theFTab.Calculate("[Shape].GetZ", z)
  end 'if
end

'Return editing state to pre-script running state
theFTab.SetEditable(edit_state)
```


Figure 8-6 (Continued). Sample Avenue Script for Calculating Node Elevations of a Water Distribution System Model from a Digital Elevation Model (DEM)

Developed by Hsu (1999), the 3D.ADD_XYZ script can be downloaded from the ESRI ArcScripts Website (*www.esri.com/arcscripts/scripts.cfm*).

5. Select File \ Export from the ArcView pulldown menus and enter the export file name for a dBase IV (DBF) or delimited text (TXT) file.
6. Select File \ Synchronize \ Database Connections from the WaterCAD's pulldown menus and set up a link between the export file (created in the previous step) and WaterCAD.
7. Select Elevation as the WaterCAD field into which the elevation data should be imported and the name of the corresponding field in the export file (e.g., Z-coord).
8. Select the link in the Database Connection Manager and click the Synchronize In button.
9. Finally, validate the accuracy of imported elevation data by checking the elevations at a number of nodes against known elevations.

3D Case Study

Figure 8-5 shows three stacked surfaces: ground surface elevations, demands, and modeled average daily pressures for the Borough of White Haven (Pennsylvania, USA). The land area of the borough is 773 acres, the population is 1,091, and the average daily demand is 197,000 gallons per day. The distribution system consists of water mains ranging in size from 2 to 8 in. It is interesting to note the high-pressure areas coinciding with the low-elevation areas, and vice versa. It also quickly becomes apparent that the high-demand areas predominate in the lower elevations. A display such as Figure 8-5 is an effective visual aid for displaying the network model results. Most important, such graphics can be easily understood by system operators without extensive training, which is usually required to understand the tabular output from a network modeling computer program.

Summary

This chapter illustrated how GIS techniques can be particularly useful in the planning activities of a water distribution system. Remote sensing and DEM data can be integrated with a GIS to develop a hydraulic model of a water distribution system. The integrated information can be used to prepare a capital improvement program for a water utility.

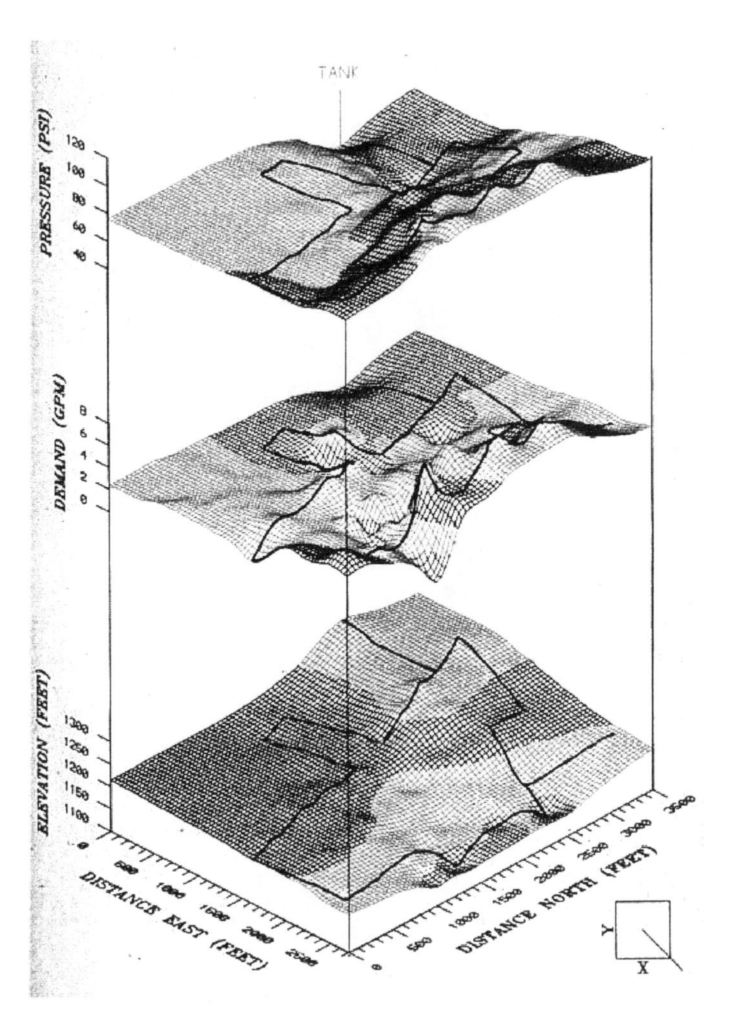

Figure 8-5. 3D Mapping of Water Distribution Network Model

Self Evaluation

1. Prepare a list of GIS applications for water distribution systems.
2. What is remote sensing and how does it help water distribution systems?
3. How is satellite imagery used to create a land use map?
4. How can you use 3D data in a GIS to support the tasks related to hydraulic modeling?
5. Does your water distribution system hydraulic model use GIS data? How?

Chapter

9 Wastewater System Applications

Ongoing advances in GIS and related technologies are providing more and more powerful and cost-effective tools for sewer system mapping.

3D Fly-Through Animation for a Sewer Construction Project in Pittsburgh (Pennsylvania, USA)

LEARNING OBJECTIVE

The learning objective of this chapter is to illustrate GIS applications for wastewater systems with a special emphasis on needs analysis and mapping. Major topics discussed in this chapter include

- ☑ Sewer system mapping
- ☑ Digital orthophotos
- ☑ Global positioning system (GPS)
- ☑ Needs analysis
- ☑ Case studies

PHILADELPHIA'S COMBINED SEWER OVERFLOW (CSO) PLAN

Application area	CSO program support
Reference	Byun and Marengo, 2001
Project Status	Ongoing in 2001
GIS software	ArcInfo and ArcView
Other software	1. RUNOFF block of EPA's Storm Water Management Model (SWMM), model component developed to simulate both the quantity and quality of runoff in a drainage basing and routing of flows to the major sewer lines; 2. Extended Transport (EXTRAN) block of SWMM, a hydraulic flow routing model for open channel and/or closed conduit systems; and 3. U.S. Army Corps of Engineers' Storage, Treatment, Overflow, Runoff Model (STORM), a planning-level model which is applied for quantity and quality analysis of urban watersheds and storage/treatment alternative screening.
GIS data	CSO regulators, sewersheds, sewers, manholes, hydrologic features, water pollution control plants, TIGER/Line data
Hardware	Dell Precision Workstation 410
Study area	City of Philadelphia (Pennsylvania, USA)
Organization	Philadelphia Water Department Office of Watersheds and Camp Dresser & McKee, Inc.

In 1994, the U.S. EPA issued the National CSO Control Policy through the National Pollution Discharge Elimination (NPDES) permit program. Since then, the Philadelphia Water Department (PWD) has developed a comprehensive CSO compliance program in an effort to improve Philadelphia's water environment and meet NPDES permit conditions.

The CSO program has produced a characterization of the CSO system, a program to implement the Nine Minimum Controls (NMC), and development of a CSO Long-Term Control Plan (LTCP).

The program's efforts have focused on the development and integration of hydrologic and hydraulic (H&H) models and a GIS for all its phases. PWD elected to develop a GIS to serve as a repository for the sewer system data essential to developing a sewer system model. The program has utilized a GIS to represent the city's water distribution and sewer collection systems. The system has served further as a basis for a comprehensive, geographically based maintenance management system. The initial characterization of capture and overflow for Philadelphia's combined collection system was developed using the STORM software. Model development focused on a detailed simulation of sewers and regulators using the RUNOFF and EXTRAN blocks of the SWMM software. Together, these resources have provided PWD with the tools necessary to successfully develop and implement Philadelphia's CSO program. These tools provide an accurate characterization of the combined sewer system components and inputs, the system condition, measurements of its behavior, and simulation of its performance now and in the future.

Data are required to describe the collection system, the combined sewersheds (drainage areas), and the meteorological conditions. The data collection sites included the interceptor sewer network, regulators, storm relief structures, trunk sewers, special junctions, and other hydraulic control points. The drainage area data collection and processing task included data describing sewershed boundaries, land use, population, housing densities, and general topography.

PWD's implementation of technically viable and cost-effective improvements and operational changes with the use of GIS-based computer models have enhanced the system performance and reduced CSO volume and frequency. The GIS component of the CSO program resources is serving as a tool for evaluating the CSO impact on a watershed basis.

Figure 9-1 shows the City of Philadelphia, its sewersheds, and the wastewater collection system network.

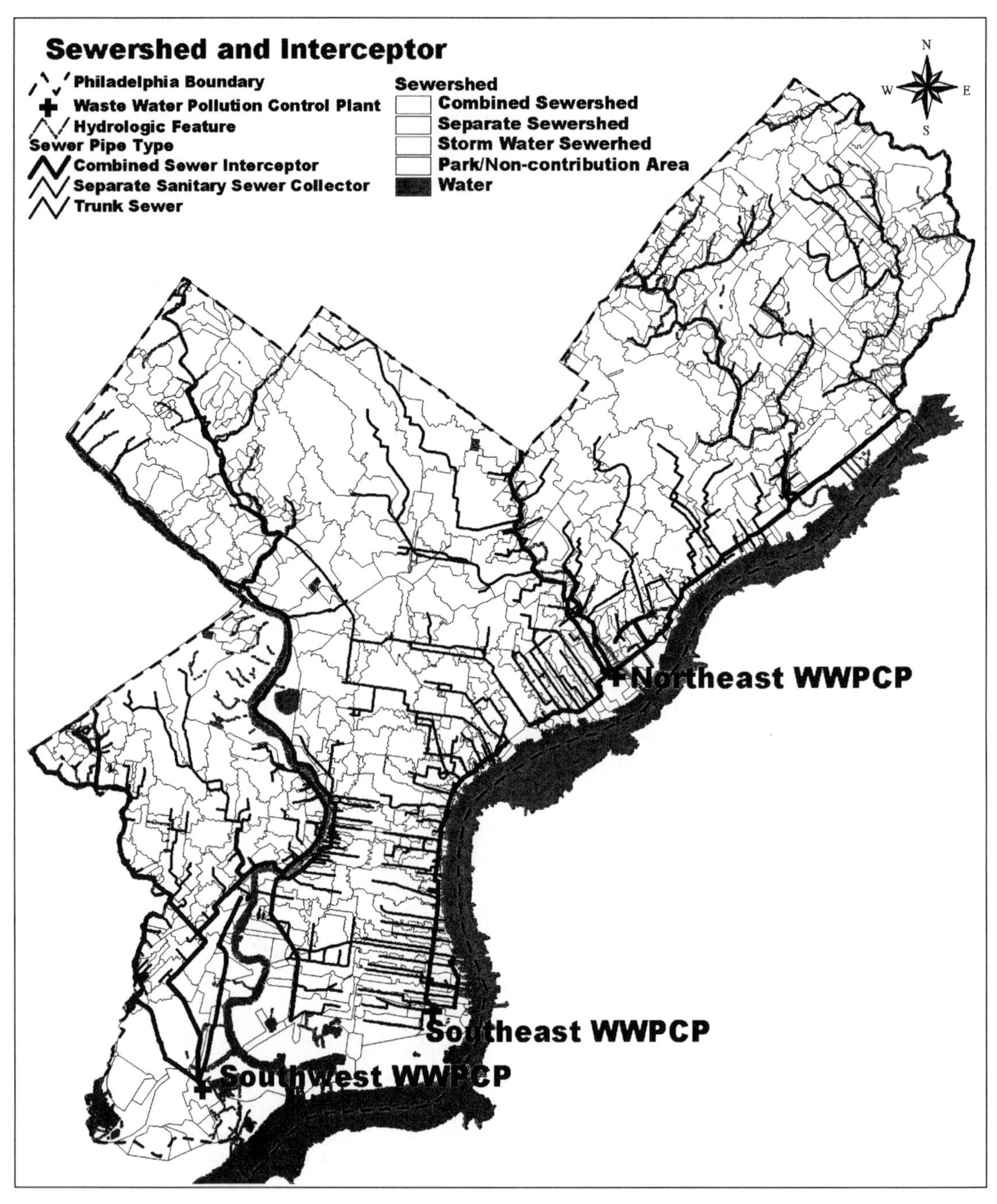

Figure 9-1. City of Philadelphia's Wastewater Collection and Conveyance System

WASTEWATER SYSTEM APPLICATION EXAMPLES

GIS and mapping applications for wastewater collection and conveyance system include:

- Mapping work for NPDES requirements:
  - EPA's CSO regulations, such as System Characterization, Nine Minimum Controls (NMC), and Long Term Control Plan (LTCP); and
  - EPA's sanitary sewer overflow (SSO) regulations, such as Capacity, Management, Operations, and Maintenance (CMOM).
- Planning: assessment of the feasibility and impact of system expansion
- Hydrologic and hydraulic (H&H) modeling of combined and sanitary sewer systems, including
  - Automatic delineation of watersheds and sewersheds;
  - Model simplification or skeletonization, that is, reducing the number of manholes and conduits to be included in the H&H model;
  - Estimating dry weather sewage flow rates from land use, census data, and billing records;
  - Estimating wet weather sewage flow rates from land use, soil, surface imperviousness, and slope; and
  - Estimating surface elevation and slope from digital elevation model (DEM) data.
- Documenting fieldwork, including
  - Work order management using a point-and-click approach;
  - Inspection and maintenance of overflow structures and manholes;
  - Television (TV) inspection of sewers;
  - Flow monitoring and sampling; and
  - Smoke testing, dye testing, and inflow/infiltration (I/I) investigations

H&H modeling and planning applications of GIS were presented in Chapter 7 (Modeling Integration) and Chapter 8 (Water System Applications), respectively. This chapter focuses on the mapping applications of GIS. Although the mapping application will be illustrated for wastewater systems, the methodology is quite general and can be applied to water and stormwater systems.

GIS-Based Mapping Program

In many wastewater collection systems, there is a backlog of revisions that are not shown on the sewer system maps and the critical information is recorded only in the memories of employees. However, there is no longer any excuse to procrastinate, because GIS-based mapping is easy and affordable.

A GIS-based mapping program (GMP) is a comprehensive package of data collection, mapping, and data management services. With a GMP, one gets new sewer infrastructure maps that are complete, accurate, up to date, and affordable. Because a GMP provides accurate and up-to-date information at your fingertips, responding to emergencies is much faster. A GMP uses the latest computer mapping technology to incorporate comprehensive information about a sewer infrastructure gleaned from interviews with staff, from an inventory of records, and from field and airborne investigations.

Three main types of computer mapping systems are available today for developing a GMP: computer-aided mapping (CAM), automated mapping and facilities management (AM/FM), and GIS. Definitions and comparisons of these technologies are provided in Chapter 1 (GIS Basics). According to that information, each of the three technologies has distinctly different characteristics and applications and no single system offers all of them. For instance, one cannot simply use a CAM or AM/FM system as a GIS. Only a GIS has the capability to relate data across layers to allow spatial analysis (Dueker, 1987). A CAM map does not have the graphical intuition of human eye and cannot, for example, tell which sewershed is adjacent to which (Berry, 1994). A GIS map does not have this limitation because the spatial relationships among its features or topology are simply a part of the map intelligence.

A commercial map atlas company may use a CAM system because its applications are primarily for cartographic products. A telephone company will use an AM/FM system to support its telephone system operations and maintenance, because it must be able to quickly trace a cable network and retrieve its attributes (Korte, 1994). For a sewer system, a GMP is most suitable because it must conduct many types of spatial analyses, asking questions like how many customers, by type (residential, commercial, industrial), are located within 1,000 ft of a proposed sewer line. Most important, in addition to sewer system management, a GMP must also support other applications of the municipality. For example, the Planning Department must be able to generate 200-ft notification lists as part of its plan review process. The Public Works Department must be able to conduct maintenance tracking and scheduling. The Public Safety Department must be able to perform crime location analysis. Thus, if the sewer system is owned, operated, and maintained by a city rather than a sewer authority, its GMP might have to be "enterprise-wide"—one that can simultaneously address the mapping needs of all the city departments. GIS-based mapping is therefore the most appropriate technology to meet all the mapping needs of a municipality.

GIS-Based Mapping Case Study

This case study (Shamsi et al., 1996) shows how a suite of advanced technologies—including GIS, digital imaging, digital orthophotography, and GPS surveys—were applied to develop a GMP and create an affordable and proactive management tool for the Public Works Department of the Borough of Ramsey (New Jersey, USA).

The Ramsey Board of Public Works is responsible for the operation and maintenance of water distribution and sanitary sewer collection systems throughout the borough. These operations necessitate the daily use of a variety of map products and associated geographically referenced (or "georeferenced") information resources, which describe or are related to a specific location, such as a land parcel, manhole, sewer segment, or building. More than 80% of all the information processed by local governments is georeferenced. In an effort to more efficiently manage its geographically referenced data, the board started to explore the benefits and applications of GIS technology and computer mapping in early 1993. As a first step, with the assistance of Chester Engineers, Inc. (Pittsburgh, Pennsylvania, USA), the board started a GIS pilot project. The goals of the pilot project were to:

- Thoroughly evaluate the benefits and costs of a GIS,
- Develop specifications for GIS implementation,
- Confirm the suitability of tasks selected for GIS automation,
- Demonstrate GIS functional capabilities,
- Firmly quantify the unit costs of GIS implementation,
- Demonstrate GIS benefits to borough departments not participating in the pilot project,
- Identify any technical problems,
- Provide immediate tangible benefits, and
- Assess the quality of existing records and procedures.

The GIS pilot project produced a GIS Needs Assessment Report, a GIS Implementation Plan, and a functioning GIS demonstration system for a selected portion of the borough. With the knowledge and experience gained from the pilot project, the board was prepared to pursue a broader implementation of GIS technology. The geographic limits of the pilot project area were selected on the basis of two criteria:

- Size: large enough to allow realistic evaluation, small enough to be affordable.
- Location: encompassing a portion of the borough where activities supported by priority applications are likely to occur.

The selected pilot area measuring approximately 2,600 ft by 2,600 ft is located in the central portion of Ramsey. Figure 9-2 shows the pilot area. The pilot project was completed in one year for approximately $40,000 (in 1994 dollars).

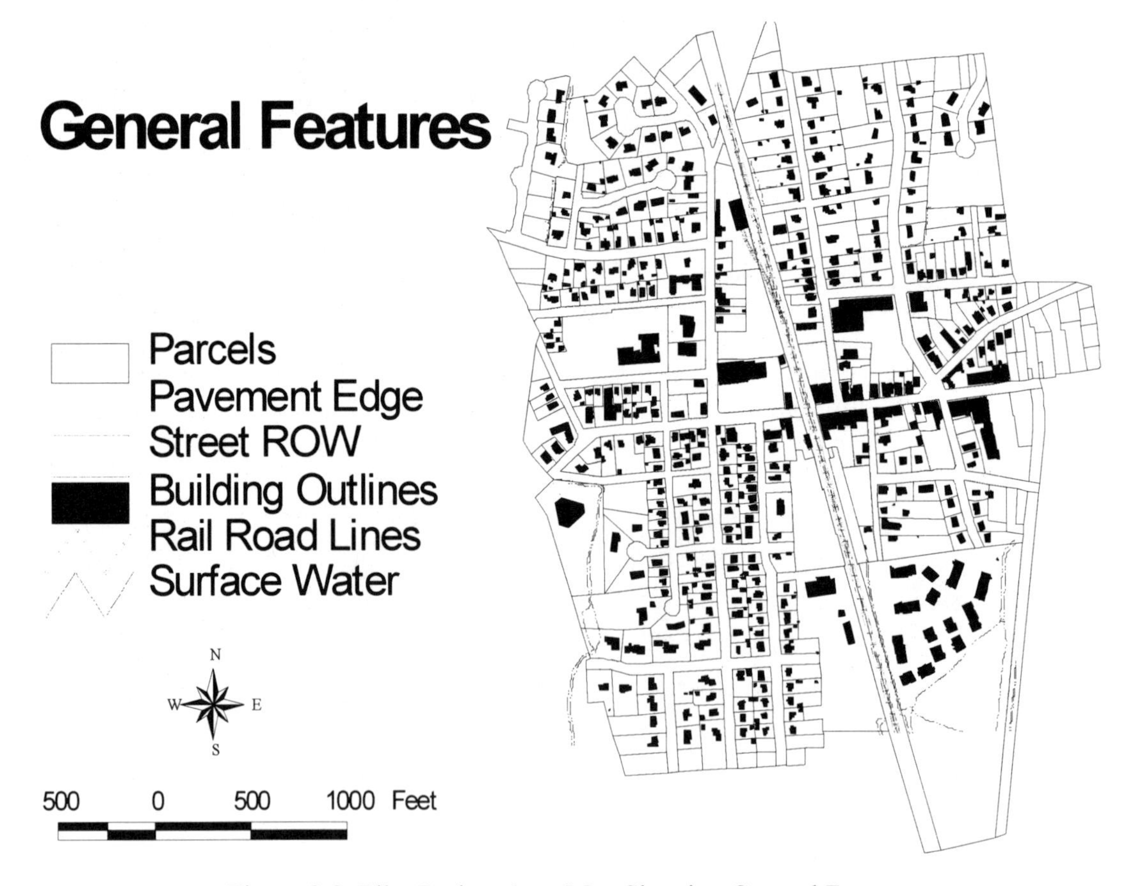

Figure 9-2. Pilot Project Area Map Showing General Features

NEEDS ASSESSMENT

The crucial first step to a successful GIS project is performing a needs analysis (Wells, 1991). The needs assessment process includes two fundamental tasks: identifying application needs and system design.

IDENTIFYING APPLICATION NEEDS

This step identifies potential applications that can be performed more efficiently using GIS technology. The applications are identified through interviews with key managerial and technical personnel and through an inventory of potential data resources, for example, existing maps, tabular data files, relevant maps and data available from external sources, and previously automated databases. Table 9-1 lists the potential applications for Ramsey identified by the needs assessment.

Table 9-1. Potential Applications of GIS Technology

Department	*Representative Applications*
Public Works (responsible for operation and maintenance of the sewer system)	Emergency repair Utility markout Asset management Maintenance tracking and scheduling One Call support Critical water user location Wellhead protection
Engineering and Planning	Generation of 200-ft notification lists Demographic analysis Planning review Permit approval Green space analysis Floodplain mapping Drainage management (storm sewer maps)
Tax Office	Parcel location and data retrieval Tax map production Property photo storage and retrieval Automated public access
Public Safety	Fire hazard locations Hazardous materials inventory Hydrant flow data retrieval Preferred hydrant location Crime location analysis

From the applications listed in Table 9-1, the following two priority applications were selected for pilot testing:

1. Public Works Department: Management of infrastructure asset location information in support of emergency repair and utility markout functions. The GIS will automate the storage and retrieval of location and maintenance data for water and sewer infrastructure. Expected benefits include more rapid emergency repair response, more efficient utility markouts, and more efficient scheduling of routine maintenance.

2. Planning Department: Generation of 200-ft notification lists as part of plan review process. GIS automation will generate a 200-ft buffer, identify affected parcels, retrieve lot and block numbers, retrieve property owner names and addresses, and print a notification list. GIS benefits are expected to reduce the time required to complete these tasks dramatically.

SYSTEM DESIGN

This step recommended a GIS to support the applications identified in the previous step. System design includes determining specifications for the mapping, database, hardware, software, data communications, data maintenance, and organizational issues. Needs-driven design is particularly critical for mapping and database components. The design and content of the mapping and database components determine the fundamental capabilities of the entire GIS. These components can be considered to be the foundation of the system. Once actual implementation has begun, extensive changes to either the basemap or the database design are very costly. System design recommendations for Ramsey are presented below.

MAPPING REQUIREMENTS

The foundation for a successful GIS is an appropriately designed basemap. The basemap is the underlying common geographic reference for all other map layers. Basemap design addresses two fundamental map characteristics: accuracy and depicted feature types. Both characteristics vary with map scale. Generally, larger scale maps are more accurate and depict more detailed feature types. Smaller scale maps, such as USGS quadrangle maps, generally show only selected or generalized features. Table 9-2 summarizes the relationships among map scale, accuracy, and feature detail.

Table 9-2. Relationships Among Map Scale, Accuracy, and Feature Detail

Map Scale	*Minimum Horizontal Accuracy, per National Map Accuracy Standards*	*Examples of Smallest Features Depicted*
1 in. = 50 ft	± 1.25 ft	Manholes, catch basins
1 in. = 100 ft	± 2.50 ft	Utility poles, fence lines
1 in. = 200 ft	± 5.00 ft	Buildings, edge of pavement
1 in. = 2,000 ft	± 40.00 ft	Transportation, developed areas, watersheds

Selection of an appropriate basemap scale is largely determined by the earlier choice of GIS applications. Each application inherently requires a certain minimum basemap accuracy and certain map features. For engineering and public works applications, the required map accuracy is in the range of ±1 ft, as dictated by the need to accurately locate specific physical features, such as manholes and catch basins. Planning applications, which most often deal with area-wide themes, do not generally require precise positioning. Accuracies of ±5 ft, or perhaps as much as ± 40 ft, are often acceptable. Less detailed maps, showing nothing smaller than roads and buildings, for example, may be adequate for many planning applications.

Whatever the range of mapping requirements, the GMP basemap must be suitably accurate and detailed to support the application with the most demanding application, requiring map accuracies of better than ± 2 ft. Utility asset location also requires mapping that depicts specific small features such as manholes and catch basins. As shown in Table 9-3, these requirements are met by a map scale of 1 in. = 50 ft.

Table 9-3. Minimum Map Scale Required to Support Pilot Applications

Application	*1 in. = 50 ft*	*1 in. = 100 ft*	*1 in. =200 ft or less*
Utility asset location	☑		
200-ft parcel buffers		☑	
Census demographics			☑

The Borough of Ramsey did not have any complete large-scale mapping resources suitable for utility asset location. System design recommended the following GMP specifications:

- A new digital orthophoto basemap should be developed from aerial mapping photography obtained in April 1993. This basemapping should be accurate to the tolerances normally associated with mapping at a scale of 1 in. = 50 ft.

- To facilitate updating and maintenance of the basemap, the borough has a need for a common coordinate system for all subsequent mapping efforts. Orthophoto base mapping should be established on the New Jersey State Plane Coordinate System. Survey monuments should be established to enable subsequent mapping efforts to be tied to the basemap coordinate system.

- Lot and block boundaries should be included as a unique layer in the basemap. Lot and block geography is the fundamental map base for all property-related functions within the borough.

- Additional map layers for the water distribution system, the sanitary sewer collection system, the stormwater collection system, and census geography should be developed to support the priority applications.

Table 9-4 summarizes the required map layers and the features depicted in each layer.

Table 9-4. Map Layers and Depicted Features

Layer	*Features*
New Jersey State Plane Grid System	Grid tics, survey monuments
Photogrammetric basemap	Building outlines, street outlines and centerlines, vegetation, water courses, topography
Parcels	Lot and block boundaries
Sanitary sewer collection system	Sanitary sewers, manholes, cleanouts
Stormwater collection system	Storm sewers, manholes, catch basins
Census geography	Blocks, block groups, tracts

DIGITAL ORTHOPHOTO BASEMAP

To meet the recommendations of the system design, a digital orthophoto produced from stereo aerial mapping photography was used as the basemap. Digital orthophotos have been widely accepted as an optimal land base for GIS. A digital orthophoto is a computer-compatible raster image derived from aerial photography that has been scanned at very high resolution. The image is corrected to remove distortions due to the attitude of the aircraft at the time of exposure, image displacement due to topographic relief, and the distortion introduced by the camera (Michael, 1994). Digital orthophotos are very detailed, can be easily interpreted, and provide highly accurate data that can be easily quantified and verified.

Global Positioning System (GPS)

Paper-to-digital map transfer is an important first step in a computer mapping project. An equally critical step is to identify and georeference features on digitized map layers and keep them up to date. Most municipalities cannot afford traditional survey methods for field verification of map features. The GPS technology offers a promising solution to placing current data on maps (Lewis, 1993).

A GPS represents a space-age revolution in GIS data collection. Satellite-based GPS technology has been used since the early 1980s. A GPS consists of a constellation of satellites that orbit the earth twice daily and transmit precise time and position signals. GPS receivers read signals from orbiting satellites to calculate the exact spot of the receiver on the Earth.

Previously, the GPS technology was utilized for high-accuracy field surveys, basemap control, or positioning the military assets. In addition to surveying professionals, the traditional users of GPS, the new user community of GPS includes GIS data collectors, environmental specialists, and utility engineers (Leick, 1992). The new line of GPS brings technology to GIS practitioners, who can populate maps with the location of features such as manhole covers, catch basins, and overflow points.

GPS technology was used in the pilot project to verify the locations of valves, hydrants, and manholes.

Database Requirements

A GMP database, though not as visible as the basemap, is an equally critical system component. To successfully support GMP applications, the database must provide appropriate information in a useful and accessible form. The design of the database, like that of the basemap, is driven by applications needs. The first step in designing the database is the development of a data dictionary, which defines the categories of information required to describe each map feature. Each unique category of information is called an attribute. Table 9-5 presents the data dictionary for Ramsey's pilot project.

Table 9-5. Sewer System GMP Data Dictionary

Feature Type	*Required Attributes*
Sewers	Sewer ID Pipe material Pipe size From manhole ID To manhole ID
Manholes	Manhole ID Manhole type Lid type Depth Inflow problem status
Catch basins	Catch basin ID

Other database design issues include methods for converting maps and data to digital form, and procedures for maintaining maps and data. There are currently two methods for converting paper maps to digital form: scanning and line digitization. Scanning is generally most efficient when converting maps for archival purposes. Digitized files, when properly prepared, are ready for immediate use in a GMP. Scanned files generally require extensive post-processing to convert raster images to the vector format required by the priority applications. Having reviewed the borough's available maps, and considered the mapping requirements of the priority applications, the use of digitization was recommended for map conversion. All of the borough's sewer system data were stored in a non-digital form. This information was manually entered into the GIS database. Map features and attribute information were gleaned from a variety of sources. Table 9-6 lists recommended sources for specific map features and attributes specific to the sewer system.

Table 9-6. Recommended Source Documents for Map Features and Attributes

Features Types	*Source Documents*			
	Basemap	Utility Markouts	Sewer System As Built, 1975, 50-Scale	Sewer System As Built, 1970, 100-Scale
Sewers	☑	☑	☑	☑
Manholes	☑	☑	☑	☑
Catch basins	☑	☑		☑

HARDWARE AND SOFTWARE

The selection of computer system components for a GMP is driven by applications needs and software functional requirements. Additional selection criteria include:

- Low cost
- Ease of use
- Expandability
- Compatibility

A single PC-based system was recommended for evaluating the demonstration GIS datasets developed in the pilot project. The hardware specifications included a 500-MB Pentium I PC, 17-in. VGA color monitor, 14,400-bps fax modem, and Hewlett-Packard DesignJet 650C full-size color plotter. Although some of these items are obsolete today, they were the state-of-the-art in 1994.

ESRI's ArcInfo (version 6) and ArcView (version 2) were the two main software programs used in the Ramsey pilot project. ArcInfo is a complex program and requires substantial technical expertise. Because the borough did not have the technical staff to run and maintain an ArcInfo-based GMP, data conversion work was done by Chester Engineers using ArcInfo. Due to its low cost, ease of use, and compatibility with the ArcInfo file format, ArcView was installed in the borough for routine display and plotting of maps and querying the GMP database.

MAPPING RESULTS

To accomplish the pilot project task of presenting a functioning GIS demonstration system, several maps were produced. Some sample sewer system maps are portrayed in Figures 9-2 through 9-8.

- Figure 9-2 shows the general features of the pilot project area. This map also shows parcels, building outlines, pavement edges, street rights of ways, railroad tracks, and surface water.

- Figure 9-3 shows the sanitary sewers and manholes overlayed on the parcel map. The sewer lines have been depicted according to pipe diameter. This map was created by classifying the "sanitary sewers" theme by the "diameter" attribute.

- Figure 9-4 shows three adjacent tiles of the digital orthophoto basemap created from low-level vertical mapping photography. The digital orthophoto has been processed to remove scale distortion, and it creates a very accurate and true-to-scale basemap at an accuracy of ± 1.25 ft. Sanitary sewer lines showing flow direction have also been plotted on the digital orthophoto. The sewer features were taken from as-built drawings and adjusted to locations of surface features that have been field verified.

- Figure 9-5 shows an ArcView screenshot displaying sanitary sewer lines, manholes, catch basins, cleanouts, and problem areas plotted on the digital orthophoto. Figure 9-5 also shows a problem area still (roots entering from the right side of the sewer) captured from a video of the internal TV inspection of the sewers. The stills are displayed using ArcView's "hot link" function. Hot link allows a user to point and click on a map feature to view other documents or to launch other applications from within ArcView.

- Figure 9-6 shows an ArcView screenshot displaying a 200-ft parcel buffer. The circle drawn has a 200-ft radius. All lots that fall within this radius are automatically picked from the database so that owners, listed in the query table, may be contacted. Selecting polygons within a given radius was employed to perform this analysis. Using this method, the Planning Department can generate 200-ft notification list for its plan review process. Form letters can be created using the mail-merge function of a word processor.

- Figure 9-7 shows an ArcView screenshot displaying parcels according to their assessed value plus improvements. This display also utilizes the ArcView hot link function, which allows the display of scanned property photographs stored within the database.

- Figure 9-8 shows an ArcView screenshot illustrating how the GMP database can be queried simply by using a point-and-click method. The query results for a manhole are displayed in a pop-up window.

CASE STUDY CONCLUSIONS

Conducting a pilot project is an effective means of performing a needs analysis study. Several tangible benefits of a GMP were identified during the pilot project. They included rapid, more cost-effective responses to emergency repairs, more efficient utility markouts, compilation of an infrastructure inventory, creation of asset valuation for insurance, and development of annual preventive maintenance plans. The pilot project demonstrates that, in addition to meeting the mapping needs of a sewer system, a GIS-based mapping program also offers extensive benefits to other departments of a local government.

Figure 9-3. Sewer Map

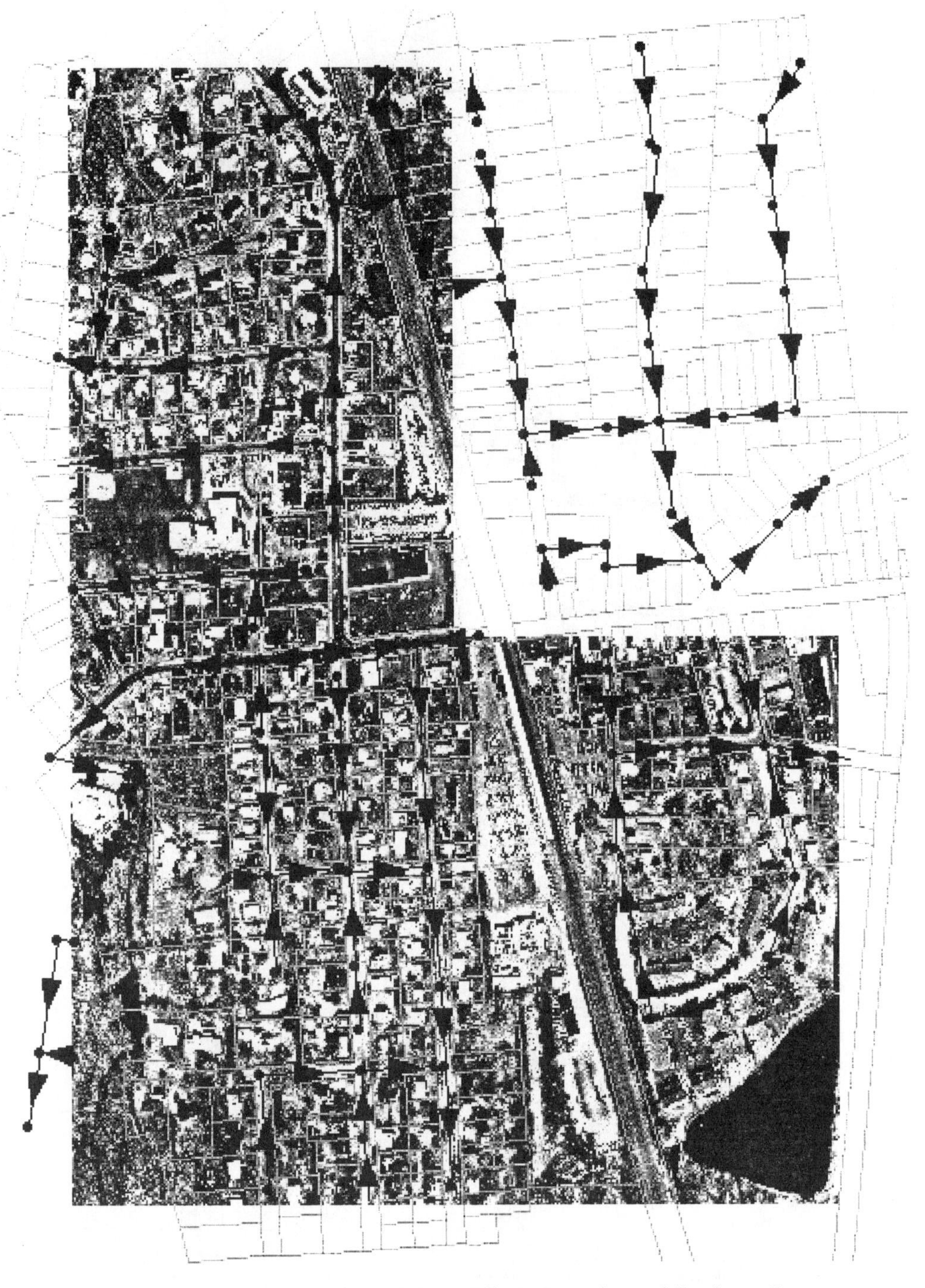

Figure 9-4. Digital Orthophoto (Three Tiles), Parcels, and Sanitary Sewers

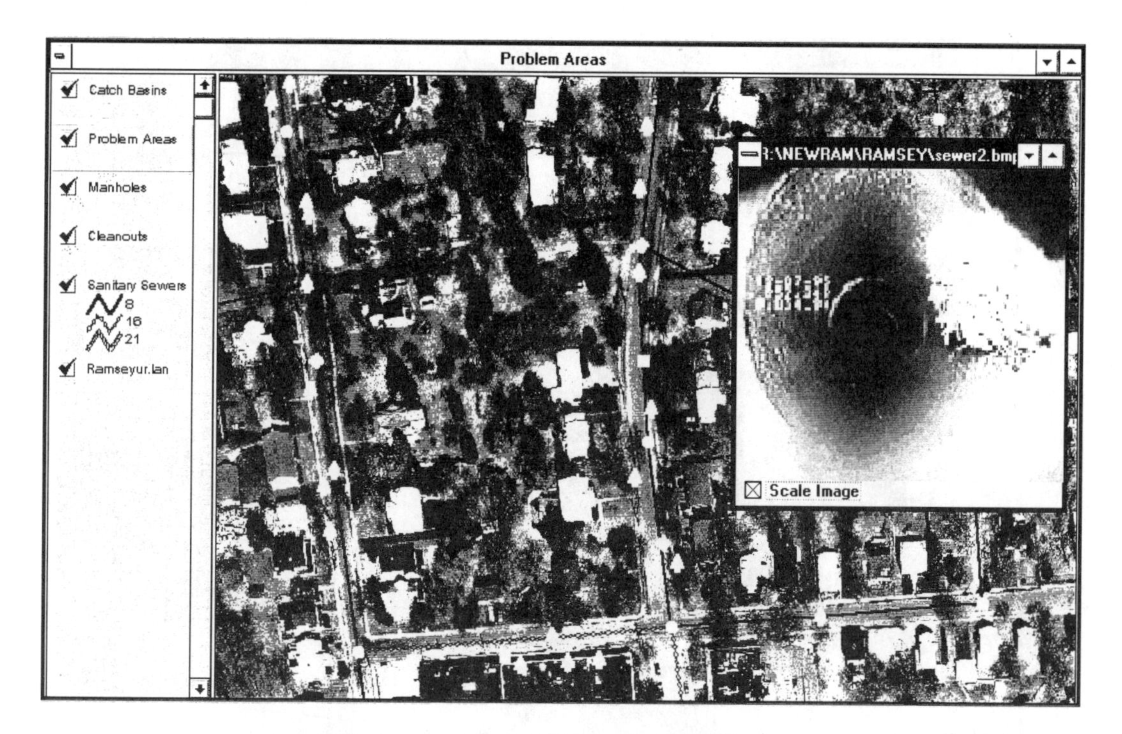

Figure 9-5. Problem Areas and TV Inspection Video

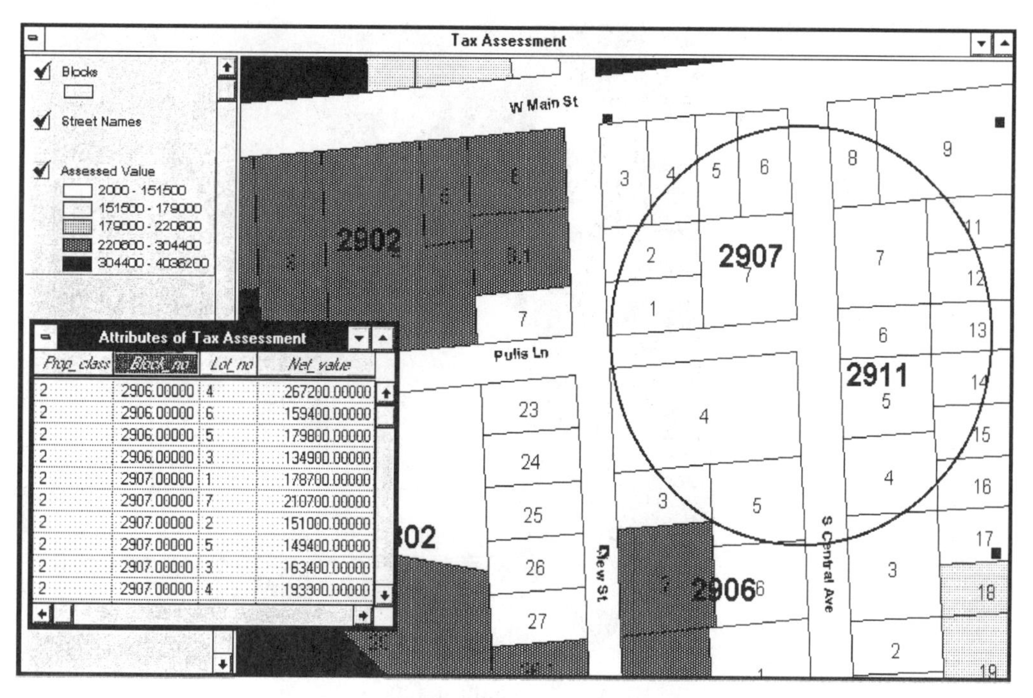

Figure 9-6. Parcel Buffer

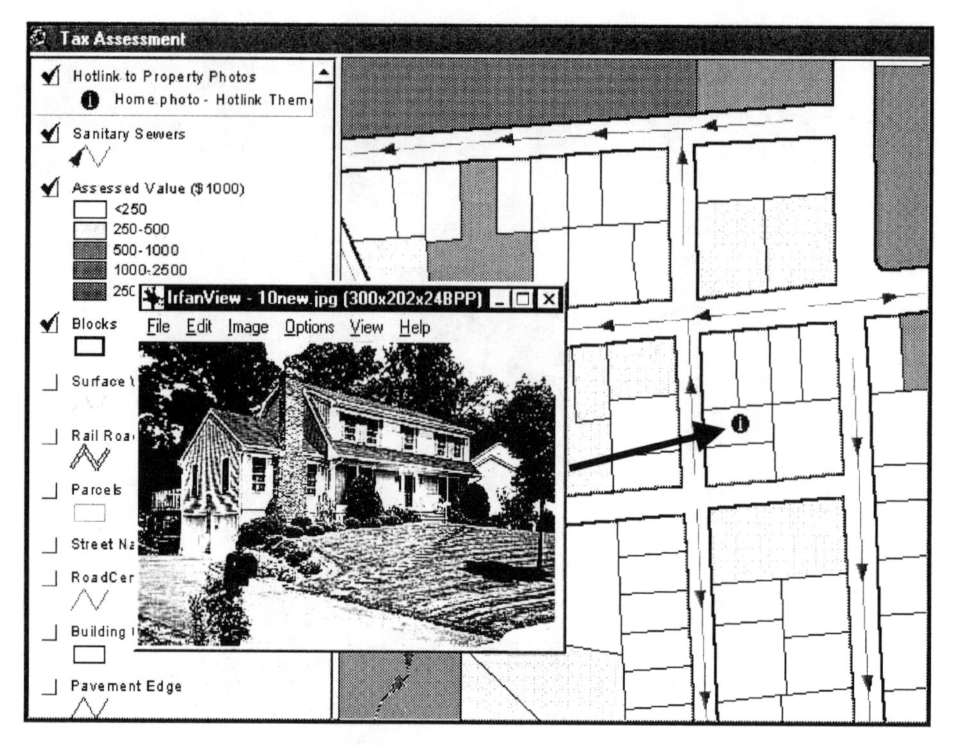

Figure 9-7. Property Assessment

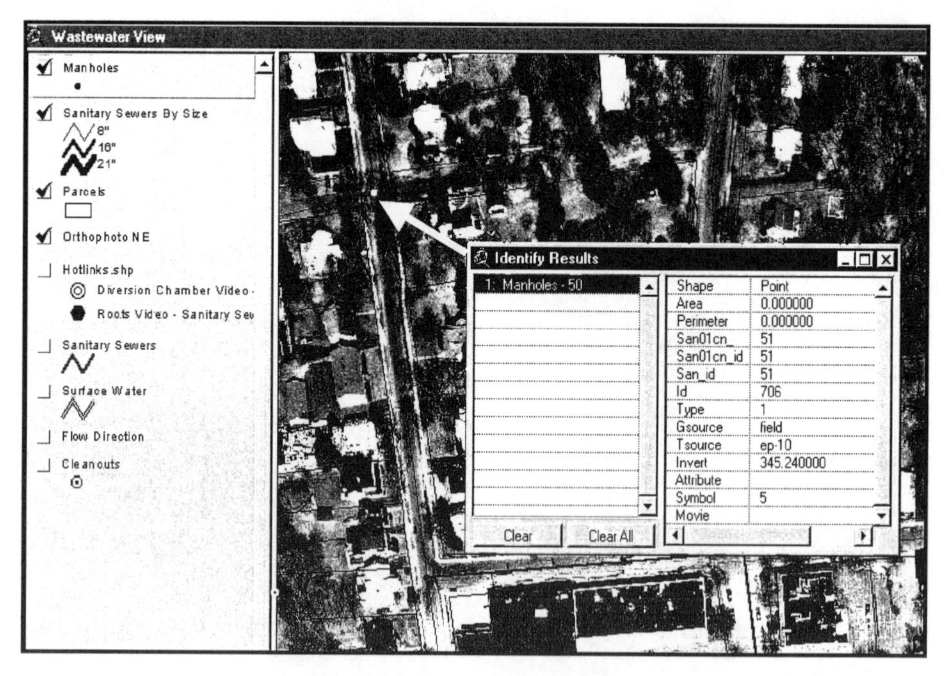

Figure 9-8. Database Query

SUMMARY

This chapter described an application of a suite of advanced technologies, including GIS, digital imaging, digital orthophotography, and GPS surveys to develop a GMP for the sewer system of a local government. Mapping, database, and computer hardware and software requirements were described with the help of a case study for a small city. The importance and benefits of a needs analysis prior to implementing a full-scale GMP were demonstrated. Various tasks of the needs analysis were explained. When carefully done, a needs analysis accomplishes several objectives of a GPM project. It clarifies the project's specific needs, provides a process for arriving at interdepartmental understandings about the objectives and limits of a project, and establishes a detailed basis for designing the maps and the database.

SELF-EVALUATION

1. What is a GMP?
2. How is a GMP map different from a CAD drawing?
3. What is the relationship between map accuracy and scale?
4. What is needs analysis, and how is it conducted?
5. What are the benefits of a needs analysis and of a pilot project?

Chapter

10 STORMWATER SYSTEM APPLICATIONS

The objective of stormwater management is to prevent or mitigate the adverse effects of conveying an excessive quantity and poor quality of stormwater runoff. A GIS can be used to identify appropriate watershed-wide stormwater management control standards and criteria.

Flooded Streets of Lewisburg Borough, Pennsylvania, Located in the Bull Run Watershed, After the Flood of September 7, 1999.

LEARNING OBJECTIVE

The learning objective of this chapter is to illustrate GIS applications for stormwater systems with a special emphasis on watershed stormwater management. Major topics discussed in this chapter include

- Stormwater management concepts
- Hydrologic modeling, using the Penn State Runoff Model
- Using a GIS to prepare hydrologic model input data
- Watershed-wide stormwater management implementation
- Case studies

GIS APPLICATIONS IN NORTH CAROLINA AND FLORIDA

Many communities are using the latest Information Technology (IT) tools to develop "Watershed Information Systems." For example, the City of Charlotte and surrounding Mecklenburg County in North Carolina—which make up one of the fastest growing metropolitan areas of the United States—have developed a Watershed Information System called WISE that integrates data management, GIS, and standard stormwater analysis programs like the Hydrologic Engineering Center's (HEC) HEC-1, HEC-2, HEC-HMS, and HEC-RAS. This integration approach provides them with a "living" watershed model: one that never outdates and can be easily updated if the input data (such as zoning or land use) or modeling software are changed in the future. Using this method, the existing hydrology and hydraulics (H&H) models can be updated at a fraction (less than $100,000) of the cost of developing a new model (more than $1 million). WISE data management system emphasizes data storage and access using pre- and post-processing techniques. Pre-processors funnel appropriate data in the correct format to industry-standard H&H models. Post-processors extract and assimilate model results inside a GIS (Edelman et al., 2001).

Charlotte County, Florida, is using GIS to monitor and direct stormwater maintenance activities based on performance. They have integrated engineering, environmental, and socioeconomic objectives into a stormwater management tool. The management tool was developed by analyzing existing expenditures and generating replacement schedules based on performance data about flooding level of service, biological indicators, and life-cycle analysis. Taxing districts were delineated based on watershed boundaries, land use, water quality, and flood management objectives. ArcInfo GIS was used to delineate Municipal Service Benefit Units, which are geographically based assessment districts that collect fees for stormwater management (Walter, 1998).

Stormwater System Application Examples

Typical applications of GIS for stormwater systems include:

- Watershed stormwater management
- Planning: assessment of the feasibility and impact of system expansion
- Floodplain mapping and flood hazard management
- Mapping work for Stormwater National Pollution and Discharge Elimination System (NPDES) permit requirements
- Hydrologic and hydraulic (H&H) modeling of combined and storm sewer systems, including
    - Automatic delineation of watersheds and sewersheds;
    - Model simplification or skeletonization (i.e., reducing the number of manholes and conduits to be included in the H&H model);
    - Estimating stormwater runoff from the physical characteristics of the watershed (e.g., land use, soil, surface imperviousness, and slope); and
    - Estimating surface elevation and slope from digital elevation model (DEM) data.
- Documenting field work, including
    - Work order management using a point-and-click approach;
    - Inspection and maintenance of stormwater system infrastructure, such as pipes, manholes, culverts, catch basins, inlets, outfall structures, and headwalls;
    - TV inspection of sewers;
    - Sewer cleaning;
    - Flow monitoring and sampling; and
    - Smoke and dye testing.

H&H modeling, planning, and mapping applications of GIS were presented in, Chapter 7 (Modeling Integration), Chapter 8 (Water System Applications), and Chapter 9 (Wastewater System Applications), respectively. This chapter will focus on the stormwater management applications of GIS.

STORMWATER MANAGEMENT REGULATIONS

A stormwater management practice is one that reduces the quantity and/or quality of water being discharged into a land or water area (Wanielista, 1978). The objective of stormwater management is to prevent or mitigate the adverse effects related to the conveyance of excessive quantity (rates and volumes) and poor quality of stormwater runoff. In this chapter, we will focus on the stormwater quantity aspects. There are two types of stormwater management: *local* and *watershed-wide*. In local stormwater management, different areas of a watershed implement different stormwater management plans locally to solve their individual stormwater problems (e.g., flooding, sedimentation, or erosion) without considering the adverse downstream effects. Early efforts to mitigate storm flows followed this approach, which consisted of simple routing of stormwater through gutters and sewer systems (e.g., using the Rational Method) with the objective of removing the stormwater as quickly as possible. This approach is totally consistent with the Common Enemy Rule in effect in most states until recently (Kibler and Aron, 1980). It has been recognized for sometime that simply bypassing storm flows really shifts the location of the problem and very often aggravates the problem by compounding flows downstream. The end result is an increase in total flow, peak-flow rates, stream velocity, and stream stage in the downstream channels. Most of the physical, social, and economic problems associated with stormwater are attributable to unwise land use, insufficient attention to land drainage in urban planning, and ineffective updating of existing stormwater control systems. The most pressing institutional problems of public agencies are the lack of watershed-wide policy, criteria, laws, and guidelines for the development of stormwater programs and facilities (Poertner, 1988).

A more effective solution to stormwater management is the watershed-wide approach. This approach implements a comprehensive stormwater plan throughout the watershed to prevent the adverse effects of stormwater, both at a particular site and anywhere downstream where the potential for harm can be reasonably identified. This approach maintains, as nearly as possible, the natural runoff flow characteristics throughout the watershed. This can be accomplished either by augmenting the infiltration process or by temporarily storing stormwater for release at controlled rates of discharge. This recent trend represents the Reasonable Use Doctrine (Kibler and Aron, 1980).

A significant change in the approach to stormwater management in Pennsylvania occurred with the passage of the Storm Water Management Act (Act 167). This legislation, passed by the Pennsylvania General Assembly on October 4, 1978, requires a comprehensive watershed-wide approach to planning and managing excess stormwater runoff (Department of Environmental Protection, 1985). The Act provides for the preparation of "stormwater plans" for all the state watersheds by counties and implementation of such plans by municipalities. The Pennsylvania Department of Environmental Protection (DEP) provides 75% financial assistance to counties for the preparation of the stormwater plans for the county watersheds. To implement the act, DEP has divided the entire state into 356 watersheds, which are shown in Figure 10-1. During the 13 years from 1984 to 1997, 56 stormwater management plans were completed and 23 were in progress.

TECHNICAL APPROACH

Effective stormwater management requires the linking of specialized computer models to the GIS (Pearson and Wheaton, 1993). The technical approach used in the development of stormwater management plans for the Pennsylvania watersheds is based on GIS and computer modeling.

The use of GIS in modeling urban stormwater systems had been more limited due to the need for large, expensive and detailed spatial and temporal databases, along with the fact that many computer tools used in urban stormwater modeling were not easily amenable to integration with GIS. However, as local data-gathering efforts have increased and software integration has evolved, the use of GIS in urban stormwater is now widespread (Heaney et al., 1999). Furthermore, rapidly developing computer technology has continued to improve the modeling necessary for watershed stormwater management planning. The use of GIS provides an accurate, manageable way of compiling and evaluating modeling parameters such as land use, slope, and soil types. Each parameter can be digitized from existing maps, aerial photographs, or satellite images and stored in "layers" on a computer. Each layer can then be easily combined for input into a hydrologic model. The use of GIS, as a side benefit, provides counties with an easily updated database for non-stormwater-related planning activities. DEP encourages the development and use of new techniques to improve and lower the cost of stormwater plan preparation (Jostenski, 1988).

MODEL SELECTION

Hydrologic modeling plays an important role in developing stormwater management standards. For example, an important part of the stormwater management plans required by Act 167 is the development of a hydrologic model for each watershed. The model must be used to simulate stormwater runoff hydrographs corresponding to design storms of various durations and frequencies. The hydrograph information must be subsequently used to estimate the relative effects of subbasin discharges

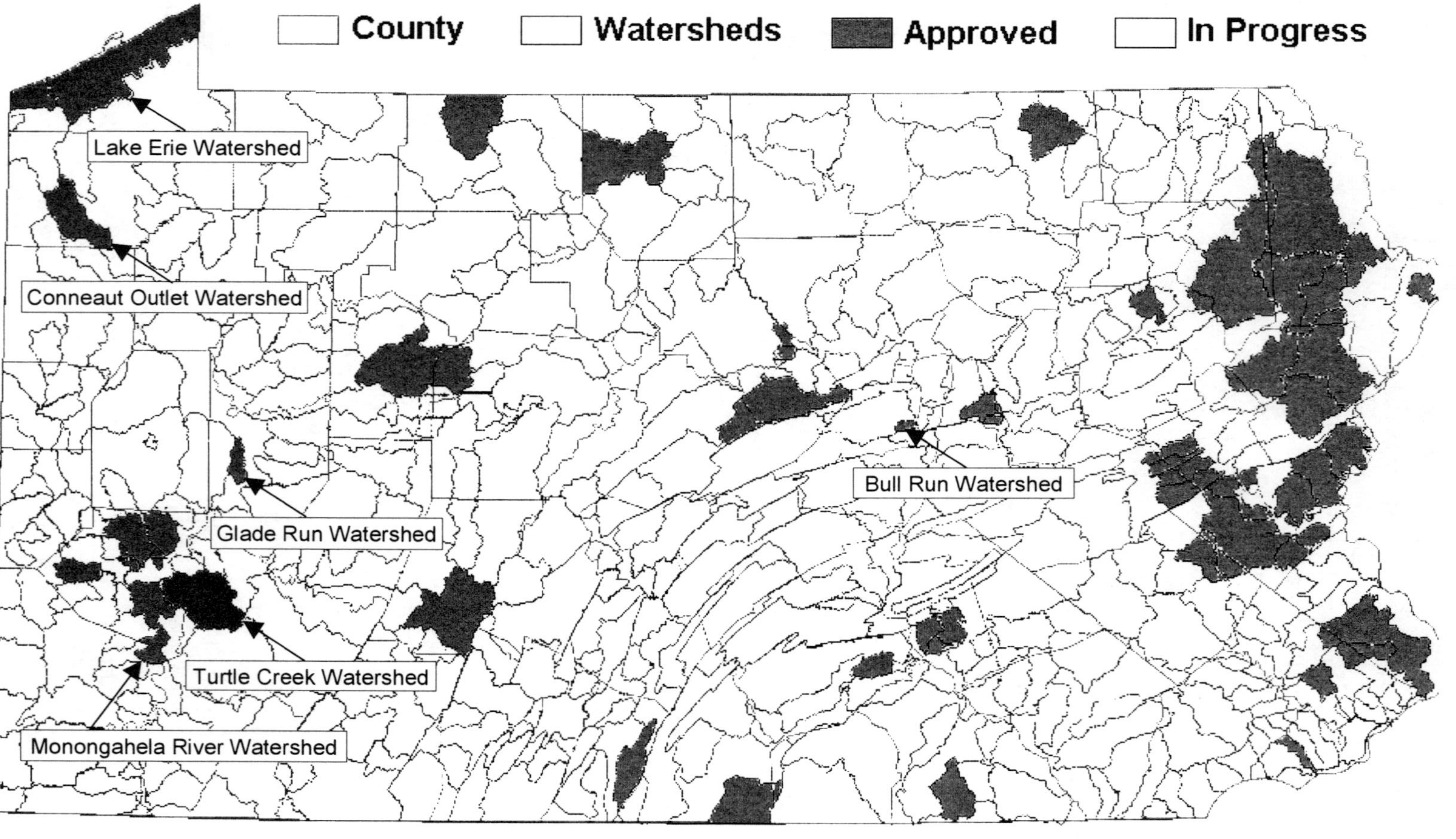

Figure 10-1. The 356 Pennsylvania Watersheds Designated for Stormwater Management

on downstream locations, which in turn helps to identify appropriate watershed-wide stormwater management control standards and criteria.

Most hydrologic models in current use are lumped-parameter conceptual models, such as U.S. EPA's Storm Water Management Model (SWMM) (Huber and Dickinson, 1988), the Storage Treatment Overflow Runoff Model (STORM) (U.S. Army Corps of Engineers, 1977), HEC-1 (U.S. Army Corps of Engineers, 1990), TR-20 (Soil Conservation Service, 1983), the Hydrologic Simulation Program Fortran (HSPF) (Johnson et al., 1980), and the Penn State Runoff Model (PSRM) (Aron and Lakatos, 1990). The lumped models that have utilized GIS in model development are HEC-1 (Shea et al., 1993; Phipps, 1995), HEC-2 (Djokic et al., 1993; Thomas, 1995; Phipps, 1995), TR-20 (Hydrologic, 1974; Thomas, 1995), TR-55 (Thomas, 1995), HSPF (Ross and Tara,1993), and SWMM (Cowden 1991; Shamsi, 1993a; Shamsi and Fletcher, 1994; Curtis, 1994). Shamsi (1996) published the first application of PSRM and GIS integration.

At the modeling time during the late 1980s, the famous *distributed* models, such as SHE (European Hydrological System—Système Hydrologique Européen) (Abbott et al., 1986), DROTEL (Fortin et al., 1986), and WATFLOOD (Kouwen, 1988), were still in developing stages (Kouwen et al., 1993). The large quantity of required input data is the very characteristic of distributed models that often renders them inefficient for everyday operational hydrology (Smith and Vidmar, 1994). For instance, the most renowned distributed model, SHE, had performed well on mainframe computers, but its PC applications were limited because of the large number of computations that had to be made (DeVries and Hromadka, 1993). Distributed models had not been shown to be intrinsically better than lumped models for runoff prediction (DeVantier and Feldman, 1993). Finally, the selected model should also serve as an everyday computational tool for counties, municipalities, and developers to control the development and to aid the design of appropriate stormwater management techniques. These reasons indicated that a lumped parameter model was more appropriate in implementing the modeling objectives of Act 167.

PSRM (Aron and Lakatos, 1990) was selected as the model of choice for developing the Act 167 watershed hydrologic models in Pennsylvania. Additional reasons for selecting PSRM as the model of choice were:

1. The overall model complexity, data, and personnel requirements for PSRM are low relative to similar models, such as the Distributed Routing Rainfall Runoff Model (DR3M), HSPF, SWMM, and TR-55 (WEF, 1989).

2. At the modeling time during the late 1980s, PSRM computer run times were considerably lower than for other models, such as HEC-1 and SWMM (Aron et al., 1979).

3. Other models would require post-processing of the model output to derive stormwater control criteria. PSRM provides a release rates table, which can be used to control the post-development flows according to Act 167 requirements.

4. PSRM input parameters can be computed from a planning-level GIS. The literature and the author's own experience indicate that the GIS integration provides for an increased detail of evaluation, minimizes user subjectivity in estimating physical model input parameters of the watershed, and reduces cost of analysis due to significant time savings (Ross and Tara, 1993; Pearson and Wheaton, 1993; Shamsi, 1996).

5. GIS integration techniques are flexible enough to fit different projects and budgets. For example, manual digitization and vector polygon overlay is more cost-effective for small watersheds, whereas automated scan digitization and raster polygon overlay is more appropriate for large watersheds.

6. DEP, Pennsylvania's stormwater management regulatory agency, recommends PSRM as a preferred model.

7. All modeling and GIS analysis can be performed on PCs, and no workstations or mainframes are required.

8. The model should be simple to serve as an everyday computational tool for counties, municipalities, and developers to control development and to aid the design of appropriate stormwater management techniques.

Penn State Runoff Model (PSRM)

PSRM is a single-event simulation model based on the U.S. Natural Resources Conservation Service (NRCS) (formerly known as the Soil Conservation Service or SCS) techniques for infiltration, kinematic wave method for overland flow, and nonlinear reservoir routing for storage. PSRM was developed in 1976 at Pennsylvania State University as a simplified, quantity-only, mainframe FORTRAN program for evaluating subbasin runoff effects on downstream flow conditions. A PC Basic version was released in 1987 (Aron, 1987). Since then, stormwater quality modeling has been added. Best Management Practices (BMP) modeling is anticipated in one of the future releases. The stormwater quality version of PSRM is called PSRM-QUAL (Aron et al., 1996).

Shamsi (1988) developed an enhanced PC FORTRAN version of PSRM, called CPSRM (Chester Engineers' PSRM). The major CPSRM enhancements included database and GIS interface, watershed schematic plots, release rate tables, and expanded capacity to model a larger number of subbasins and time steps compared with PSRM (Shamsi, 1988). A flowchart illustrating CPSRM operation is shown in Figure 10-2. The latest Penn State release of PSRM has a more flexible model capacity because a larger number of subbasins can be modeled if the number of time steps is kept smaller, or a larger number of time steps can be modeled if the number of subbasins is kept smaller (Aron and Lakatos, 1990). In 1995, Shamsi developed the first Windows version of PSRM called CPSRMW capable of modeling a virtually unlimited number of subbasins and time steps, limited only by the available computer memory. All the case studies presented in this chapter used CPSRM or CPSRMW versions of PSRM.

The *physical* input parameters of PSRM are:

- Subbasin area,
- Subbasin overland flow width,
- Subbasin mean overland flow slope,
- Subbasin percentage imperviousness,
- Subbasin SCS runoff curve number,
- Subbasin centroid coordinates,
- Stream capacity, and
- Stream travel time.

The hydraulic input parameters of PSRM are:

- Manning's "*n*" (roughness coefficient),
- Depression storage for impervious areas,
- Depression storage for pervious areas,
- SCS initial abstraction factor,
- Residual infiltration time parameter, and
- Ratio of in-bank to over-bank flow velocities.

The *operational* parameters of PSRM are:

- Rainfall input hyetographs,
- Reservoir stage–storage–outflow curves,
- Length of simulation,
- Simulation and printing time intervals, and
- Output instructions.

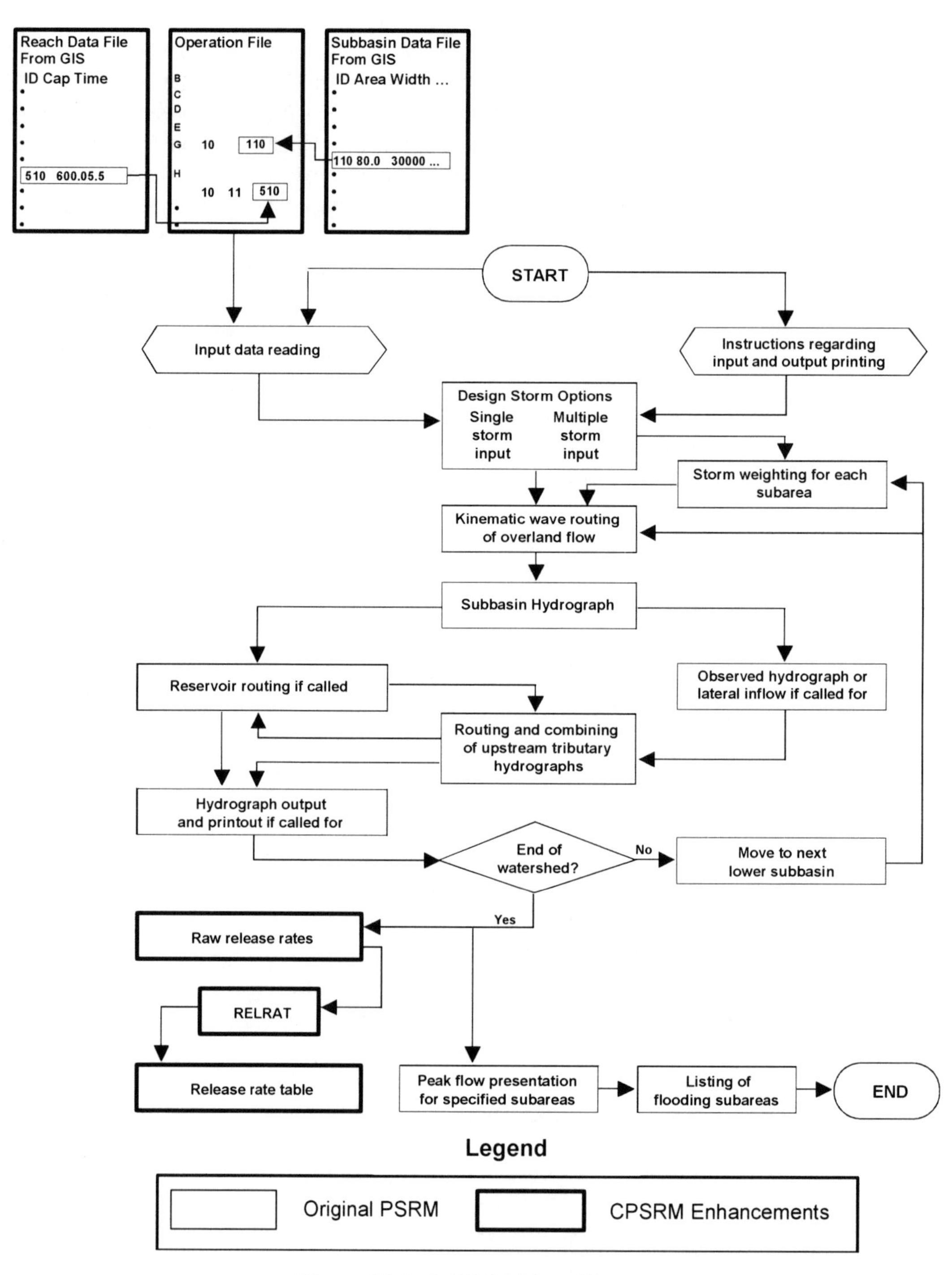

Figure 10-2. CPSRM Flow Chart

The most common application of a model is to simulate multiple "what-if" scenarios. To do this, a calibrated model requires changing only a few operational parameters without changing the physical parameters. Thus, the physical parameters can be stored separately from the operational data. As shown in Figure 10-2, CPSRM input consists of three data files: a subbasin file, a reach (stream) file, and an operation file. The first two files contain model physical parameters in a tabular (row-and-column) format. The operation file contains model operational parameters in a text format. The GIS linkage is accomplished by copying GIS data in the subbasin and reach data files, which reflects the interchange method of GIS integration described in Chapter 7 (Modeling Integration). This data structure allows direct data import from the GIS database and data editing in a spreadsheet, such as Microsoft Excel or Lotus 1-2-3. It also eliminates inclusion of large amount of subbasin and reach data in each input file. Operation file size is minimized as well. Each new scenario can be modeled by a simple edit of the small operation file or by creating a new scenario-specific operation file. Finally, there would be no need to edit the operation file if the physical parameters change in the future.

Stormwater Management Case Studies

Several case studies are presented below to illustrate the application of the GIS and modeling integration approach adopted for developing stormwater management plans for Pennsylvania watersheds. Examples are presented for both the small and large watersheds. These case studies demonstrate that PSRM-GIS linkage successfully implements the Act 167 requirements. The model is used to simulate runoff hydrographs for various durations and frequencies. Modeled hydrographs are processed to create peak-flow presentation and release rate tables. These tables are used to create a watershed release rate map, which is a practical tool to implement a stormwater management plan.

Table 10-1. List of Case Study Watersheds

Watershed	*Area in mi^2 (km^2)*	*Subbasins*
Turtle Creek	146.0 (378.0)	696
Glade Run	25.5 (66.1)	120
Bull Run	8.4 (21.8)	48
Conneaut Outlet	101.0 (261.8)	385
Monongahela River	98.0 (254.0)	432
Lake Erie	400 (1,036.8)	1,811

Table 10-1 lists the case study watersheds for which stormwater management plans were developed using the GIS-based modeling techniques described above (Shamsi, 1996; Chester Engineers, 1993; Chester Engineers, 1990). These watersheds are shown in Figure 10-1. The Bull Run and Turtle Creek watersheds will be discussed in more detail.

A small watershed application will be illustrated for the Bull Run watershed located in Union County in north-central Pennsylvania (Shamsi, 1993b; Shamsi and Fletcher 1994). This watershed is selected because it is the smallest studied watershed and the project results, especially the GIS maps, can be easily included in the book. Bull Run watershed's 8.4-mi^2 (21.8-km^2) drainage area is tributary to the West Branch Susquehanna River at the eastern boundary of Lewisburg Borough. A topographic map of the watershed is shown in Figure 10-3. The predominant land use in the watershed is open space and agriculture. Only 20% of the watershed has residential, commercial, and industrial land uses.

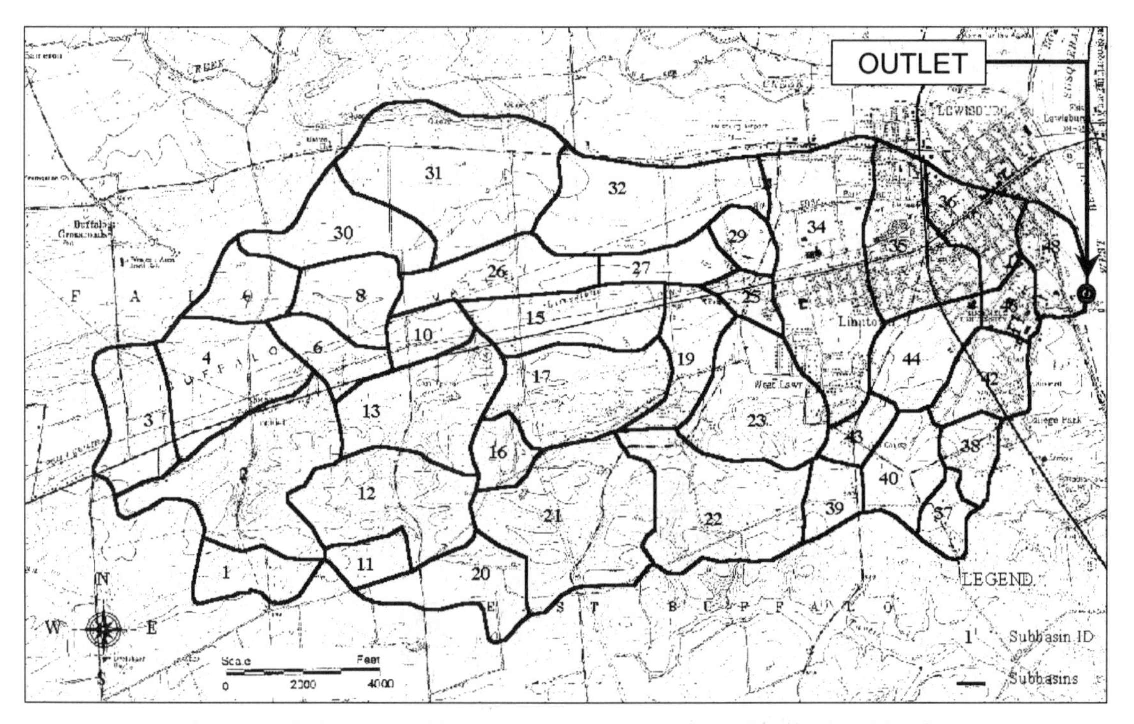

Figure 10-3. Topographic Map of Bull Run Watershed and Subbasins

A large watershed application will be illustrated for the Turtle Creek watershed (Chester Engineers, 1990) shown in Figure 10-4. The Turtle Creek watershed is a tributary of the Monongahela River at a point approximately 11.4 river mi (18.34 km) south of the City of Pittsburgh

(Pennsylvania, USA). The watershed drains approximately 146 mi^2 (378 km^2) of land with a terrain typified by rolling hills. The watershed is jointly located in Allegheny and Westmoreland counties in southwestern Pennsylvania approximately 10 mi (16 km) east of Pittsburgh. The degree of development ranges from very dense commercial, residential and industrial urban development, residential and commercial suburban development, and sparsely developed rural and agricultural areas, to undeveloped forested areas.

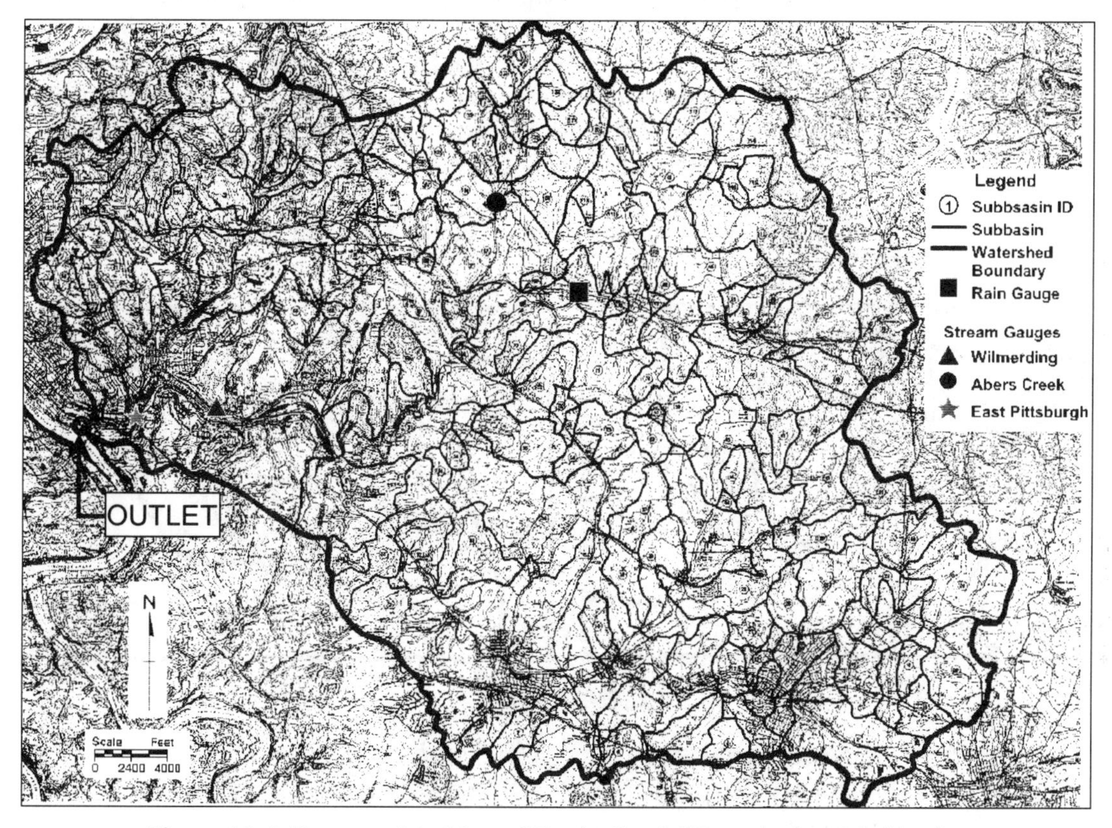

Figure 10-4. Topographic Map of Turtle Creek Watershed and Subbasins

Model Development

Watershed subdivision is the first step in model development. Traditionally, this is done manually by outlining the watershed boundary on a topographic or drainage map, identifying the major drainage paths (e.g., rivers, streams, sewers), subdividing the drainage paths into smaller segments (e.g., trunk sewers, swales) called *reaches,* and finally subdividing the watershed into smaller drainage areas called *subbasins.* PSRM requires that all but the uppermost subbasins (subbasins without any upstream subbasins) must enclose one or more reaches. Each reach must correspond to one and only one subbasin immediately upstream. There are two methods of watershed subdivision: (1) *multi-reach* subdivision, in which a subbasin may contain several reaches; and (2) *single-reach* subdivision, in which only one reach per subbasin is allowed. The first approach creates large subbasins, such as Bull Run Subbasin No. 17 containing Reach Nos. 14 and 16, shown in Figure 10-5(a). For multi-reach subbasins, computation of overland flow width and slope is difficult. The single-reach subdivision requires arbitrarily small (say 0.1 acre), multi-reach, artificial or *dummy* subbasins at each confluence point of two or more reaches, such as dummy Subbasin No. 18, shown in Figure 10-5(b). Although this approach increases the model size due to the addition of dummy subbasins, the smallest natural subbasins are created from the available topographic or drainage maps. Single-reach subbasins are consistent with many modeling packages, such as HEC-1 and allow intuitive numbering of subbasins and streams. For example, a reach and its tributary subbasin can have the same ID.

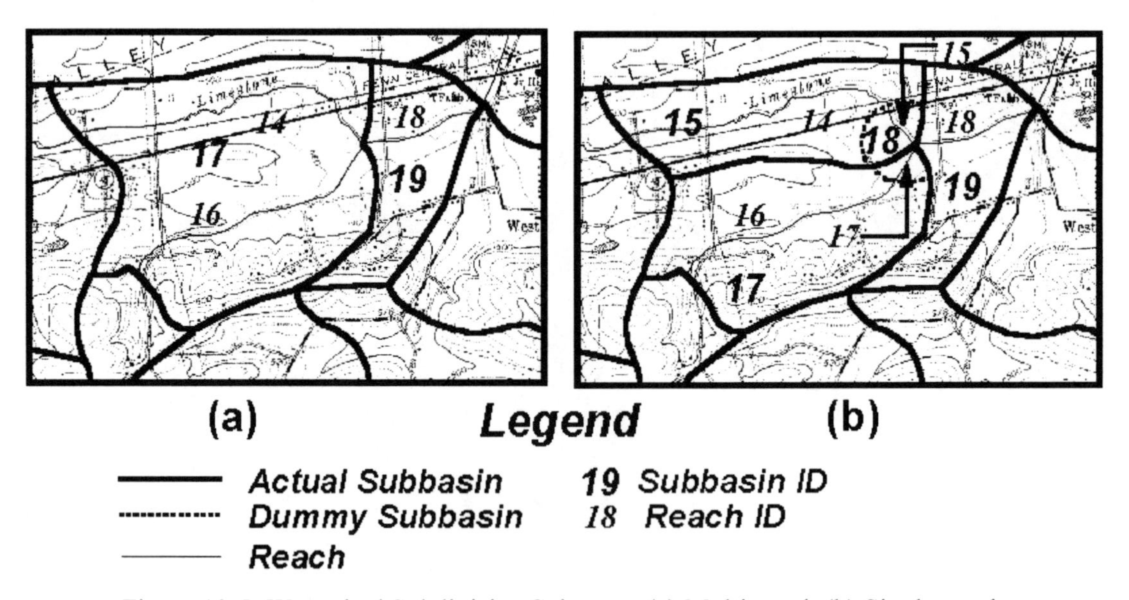

Figure 10-5. Watershed Subdivision Schemes: (a) Multi-reach (b) Single-reach

DeVantier and Feldman (1993) indicated that the difference between the lumped and the distributed models can be minimized by creating arbitrarily small subbasins. Smaller subbasins can minimize the inherent inaccuracies of the grossly lumped models. Therefore, the small size of single-reach subbasins increases the model accuracy. Due to its many advantages described above, the single-reach subdivision scheme was used in this study, which resulted in 38 actual and 10 dummy subbasins for the Bull Run watershed (Figure 10-3), and 533 actual and 163 dummy subbasins for the Turtle Creek watershed (Figure 10-4). To prevent cluttering the maps with too many subbasins, dummy subbasins are not shown in Figures 10-3 and 10-4.

Traditionally, model input parameters are estimated manually. Manual estimations require planimetering the aerial photographs, topographic maps, and/or thematic maps (e.g., land use, soils). Reach cross-section data are collected by field measurements or estimated from topographic or ground surface elevation contour maps. Initial estimates of hydraulic parameters are generally defaulted to the literature values and fine-tuned during calibration.

Manual estimation of physical parameters for large watersheds is not only tedious but also highly subjective. Sometimes, due to small budgets, no planimetering is conducted and the quantities are simply eye-balled or "guesstimated" and fine-tuned during model calibration. For instance, consider subbasin 34 shown in Figure 10-6: a 352-acre (1.42-km^2) subbasin apparently covered by residential, commercial, and perhaps industrial land use classes. It is apparent that one cannot accurately eye-ball the percent imperviousness for this subbasin, or guesstimate an SCS curve number without actually measuring the land use classes and corresponding soil types. Due to the small scale of the map, planimetering the land use classes (especially the residential type) is not feasible. Even if manual measurements were possible, they are always influenced by modelers subjectivity in delineating and measuring the areas of interest. A watershed GIS offers a bird's-eye view to instantly estimate the needed physical model parameters without any subjectivity.

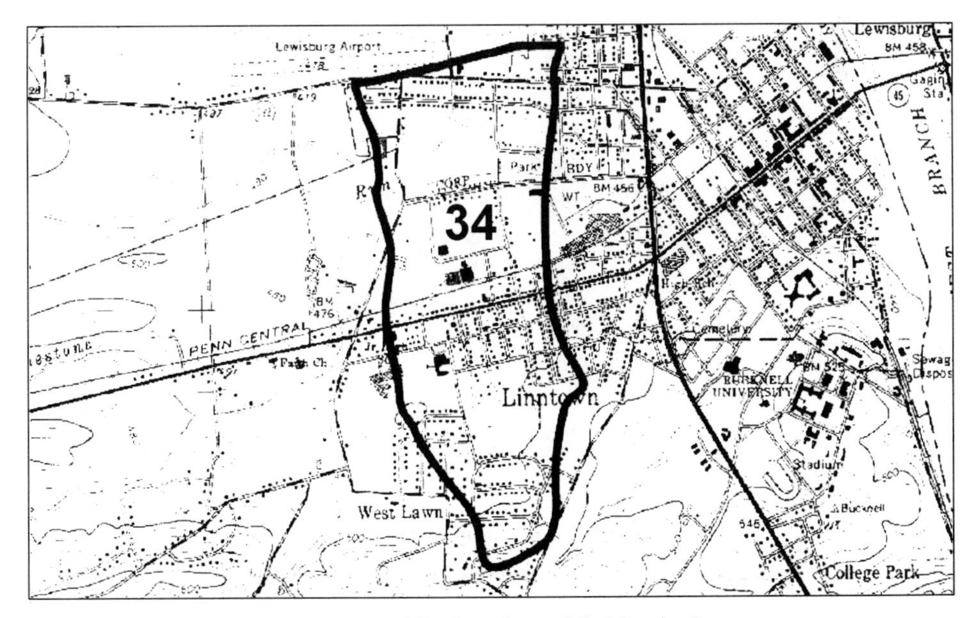

Figure 10-6. Manual Estimation of Subbasin Parameters

GIS Data Development

This section shows the front-end applications of a planning-level GIS using the interchange-type integration method described in Chapter 7, (Modeling Integration). At the study time, all-raster or all-vector approaches were quite common. In this study, both the vector and the raster systems were used to take advantage of the best features of each. For small watersheds (e.g., the Bull Run watershed), a manual-digitization and vector-polygon overlay process was used. For large watersheds (e.g., the Turtle Creek watershed), an automated scan-digitization and raster-polygon overlay process was used. The cost-effectiveness of the latter process increases as the complexity of the maps increases or as the ratio of line capture to setup time increases. Satellite imagery and DEM data are appropriately dealt with in raster format simply because it is the fundamental form of the data. Although no improvement in the quality of spatial data is possible by a raster-to-vector conversion, it is sometimes done to promote uniformity within the GIS when other layers are predominantly vector. A raster-to-vector conversion of the imagery layer was not performed due to a large number of features that would have produced an excessively complex and large vector file.

The vector-based GIS analysis was performed using ESRI's PC ArcInfo software. ArcInfo offers versatility, efficiency in assisting digitization and attribute entry, automatic editing and database creation, interactive nature, and easy spatial data display and management (Bhaskar et al., 1992). The raster-based GIS analysis was conducted in ERDAS, a digital image processing program from ERDAS Inc. Although many GIS packages offer the proposed model integration capability, ArcInfo and ERDAS were used in this study due to their mutual data exchange capability (Terstriep and Lee, 1989), leading-edge recognition (Dodson, 1993), a strong user base throughout the industry and government, and a history of frequent software updates and enhancements. More ArcInfo and ERDAS information is provided in Chapter 2 (GIS Development Software) and Chapter 3 (GIS Applications Software).

The primary objective of all the GIS analyses was the derivation of subbasin physical parameters. The central database for the storage, manipulation, and display of the collective analytical products was the ArcInfo polygon attribute file associated with the subbasin information layer. Products of vector and raster overlay analyses were merged into this common database. In turn, an ArcInfo attribute file, compatible with dBASE database handling software, was used to refine and display the pertinent information.

The basemaps for all the watersheds consisted of 7.5-min. 1:24,000-scale USGS topographic maps. All analyses were conducted on IBM-compatible PCs. Information about the GIS data layers is presented below.

PRIMARY LAYERS

Primary GIS layers are those that are digitized or scanned directly from the paper (source) maps. They are not derived from a combination or overlay of other layers. The following five primary layers were created for the Bull Run watershed.

1. Figure 10-7 shows the vector layer for the Bull Run watershed subbasins. This layer was created by digitization of subbasin boundaries, which were manually delineated on the 7.5-min. USGS topographic map of Lewisburg. This layer provided subbasin area, centroid coordinates, and overland flow width for model input.

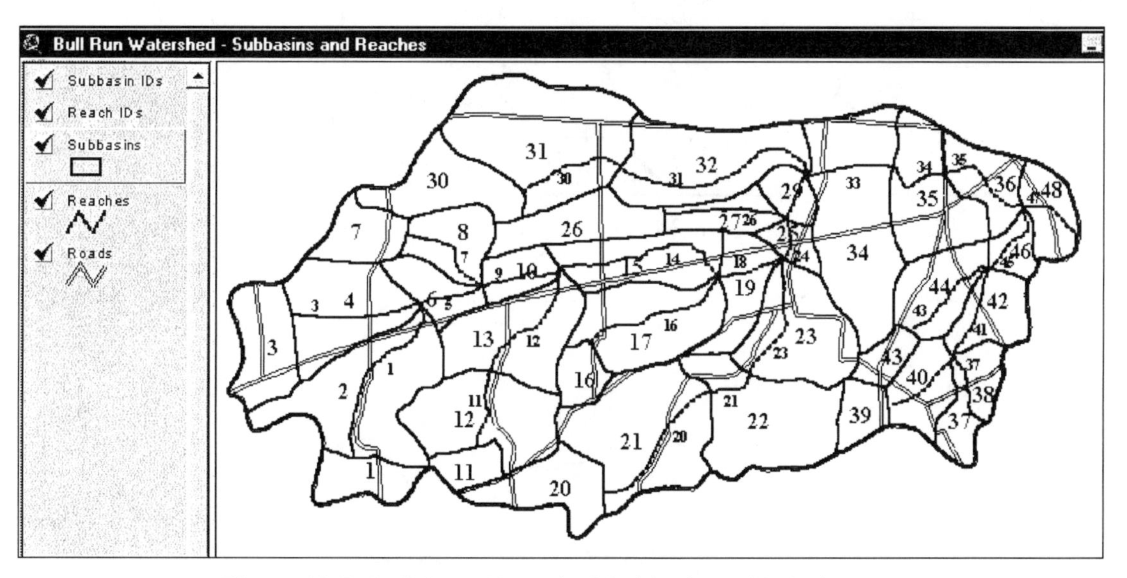

Figure 10-7. Bull Run Watershed Subbasin and Reach Layer

2. Figure 10-7 also shows the vector layer for the Bull Run streams (reaches). This layer was created by manual digitization from the USGS topographic map. This layer produced stream length and slope. Stream length, slope, and cross-section dimensions were used to compute stream capacity and travel time for model input.

   As an alternative to manual digitization of watershed boundaries and streams described above, watersheds and streams can be automatically delineated more efficiently from DEMs. Figures 10-8 and 10-9 show the DEM-based subbasins and streams for the Bull Run watershed (Shamsi, 2000). For more information about DEMs, please refer to Chapter 5 (Internet GIS).

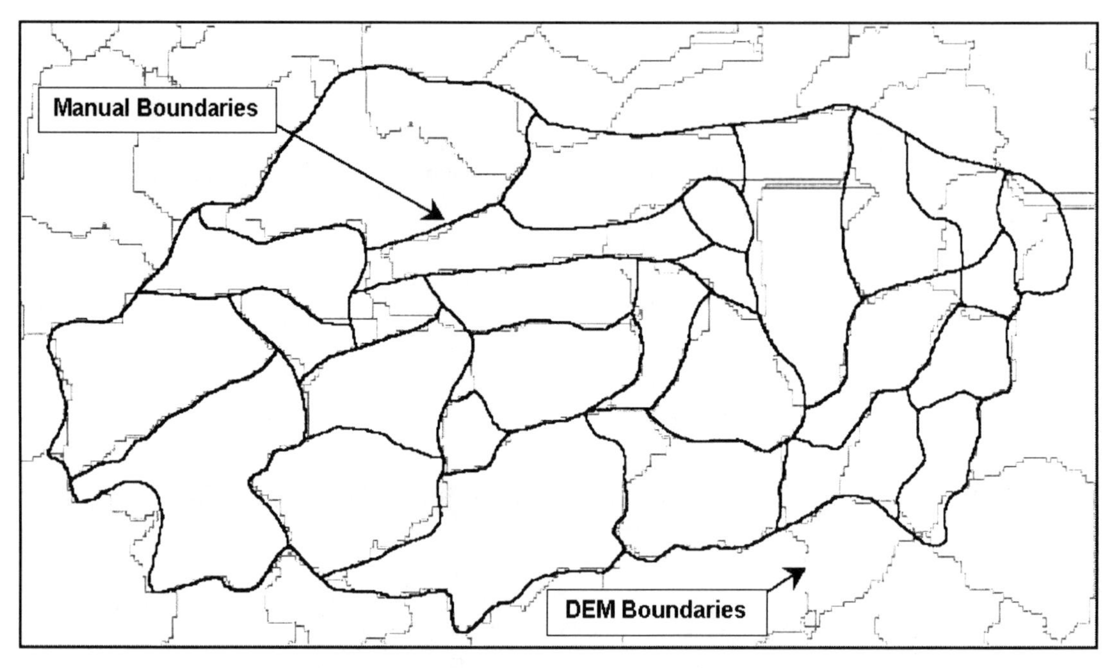

Figure 10-8. Manual Versus DEM Subbasins

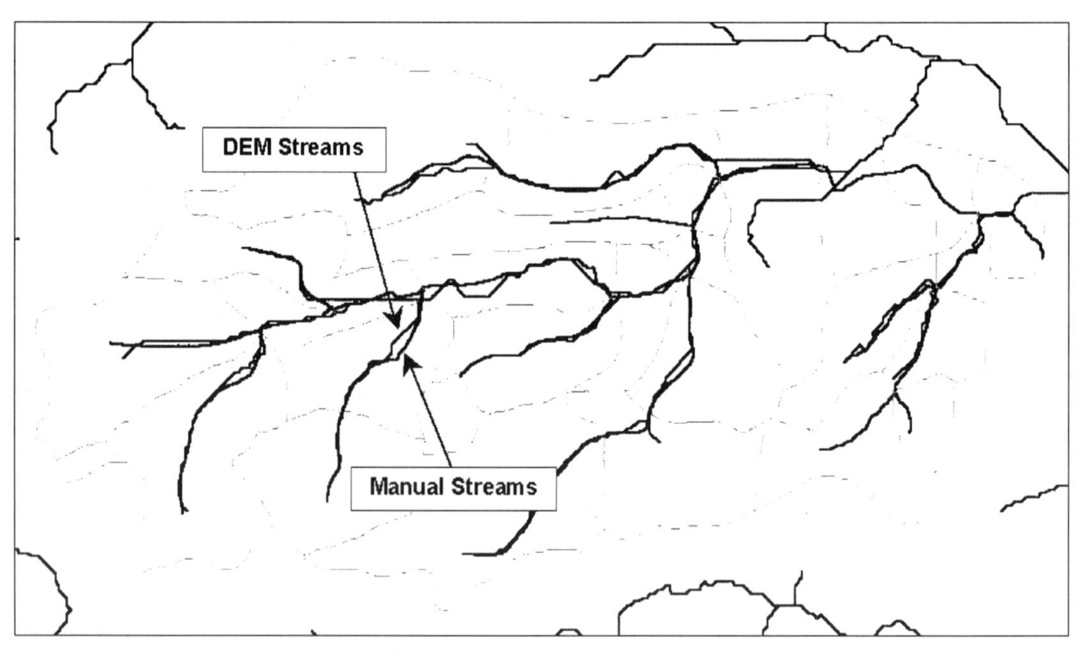

Figure 10-9. Manual Versus DEM Streams

3. Thirty-six soil types were digitized from the Union County SCS Soil Survey maps to develop the Bull Run watershed soils layer. SCS has assigned hydrologic soil groups (HSG) A, B, C, or D to each of approximately 16,000 soil types found in the United States (U.S. Department of Agriculture, 1986). More information about SCS soils data is provided in Chapter 7 (Modeling Integration). These data were used to merge the 36 digitized soil types into the four HSGs and create a vector layer shown in Figure 10-10. As an alternative to manual digitization of soil types, HSGs can also be extracted from the new STATSGO and SSURGO digital soil data from NRCS described in Chapter 5 (Internet GIS).

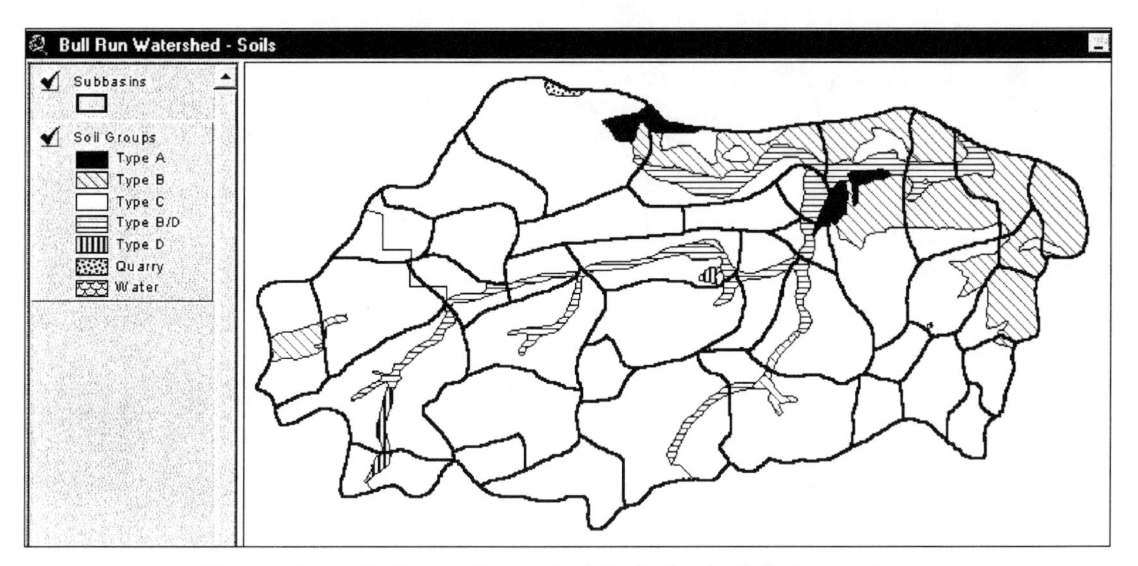

Figure 10-10. Bull Run Watershed Hydrologic Soil Groups Layer

4. Vector layer for 14 land use classes was created by digitization of 1:40,000 scale, 1990 color infrared transparencies from the National Aerial Photography Program (NAPP). The NAPP imagery and land use layer are shown in Figures 10-11 and 10-12, respectively. An example of GIS versatility: Whereas the visual inspection of USGS topographic map (Figure 10-6) could not reveal more than three land use classes for subbasin 34, the GIS-based land use map of Figure 10-12 identified nine:

- High-density residential, 54.9%;
- Medium-density residential, 7.7%;
- Low-density residential, 3.1%;
- Commercial, 4.2%;
- Industrial, 12.1%;
- Parks, 4.3%;
- Schools, 4.0%;
- Woods, 2.2%; and
- Brush, 7.4%.

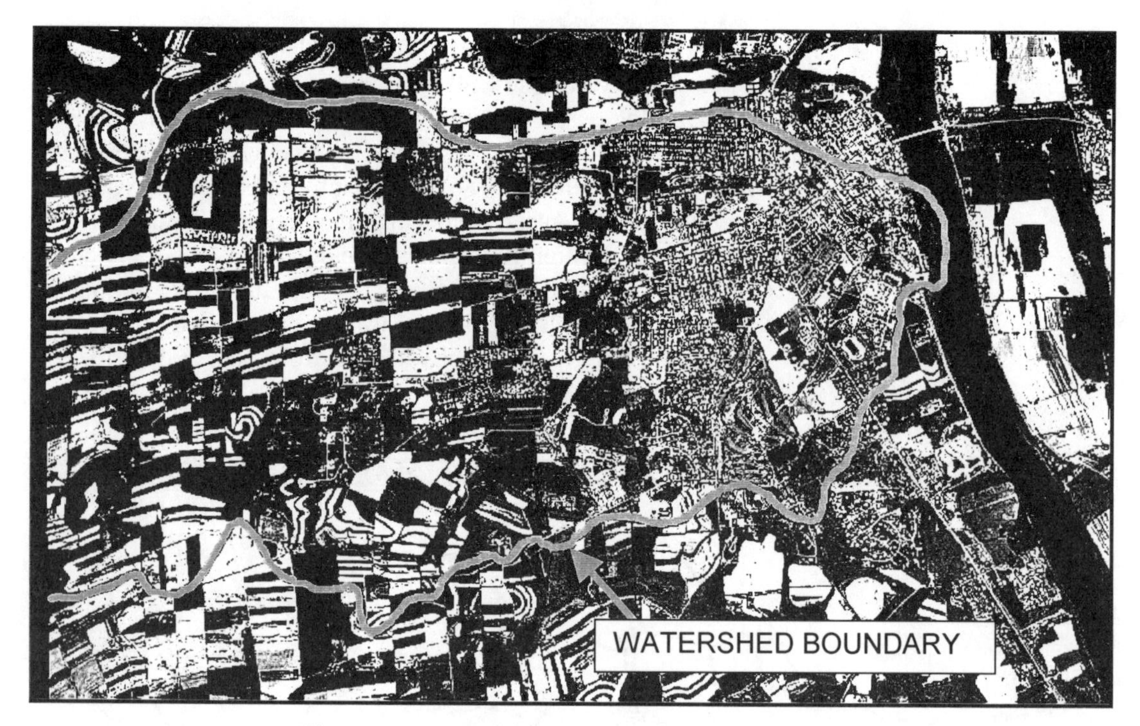

Figure 10-11. Bull Run Watershed NAPP Image

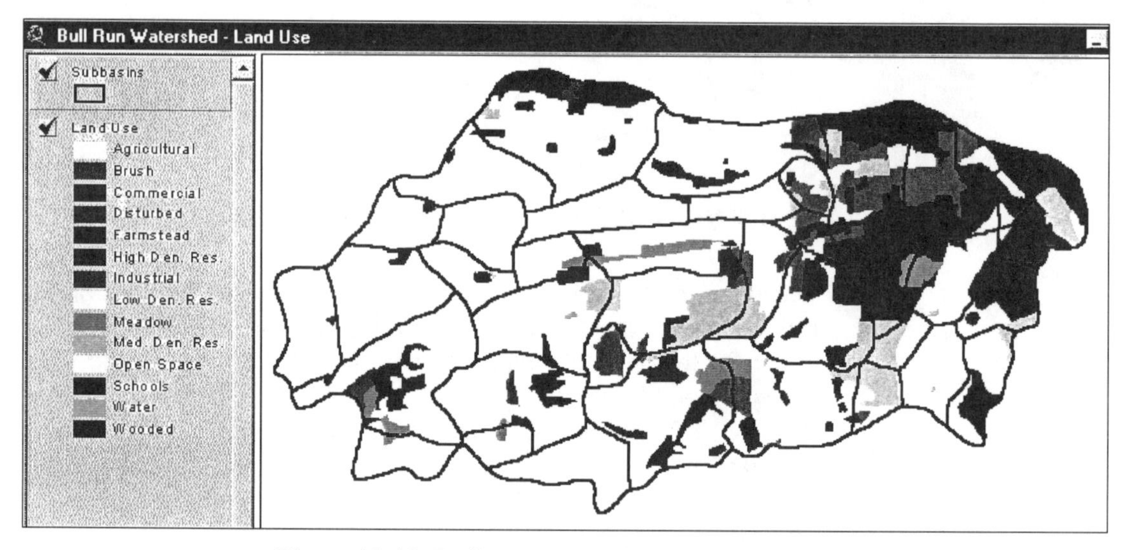

Figure 10-12. Bull Run Watershed Land Use Layer

Such a microscopic classification is the reason for the high accuracy of GIS-based estimates of subbasin physical parameters. Derivation of land use classes from the low-level aerial photography is referred to as the "conventional" method, in comparison with the remote sensing techniques described in Chapter 8 (Water System Applications) which employ automatic classification of satellite imagery. Studies have shown that the remote sensing techniques for land use classification are more cost-effective. The cost benefits have been estimated on the order of 6 to 1 in favor of the satellite imagery approach. Although remote sensing–based land use statistics may not be as detailed as those derived using the conventional method, the computed runoff curve numbers and discharges are nearly the same (Engman, 1993).

5. Raster layer of subbasin elevation was created from a 30-m USGS DEM of the watershed shown in Figure 10-13. Elevation increases from white to black pixels, and therefore the white areas depict the streams and valleys and the black areas depict the watershed divides and ridges. Note that, generally, the low (white) and high (black) areas coincide with the streams and watershed boundaries, respectively.

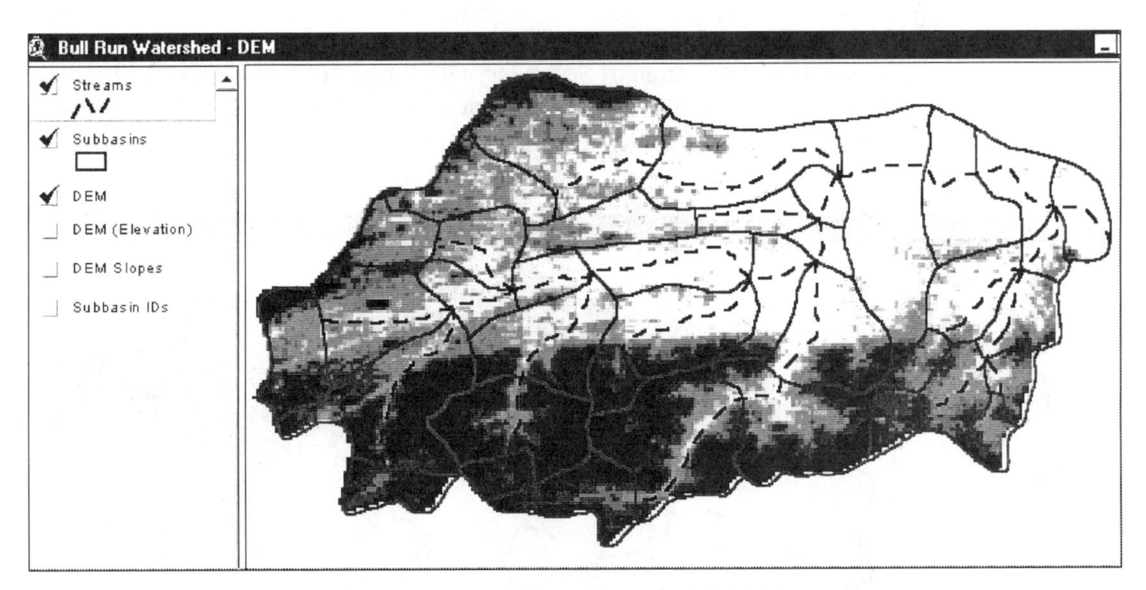

Figure 10-13. Bull Run Watershed DEM Layer

Secondary Layers

The remaining physical parameters of the Bull Run watershed model were computed from *secondary* layers. These layers are derived from post-processing (GIS analysis) of the primary layers. The following five secondary layers were created for the Bull Run watershed.

1. Figure 10-14(a) shows how the primary layers of subbasins, land use, and hydrologic soil groups were overlayed to delineate the percent imperviousness and runoff curve number polygons for Bull Run

Subbasin No. 34. Vector layer for subbasin percent imperviousness was created by overlaying the primary layers for subbasins and land use to delineate the percent imperviousness polygons shown in Figure 10-14(b). The SCS land use–percent imperviousness matrix (Table 7-1, Chapter 7) was used in PC ArcInfo to assign percent impervious values to each polygon. Bull Run watershed subbasin percent imperviousness layer is shown in Figure 10-15.

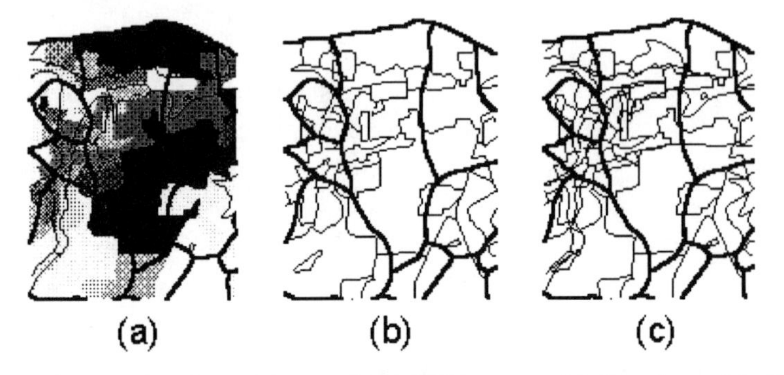

Figure 10-14. Derivation of Secondary GIS Layers: (a) Overlay of Subbasins (Thick Lines), Land Use (Shaded Areas) and soil Groups (Thin Lines); (b) Percent Imperviousness Polygons; (c) Runoff Curve Number Polygons

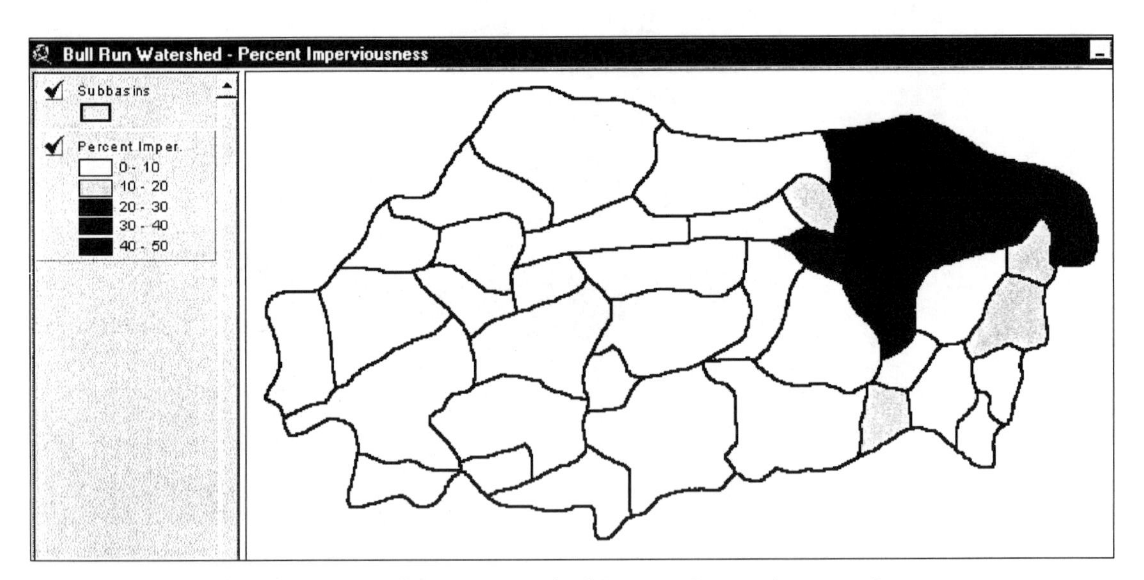

Figure 10-15. Bull Run Watershed Percent Imperviousness Layer

2. Vector layer for subbasin runoff curve numbers was created by overlaying the primary layers for subbasins, soil types, and land use to delineate the runoff curve number polygons shown in Figure 10-14(c). The SCS land use–soil–curve number matrix (Table 7-1) was used in

PC ArcInfo to assign runoff curve numbers to each polygon according to its land use and hydrologic soil group. Bull Run watershed runoff curve number layer is shown in Figure 10-16.

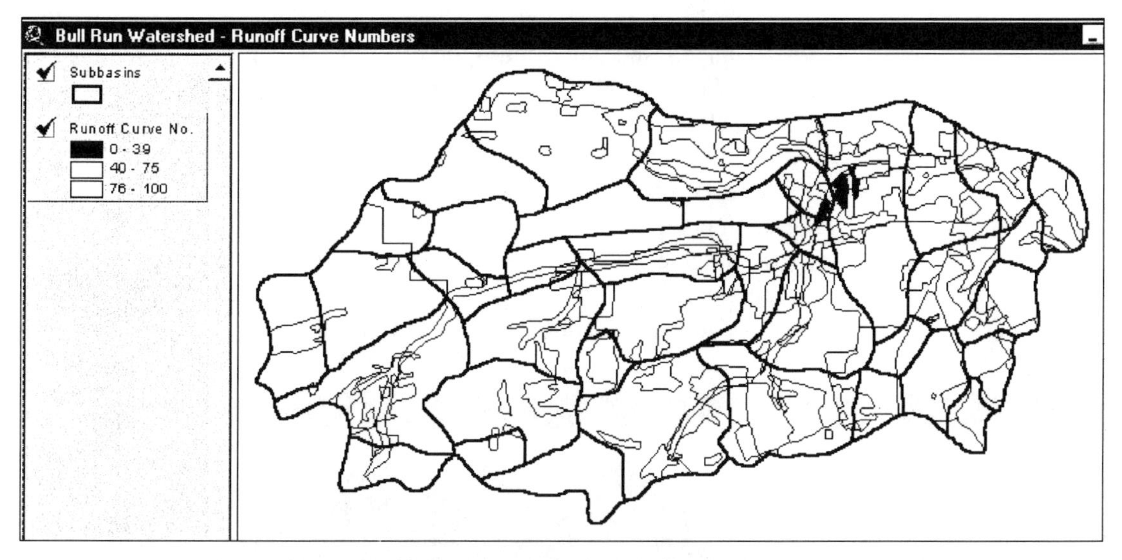

Figure 10-16. Bull Run Watershed Runoff Curve Number Layer

3. A hillshade layer clearly shows the relief of a watershed and its natural drainage pattern. A hillshading analysis can be performed on an input elevation grid theme, such as a DEM layer. The raster layer of hillshade, shown in Figure 10-17, was created by post-processing the DEM grid layer in ArcView Spatial Analyst extension (ESRI, 1996).

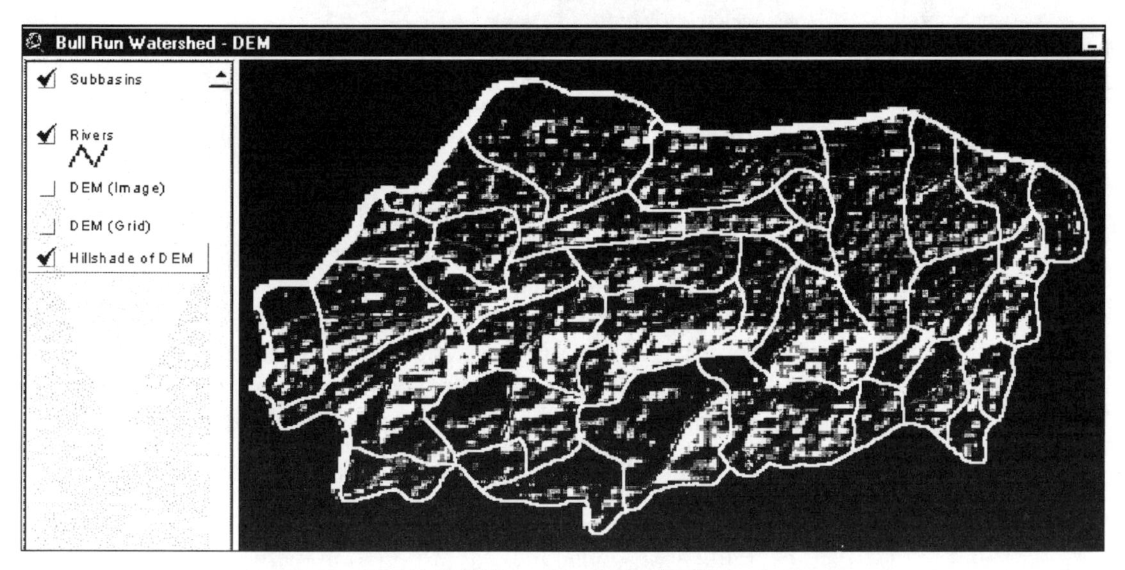

Figure 10-17. Bull Run Watershed Hillshade Layer

4. Raster layer for subbasin slope was created by overlaying the primary layers for subbasins and DEM. The ERDAS function SLOPE was used to compute the percent slope by fitting a plane to a pixel elevation and its eight neighboring pixel elevations. The difference in elevation between the low and the high points was divided by the horizontal distance, and multiplied by 100 to compute percent slope for the pixel. Pixel slope values were averaged to compute the mean percent slope of each subbasin. Figure 10-18 shows the raster slope layer. Slope decreases from blue (light gray on black and white copy) to red (dark gray/black on black and white copy) pixels. Note that, in the southern half of the watershed, most steep (blue or light gray) areas coincide with the stream locations. This observation indicates that the streams in the southern part of the watershed are steeper than those located in the northern part of the watershed

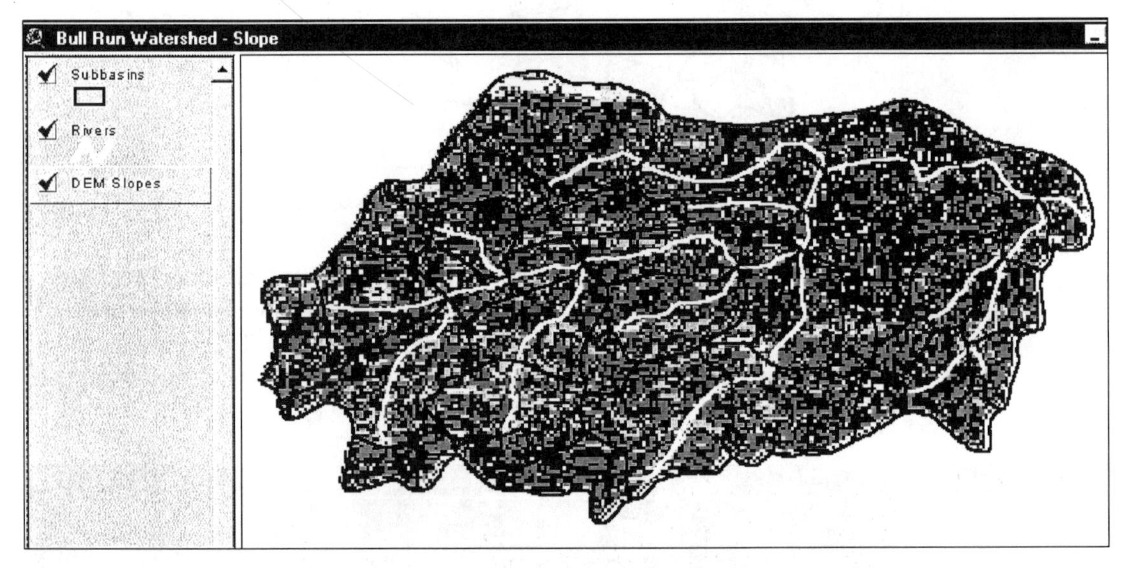

Figure 10-18. Bull Run Watershed Slope Layer

5. An overlay of roads (shown in Figure 10-7) and streams layers was used to locate potential "obstruction" locations defined by crossings of roads and streams, such as culverts and bridges. The obstructions were subsequently surveyed to estimate flow capacities of the obstructions. Estimation of stream obstructions capacities and comparing them with various modeled design flows was conducted to identify potential flooding sites, which is a required task of all the Pennsylvania watershed stormwater management plans.

Other data not required by the model itself but useful for visualization, such as municipal boundaries, transportation networks, flood hazard areas, and zoning boundaries, were also included in the watershed GIS to serve as information layers in an integrated GIS suitable for a variety of planning and siting purposes.

For the skeptics of the GIS and modeling integration approach, small system examples may not be too convincing. To appreciate the advantage of the GIS approach, we shall next focus on the large watersheds. Figure 10-19 shows the 1,811 subbasins of the 400-mi^2 Lake Erie watershed. Figure 10-20 shows the hydrologic soil group layer for the 700 subbasins of the 150-mi^2 Turtle Creek watershed. It is obvious from Figures 10-19 and 10-20 that for large watersheds the manual model parameter estimation approach would be simply too laborious and expensive.

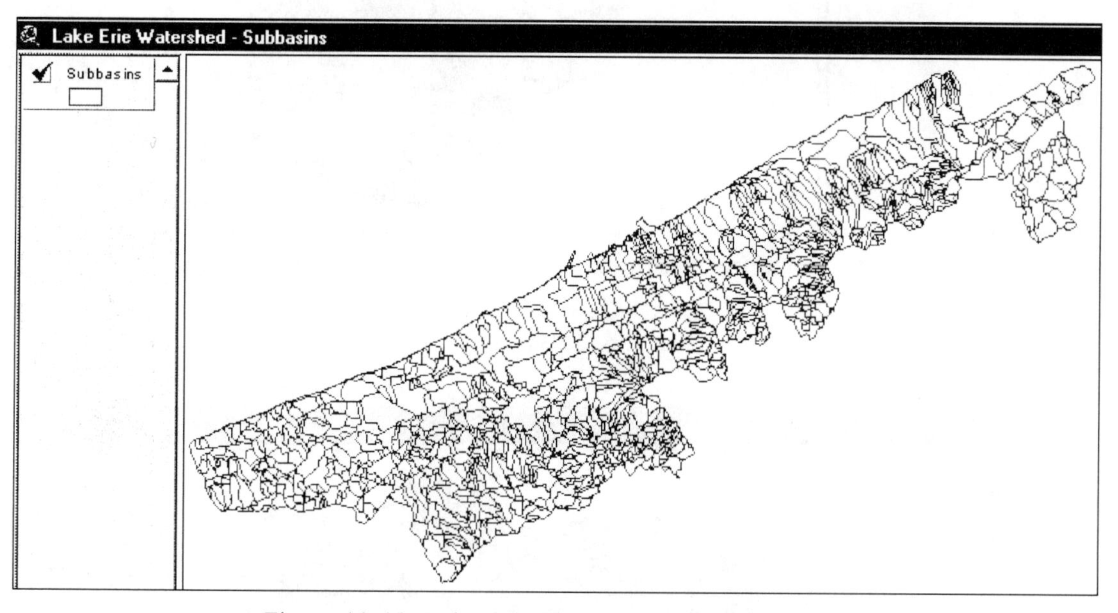

Figure 10-19. Lake Erie Watershed Subbasins Layer

The Bull Run watershed is small enough to permit land use digitization from an aerial photograph and create a vector layer. For large watersheds, land use classification is more appropriately done in a raster format by digital image processing of satellite imagery. Figure 10-21 shows the raster land use layer for the Turtle Creek watershed derived from the 30-m resolution Landsat Thematic Mapper (TM) imagery. Sixty land cover classes were identified during the unsupervised classification, but were subsequently aggregated into five broader classes for modeling, mapping, presentation, and black and white reproduction purposes. Due to a large number of polygons, manual digitization was not conducted and subbasins and soil types were scanned from the paper maps. Polygon overlay analysis was conducted using raster processing.

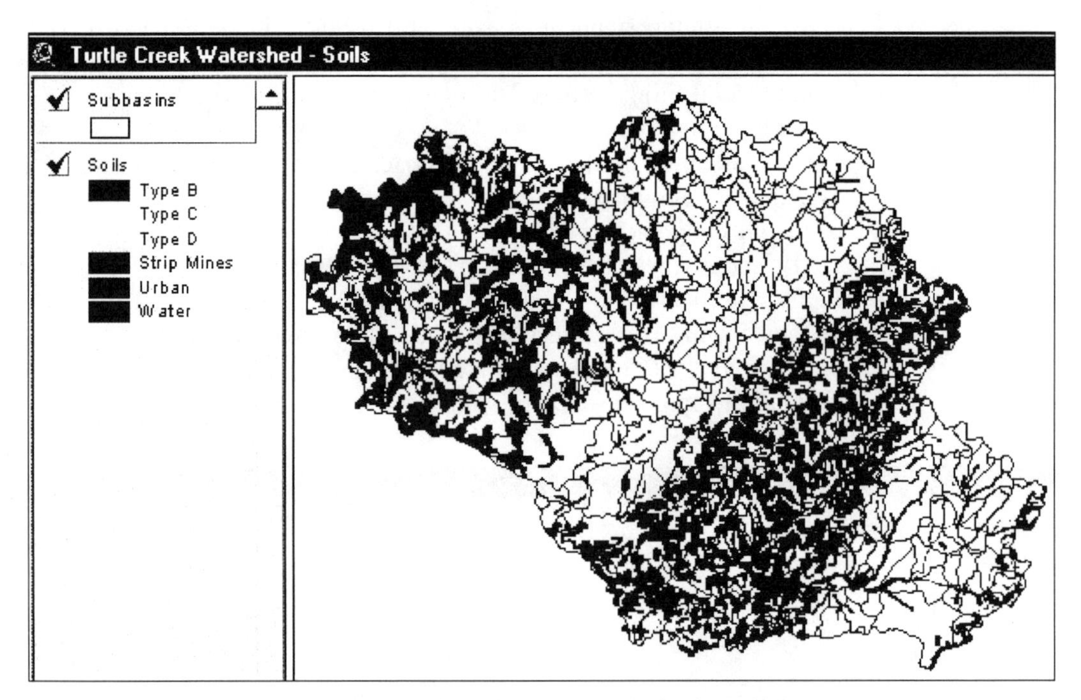

Figure 10-20. Turtle Creek Watershed Hydrologic Soil Groups Layer

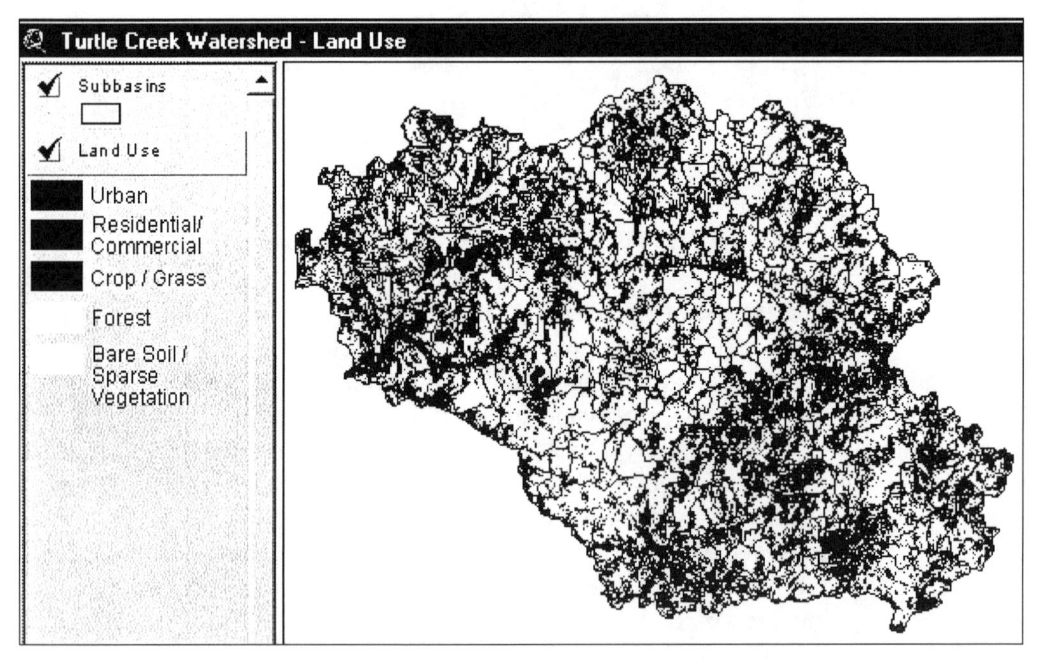

Figure 10-21. Turtle Creek Watershed Land Use Layer

The GIS integration techniques presented here are not limited to PSRM–many hydrologic models require similar input parameters. For example, the same techniques have been successfully used in H&H modeling with SWMM program (Shamsi, 1993a; Shamsi and Fletcher, 1994).

MODEL CALIBRATION AND VERIFICATION

The model input parameters derived from the above-mentioned GIS analyses were assembled to prepare a model input file. Data assembly was followed by model calibration. The objective of calibration was to change some model parameters within acceptable limits established in the literature until a reasonable agreement between some observed and modeled hydrographs was obtained. The creation of the GIS-based smallest natural subbasins provided an opportunity to—as closely and accurately as possible and practical—quantify the reach and subbasin physical parameters. The physical parameters therefore were not intended to be altered during the calibration process. Rather, to the extent possible, calibration was intended to adjust the most difficult to define hydraulic parameters and refine the default PSRM coefficients.

The integral square error (ISE), given by Equation (10-1), was used as a measure of goodness of fit between observed and modeled hydrographs (Marsalek et al., 1975):

$$\text{ISE} = \frac{\left[\sum_{i=1}^{N}(O_i - M_i)^2\right]^{1/2}}{\sum_{i=1}^{N} M_i} \times 100 \qquad (10\text{-}1)$$

where O_i = observed hydrograph value at time i, M_i = modeled hydrograph value at time i, and N = number of hydrograph values.

The calibration can be subjectively rated as excellent for $0 < \text{ISE} \leq 3$, very good for $3 < \text{ISE} \leq 6$, good for $6 < \text{ISE} \leq 10$, fair for $10 < \text{ISE} \leq 25$, and poor for $25 < \text{ISE}$ (Marsalek et al., 1975). Calibration requires continuous, preferably hourly, rainfall and stream flow data. In most study watersheds, budgetary constraints did not permit field data collection, and therefore existing historical data were utilized.

BULL RUN WATERSHED

Rainfall data were obtained from a U.S. National Oceanic and Atmospheric Administration (NOAA) rain gauge located at Selinsgrove. This was the nearest recording type rain gauge located approximately 13 mi (21 km) south of the watershed. The advantage of using a NOAA recording rain gauge is the availability of digital historical hourly precipitation data from the National Climatic Data Center (NCDC). An in-house utility Rainfall Analysis Program (RAP) was used to convert 41 years (1949–90) of raw NCDC data into meaningful hourly rainfall hyetographs (Shamsi, 1989). The only available continuous stream flow data consisted of three 1970–79 observed hydrographs collected by the Department of Civil Engineering, Bucknell University, at a point approximately 1 mi upstream of the watershed outlet. The watershed has not been developed significantly during the past decade, and the 1970–79 stream flow data were assumed to adequately represent the current flows.

Figure 10-22 compares an observed and modeled hydrograph after calibration. The calibration rainfall event occurred on May 24, 1978, and produced a total rainfall accumulation of 1.6 in. (40.64 mm) in 13 hr, which approximates a 1-year storm. The ISE value of the hydrographs is 7.1, which indicates a "good" calibration. The difference in modeled and observed peak flow and volume is 6% and 11%, respectively. Although the time to peak for the two hydrographs does not match and the modeled hydrograph shows a dual peak, calibration was considered satisfactory, especially because the rainfall input belonged to an external rain gauge.

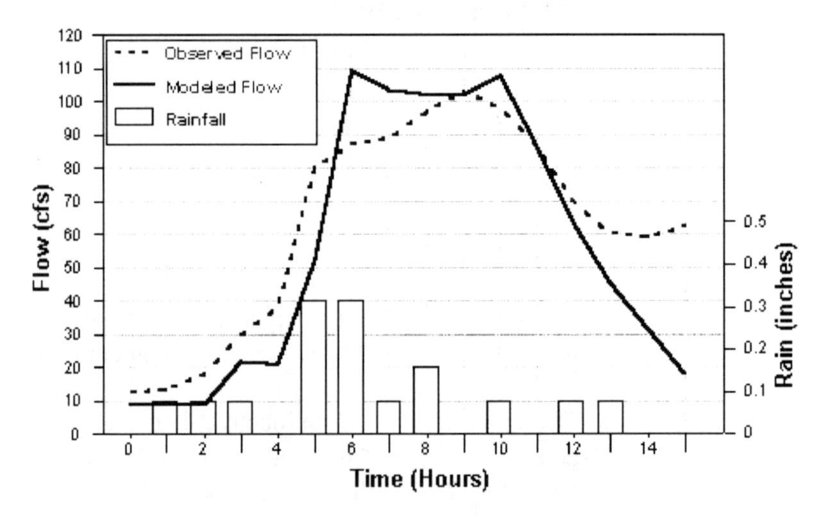

Figure 10-22. Bull Run Watershed Model Calibration

As intended, the calibration did not require modification of any physical parameters. The following hydraulic parameters were adjusted within the recommended ranges (Aron and Lakatos, 1990):

- Manning's n for pervious surfaces = 0.20,
- Manning's n for impervious surfaces = 0.02,

- Depression storage for pervious surfaces = 0.00 in.,
- Depression storage for impervious surfaces = 0.06 in. (1.524 mm), and
- The SCS initial abstraction factor = 0.10.

Model verification was performed by comparing modeled peak flows with available Federal Emergency Management Agency (FEMA) peak-flow estimates based on the regional flood-frequency method developed by the U.S. Army Corps of Engineers (COE) (Federal Emergency Management Agency, 1984). The calibrated model was run to develop 3-, 6-, 12-, and 24-hr runoff hydrographs and peak discharges for return periods of 2, 5, 10, 25, 50, and 100 years corresponding to SCS Type II rainfall distribution (U.S. Department of Agriculture, 1986). The results, shown in Figure 10-23, indicate that the modeled 12- and 24-hr peak flows reasonably agreed with the FEMA estimates.

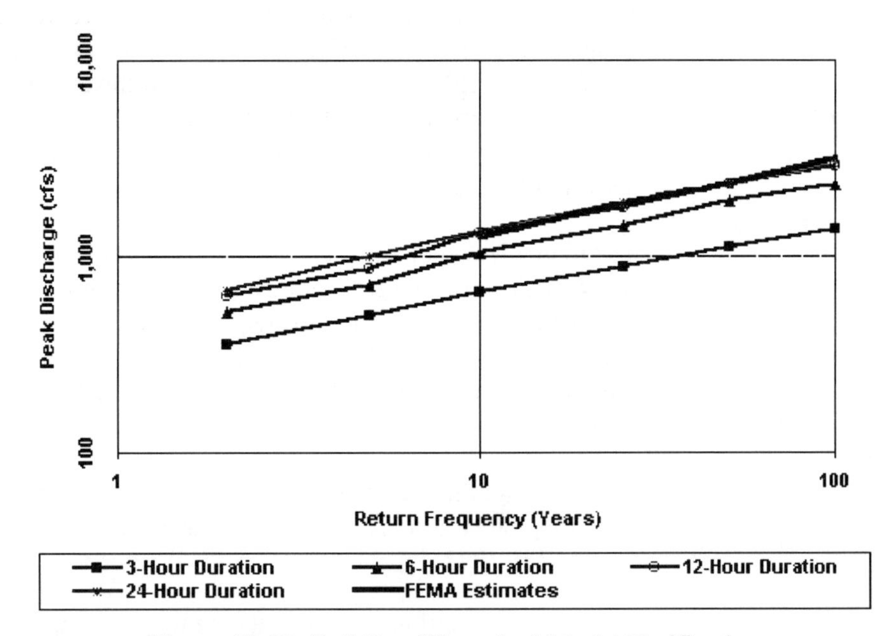

Figure 10-23. Bull Run Watershed Model Verification

TURTLE CREEK WATERSHED

Stream gaging in the Turtle Creek watershed was established as early as 1907, and at various points in time stream gauges have been operated at ten locations in the basin. In the Turtle Creek stormwater management project, stream flow data from three stream gauges were used:

1. A gauge operated by COE is situated on Turtle Creek 1.44 mi from its confluence with Monongahela River in East Pittsburgh Borough. This gauge has produced essentially continuous records since 1939. This gauge measures stream flow generated from approximately 142 mi^2 or roughly 97% of the watershed.

2. A second gauge operated by COE is situated on Turtle Creek 3.31 mi from its confluence with the Monongahela River in Wilmerding Borough. This gauge has been operating since 1968. This gauge measures stream flow generated from approximately 120.6 mi^2 or roughly 83% of the watershed.

3. A third gauge, this one operated by the USGS, is located on Abers Creek, a tributary of Turtle Creek. This gauge, which has been in operation since 1948, provides stream flow data from approximately 3,000-acre Upper Abers Creek drainage area.

Hourly precipitation data are available from a recording-type NOAA rain gauge located at Murrysville. This rain gauge is situated within the Turtle Creek watershed approximately 2.1 mi northwest of the centroid of the Turtle Creek watershed and 2.5 mi southwest of the centroid of the Upper Abers Creek watershed. The stream gauges and rain gauge locations are shown in Figure 10-4.

The model was calibrated against the Wilmerding stream flow records and Murrysville precipitation records. The calibration rainfall event which approximates a 10-year return frequency event occurred on July 8 and 9, 1985, and produced a total rainfall accumulation of 91.95 mm (3.62 in.) within a 24-hr period. This storm was selected for calibration purposes for the following reasons:

1. The magnitude of the flood which occurred on that date was quite large, the largest recorded in the watershed since the Hurricane Agnes flood in 1972.

2. Usable streamflow and rainfall data were available for the event.

3. It is a relatively recent event whose response reflects current hydrologic conditions in the watershed.

Figure 10-24 shows the modeled and observed hydrographs. The ISE value of the hydrographs is 9.9, which shows a "good" level of calibration (Marsalek et al., 1975). Calibrated modeled peak discharge and time to peak are within 4% of the observed values. Though not exact, the general shape of the modeled and observed hydrographs are quite similar. The modeled total stream discharges approximate the measured volume to within 10%. Once again, the calibration did not require modification of any subbasin physical parameters.

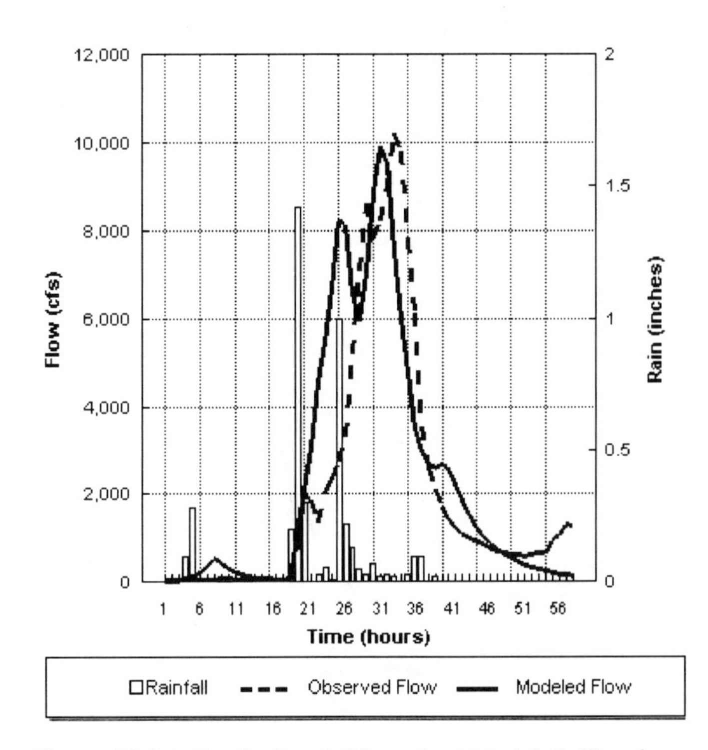

Figure 10-24. Turtle Creek Watershed Model Calibration

The experience gained through the case studies listed in Table 10-1 indicates that the GIS-based models require considerably less calibration time. This observation implies that GIS-based model parameters are more accurate than those computed from traditional manual map measurement techniques.

EFFECT OF STORM DISTRIBUTION

The existence of long-term rainfall distribution and stream flow provides an opportunity to demonstrate the effect of rainfall temporal distribution on stream discharge estimates used in stormwater management planning. The Turtle Creek watershed will be used as an example (Maslanik and Shamsi, 1991).

Various design rainfall hyetographs were input to the calibrated model. The rainfall volume–duration-frequency data reported in the Field Manual of Pennsylvania Department of Transportation Storm Intensity–Duration–Frequency Charts (Aron et al., 1986) were used to estimate 24-hr rainfall depths for 2-, 10-, 25-, 50- and 100-year return periods. Storm hyetographs were produced using two types of storm distributions: the familiar SCS Type II distribution (U.S. Department of Agriculture, 1986) and a *historical* distribution produced through the analysis of Murrysville rainfall records using RAP computer program (Shamsi, 1989). The historical distribution represents the mean temporal distribution of all the historical rainfall events. Both types of distributions are shown in Figure 10-25. The storm hyetographs were modeled to produce estimates of peak stream discharges at the Turtle Creek (East Pittsburgh) and Upper Abers Creek stream gaging locations for each of the five return frequencies.

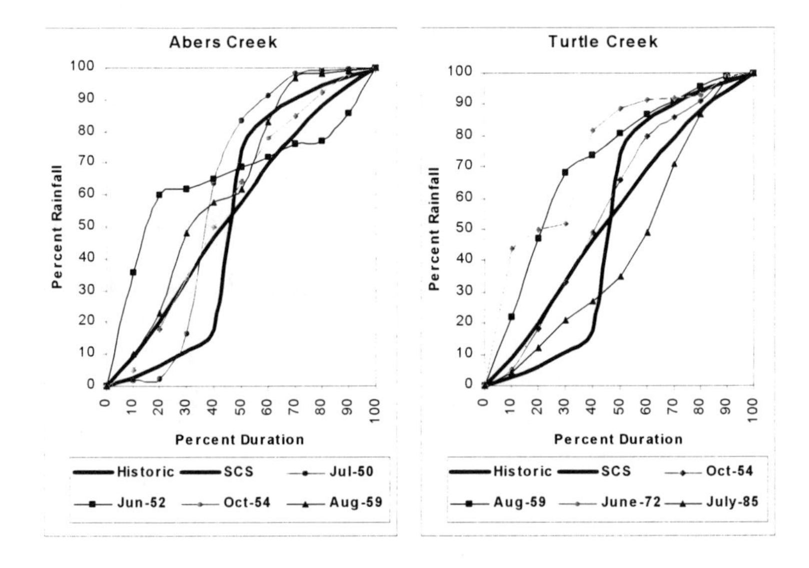

Figure 10-25. Comparison of Rainfall Distributions

It has been reported that the largest floods in the Upper Abers Creek watershed are caused by thundershower convection type storms containing high-intensity episodes (Turtle Creek Watershed Association, 1977). In contrast, the largest of the floods occurring at the Turtle Creek gauge have been attributed to hurricane-caused rainfall events, which tend to be longer-duration, relatively low-intensity events.

The relative difference in the ability of the modeled flood frequency estimates to replicate statistical estimates was investigated by comparing some observed storms distributions to the modeled SCS Type II and historical distributions. Four storms recorded at the Murrysville rain gauge that produced the four largest floods of records at each gaging station were selected for comparison. These storms were converted to dimensionless mass diagrams describing the distribution of cumulative percentage rainfall with cumulative percentage duration. Comparisons of these storms to the SCS and historical distributions are given in Figure 10-25. A visual comparison of the distribution curves indicates that the major flood-producing events in the Upper Abers Creek watershed do tend to parallel the SCS Type II distribution. The largest floods measured at the Turtle Creek gauge, conversely, are generally produced by evenly distributed storms more closely approximated by the historical distribution.

Statistical analyses of the long-term stream flow records from the Turtle Creek gauges have been conducted by various government agencies to develop flood frequency estimates (or statistical estimates). Statistical estimates for the Upper Abers Creek gauge were determined using a Log-Pearson Type III frequency analysis of annual peak-flow data (Federal Emergency Management Agency, 1979). Statistical estimates for the Turtle Creek gauge were developed using a partial series flood frequency analysis (U.S. Army Corps of Engineers, 1963). Modeled flood frequency estimates were compared with known statistical estimates. Comparisons of the various flood frequency estimates are provided in Figure 10-26. For the Upper Abers Creek watershed, the SCS Type II distribution produces a close approximation of statistical estimates throughout the range of return intervals investigated. The historic distribution, however, consistently produced a significant underestimation of discharges. In the case of the Turtle Creek watershed, the SCS Type II estimates matched the statistical estimates at the shorter return periods. However, they diverge at the longer return periods. The modeled historical estimates tend to converge with the statistical estimates at the longer return periods.

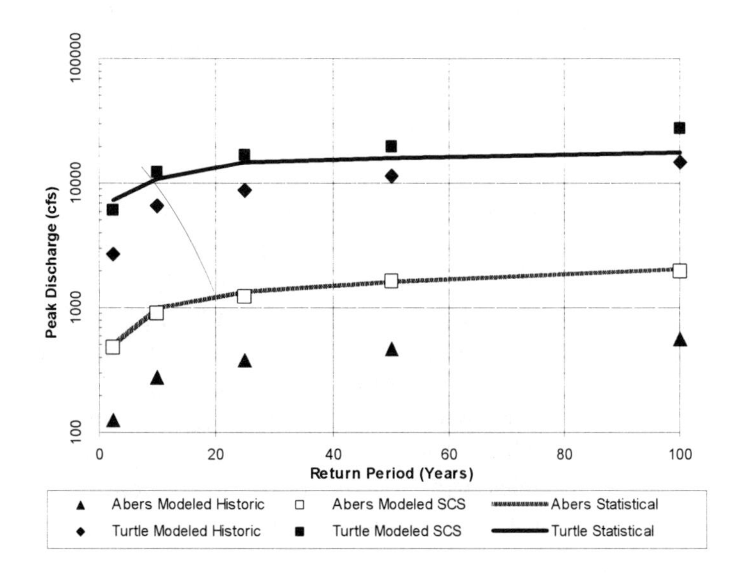

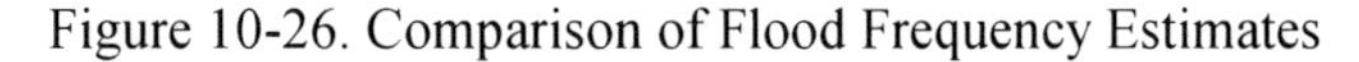

Figure 10-26. Comparison of Flood Frequency Estimates

STORMWATER MANAGEMENT PLAN IMPLEMENTATION

The most effective means of satisfying the basic requirements of Act 167 would be by controlling runoff from new land development such that both the total volume and peak rate of runoff from new development are identical to that which occurred from the site before the land was changed. In other words, the post-development runoff hydrograph would be identical to the pre-development hydrograph. This goal is usually accomplished by storing portions of the runoff and "bleeding" it at a controlled rate. This procedure tends to extend the hydrograph, increases the period of time during which the elevated flow rates are discharged, and creates a potential for the extended high flows to combine with peak discharges originating from other points in the watershed. This combination of flows can result in increases to downstream peak flows, but it can be avoided if subbasin *release rates*, defined below, are used to control the runoff from the subbasins.

A runoff hydrograph at a given point of a drainage system is the sum of individual flow contributions from all the upstream subbasins. PSRM displays this information in a *peak-flow presentation table* (PFPT). The main function of a PFPT is to display the individual runoff contributions from upstream subbasins to downstream reaches, including the timing of such peak-flow contributions. PFPTs show how subbasin runoff travels downstream and contributes to flooding at downstream reaches. This is accomplished in the model by lagging the runoff hydrograph from each subbasin by the travel time from the subbasin outlet point to the subbasin for which the PFPT has been created. The Bull Run watershed PFPT for a 100-year, 24-hr design storm is shown in Table 10-2 (Shamsi, 1996). Release rate, as a percentage, is defined as:

$$R_{ij} = \left(\frac{Q_{ij}}{Q_i}\right) \times 100 \qquad (10\text{-}2)$$

where R_{ij} = release rate of subbasin i at a downstream reach j, Q_{ij} = pre-development discharge contribution of subbasin i to the peak discharge of a downstream reach j, Q_i = pre-development peak discharge of subbasin i. Q_{ij} and Q_i are available from the PFPT. Release rates for all subbasins are computed at all downstream reaches and tabulated in a *release rate table* (RRT). RRTs are prepared for 3-, 6-, 12-, and 24-hr durations corresponding to 100-year return frequency. It has been found that the 24-hr, 100-year design storm produces the most stringent (smallest) release rates. The 100-year, 24-hr RRT for the Bull Run watershed is shown in Table 10-3 (Shamsi, 1996).

Table 10-2. Peak Flow Presentation Table at the Outlet of Bull Run Watershed (After Shamsi, 1996)

Subbasin		Flows (cfs) arriving at specified time (minutes)														
ID	TT	780	840	900	960	1,020	1,080	1,140	1,200	1,260	1,320	1,380	1,440	1,500	1,560	1,620
1	494.5	0.6	0.8	1.2	1.8	2.6	3.0	4.3	10.0	135.4	27.2	13.6	10.5	8.6	7.6	6.8
2	433.2	4.1	5.6	8.9	13.2	19.5	23.6	39.1	280.3	151.8	49.8	30.4	24.6	22.1	20.1	17.7
3	477.6	0.2	0.3	0.7	1.5	2.4	3.1	5.5	16.8	205.0	36.0	18.7	14.3	11.9	10.5	9.3
4	433.2	5.0	6.1	8.5	11.7	16.2	19.1	30.3	198.5	105.2	34.8	21.6	17.6	15.9	14.5	12.9
6	408.7	3.3	4.0	5.1	7.2	9.1	11.9	20.3	153.3	20.7	10.4	7.1	6.0	5.7	5.1	4.6
*	*	*	*	*	*	*	*	*	*	*	*	*	*	*	*	*
26	265.2	9.9	12.4	18.7	34.8	105.9	162.6	37.3	22.4	16.5	12.1	10.1	9.3	8.3	7.5	5.4
27	204.8	9.1	13.2	20.8	116.0	92.1	22.3	13.6	9.7	7.5	5.8	5.2	4.7	4.1	3.6	1.0
29	176.5	4.1	6.9	22.8	89.5	22.6	11.4	7.6	5.5	4.1	3.0	2.7	2.4	2.1	1.4	0.3
30	374.6	6.4	8.0	11.4	22.3	20.9	33.3	246.4	40.1	24.6	17.5	13.8	12.1	11.1	9.9	9.0
31	310.3	6.6	10.1	15.4	42.5	40.6	415.3	241.2	101.9	65.9	45.9	35.4	30.6	26.8	23.5	20.2
32	176.5	27.3	42.3	129.1	496.4	132.9	66.0	44.1	32.1	24.5	18.6	16.7	14.6	13.1	9.5	3.5
34	129.4	66.2	88.0	534.9	277.7	149.9	100.9	75.4	57.9	45.7	37.5	32.5	28.3	25.7	17.6	11.5
35	93.2	63.9	263.3	271.5	90.5	53.7	38.2	30.0	24.0	19.8	17.5	15.5	13.8	12.2	6.3	3.7
36	41.5	120.0	180.0	39.1	22.7	16.7	13.4	11.2	9.4	8.5	7.9	7.0	6.5	3.7	1.3	0.5
37	116.2	2.7	3.8	13.5	45.0	10.8	6.6	4.9	3.9	3.0	2.5	2.3	2.0	1.7	1.2	0.2
38	100.6	6.8	7.9	57.3	43.3	18.7	11.9	8.9	7.0	5.6	4.8	4.3	3.7	3.4	2.3	1.2
39	128.5	5.5	6.3	11.6	76.3	20.9	11.8	8.6	6.8	5.4	4.5	4.1	3.7	3.3	2.8	1.2
40	100.6	8.2	10.1	100.8	63.4	24.0	15.2	11.3	8.8	6.9	5.9	5.3	4.5	4.1	2.3	0.6
42	67.5	19.7	109.4	114.7	37.2	20.6	14.0	11.0	8.6	7.1	6.4	5.6	4.9	4.1	1.5	0.5
43	103.9	7.3	11.8	59.4	20.4	10.1	6.8	5.3	4.3	3.5	3.1	2.8	2.5	2.2	1.3	0.3
44	67.5	27.6	153.8	156.3	45.0	24.5	17.0	13.5	10.7	8.9	8.2	7.2	6.3	5.3	1.7	0.4
46	41.5	45.4	72.4	14.4	8.0	5.7	4.4	3.5	2.7	2.4	2.2	1.8	1.6	0.7	0.0	0.0
48	0.0	171.6	38.3	19.5	13.0	9.6	7.5	5.8	4.8	4.5	3.9	3.3	3.1	0.5	0.1	0.0
Total outflow		873.3	1,548.9	2,131.1	2,480.4	2,621.6	2,839.3	2,793.8	1,730.2	1,161.3	744.3	571.5	413.9	342.4	280.9	210.2

Notes: TT = Travel time in minutes, * = Data for subbasins 7 to 25 truncated to fit the table.

Table 10-3. Release Rate Table for the Bull Run Watershed (After Shamsi, 1996)

At	From Subbasin																							
Reach	1	2	3	4	6	*	26	27	29	30	31	32	34	35	36	37	38	39	40	42	43	44	46	48
1	100	0	0	0	0	*	0	0	0	0	0	0	0	0	0	0	0	0	0	0	0	0	0	0
2	11	100	0	0	0	*	0	0	0	0	0	0	0	0	0	0	0	0	0	0	0	0	0	0
3	0	0	100	0	0	*	0	0	0	0	0	0	0	0	0	0	0	0	0	0	0	0	0	0
4	0	0	8	100	0	*	0	0	0	0	0	0	0	0	0	0	0	0	0	0	0	0	0	0
6	9	81	9	81	22	*	0	0	0	0	0	0	0	0	0	0	0	0	0	0	0	0	0	0
*	*	*	*	*	*	*	*	*	*	*	*	*	*	*	*	*	*	*	*	*	*	*	*	*
26	0	0	0	0	0	*	100	0	0	0	0	0	0	0	0	0	0	0	0	0	0	0	0	0
27	0	0	0	0	0	*	0	0	0	0	0	0	0	0	0	0	0	0	0	0	0	0	0	0
29	5	18	5	19	87	*	8	6	6	0	0	0	0	0	0	0	0	0	0	0	0	0	0	0
30	0	0	0	0	0	*	0	0	0	100	0	0	0	0	0	0	0	0	0	0	0	0	0	0
31	0	0	0	0	0	*	0	0	0	11	100	0	0	0	0	0	0	0	0	0	0	0	0	0
32	0	0	0	0	0	*	0	0	0	10	79	13	0	0	0	0	0	0	0	0	0	0	0	0
34	3	7	3	8	13	*	11	7	8	42	39	8	12	0	0	0	0	0	0	0	0	0	0	0
35	4	9	3	10	53	*	10	7	7	22	30	7	11	6	0	0	0	0	0	0	0	0	0	0
36	2	3	1	4	5	*	79	15	15	11	57	16	19	10	6	0	0	0	0	0	0	0	0	0
37	0	0	0	0	0	*	0	0	0	0	0	0	0	0	0	100	0	0	0	0	0	0	0	0
38	0	0	0	0	0	*	0	0	0	0	0	0	0	0	0	59	100	0	0	0	0	0	0	0
39	0	0	0	0	0	*	0	0	0	0	0	0	0	0	0	0	0	100	0	0	0	0	0	0
40	0	0	0	0	0	*	0	0	0	0	0	0	0	0	0	0	0	30	100	0	0	0	0	0
42	0	0	0	0	0	*	0	0	0	0	0	0	0	0	0	65	96	36	95	29	0	0	0	0
43	0	0	0	0	0	*	0	0	0	0	0	0	0	0	0	0	0	0	0	0	100	0	0	0
44	0	0	0	0	0	*	0	0	0	0	0	0	0	0	0	0	0	0	0	0	13	100	0	0
46	0	0	0	0	0	*	0	0	0	0	0	0	0	0	0	9	29	10	27	85	31	84	17	0
48	2	4	1	4	7	*	55	13	12	10	84	13	17	9	5	7	12	11	8	8	7	7	4	4
TRR	2	3	1	4	5	*	8	6	6	10	30	7	11	6	5	7	12	10	8	8	7	7	4	4
PRR	80	80	80	80	50	*	100	100	100	60	60	100	100	100	100	100	100	100	100	90	80	80	100	100

Notes : TRR=Theoretical Release Rate, PRR=Practical Release Rate, * = Data for subbasins 7 to 25 truncated to fit the table.

The release rates range from 0 to 100%. Ideally, to protect all downstream reaches, a subbasin's release rate should be established as the minimum of its release rates for all downstream reaches. Thus, the *theoretical release rates* (TRR) can be computed by taking the minimum value along each column of RRT. However, implementation of theoretical release rates is not practical due to the following reasons:

1. Some theoretical release rates may be zero or near zero. Near-zero release rates would preclude any runoff from a site. A control to this extreme would essentially prohibit development and would be unacceptable to the public. It is, therefore, necessary to define an acceptable lower limit for the computed release rates from a cost–benefit point of view. It has been found that a 50% release rate represents a straight compromise between runoff control cost and absolute control effectiveness (Joint Planning Commission, 1988).

2. A release rate below 50 indicates that the subbasin peak occurs substantially before the watershed peak, such that the subbasin contributes only a small part to the watershed peak rate. In these subbasins, using release rates below 50% will result in requiring very large and costly detention facilities that retain high volumes of runoff. Instead, it is preferable for these subbasins to release the flow more quickly prior to the time when the watershed peak occurs. The release rate for these subbasins can be increased to 100%.

3. The theoretical release rate of the uppermost subbasins is generally found to be significantly different from that of the immediately downstream subbasins. This condition is difficult to implement because it is against the public perception. People tend to compare their regulatory requirements with their neighbors, and they may find it too discriminatory if significantly different release rates are implemented in adjacent subbasins.

The above discussion indicates that for the public acceptance, the theoretical release rates must be converted to *practical release rates* (PRR). PRRs are computed by following five steps: (1) Discard all values below 50 along each column of RRT; (2) take the minimum along each column; (3) if a column minimum is zero, replace it with 100 (this may happen if all the column values are below 50); (4) replace the minimums of the topmost subbasins by their immediately downstream subbasins; and (5) round off to the nearest 10s. The subareas with the same PRRs can be aggregated together to form zones of equal PRR values called *release rate areas*. These areas and their PRR values can be printed on the basemap to produce a *release rate map*. Figure 10-27 is the Bull Run watershed release rate map showing 10 manually delineated release rate areas.

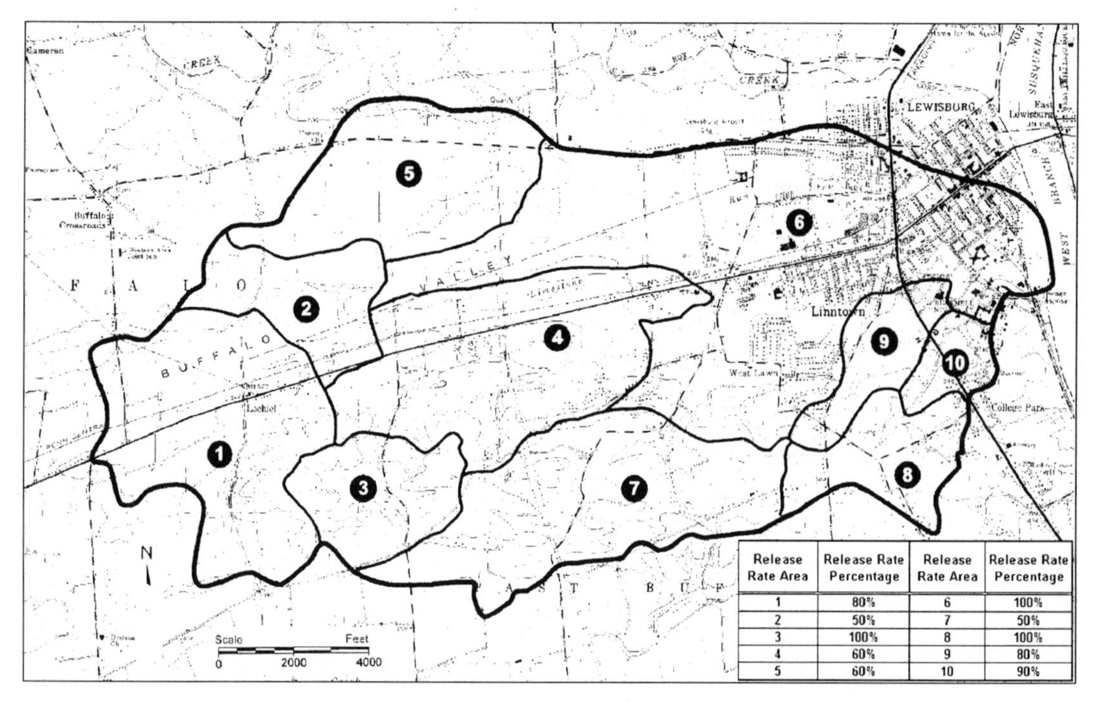

Release Rate Area	Release Rate Percentage	Release Rate Area	Release Rate Percentage
1	80%	6	100%
2	50%	7	50%
3	100%	8	100%
4	60%	9	80%
5	60%	10	90%

Figure 10-27. Bull Run Watershed Release Rate Map

Alternatively, for large watersheds, a GIS layer of the release rate areas is more suitable. Figure 10-28 shows the release rate area layer for the Conneaut Outlet watershed.

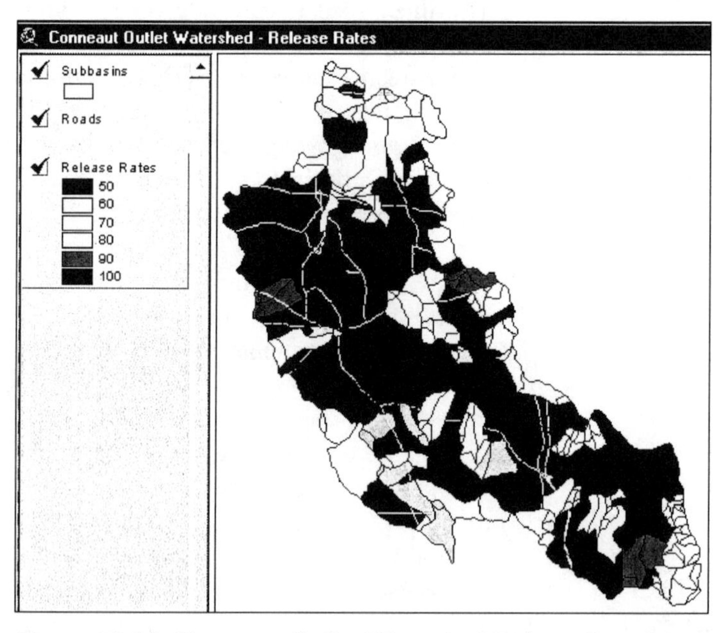

Figure 10-28. Conneaut Outlet Watershed Release Rate Layer

The release rate map is the ultimate watershed tool to implement a watershed-wide stormwater management plan based on the requirements of Act 167. A watershed stormwater management plan can be implemented by enforcing the release rate based allowable post-development peak flows calculated as:

$$Q_{pi} = (\frac{R_i}{100}) \times Q_i \qquad (10\text{-}3)$$

where Q_{pi} = the allowable post-development peak discharge in subbasin i, R_i = the release rate of subbasin i from the release rate map, and Q_i = the pre-development peak discharge of subbasin i.

For the release rate concept to work, it must be used for all development within a watershed, which results in an increase in the post-development runoff from the site. To comply with this requirement, a developer must follow four steps:

1. Identify the location of the development site on the release rate map and determine the required release rate.

2. Using PSRM, or another model, compute the pre-development and post-development runoff from the site for the 2-, 10-, 25-, and 100-year storms using no stormwater management techniques (e.g., detention basins, seepage pits, dutch drains, porous pavements). If the post-development peak rate is less than or equal to the pre-development rate, the requirements of Act 167 have been met; otherwise, proceed to step 3.

3. Apply on-site stormwater management techniques to increase infiltration and reduce impervious surfaces. Using PSRM, or another model, recompute the post-development runoff rate for the 2-, 10-, 25-, and 100-year storms. If the post-development peak rate is less than or equal to the pre-development rate, the requirements of Act 167 have been met; otherwise, proceed to step 4.

4. Using Equation (10-3), determine the allowable post-development peak discharge from the development. Using PSRM, or another model, design the necessary detention/retention facilities to meet the allowable peak runoff rate standard.

Once a watershed stormwater management plan has been adopted by a county and approved by DEP, it is the responsibility of the watershed municipalities to implement the plan. Act 167 requires each watershed municipality to adopt a stormwater management ordinance within 6 months of the plan's approval in order to regulate development within the municipality in a manner consistent with the applicable watershed stormwater plan and the provisions of Act 167.

Case Study Discussion and Conclusions

In the 1990s, watershed management reemerged as the nation's water quality strategy for the 21st century (Rubin et al., 1993). Watershed management is also one of the key provisions of the potential reauthorization of the Clean Water Act (Jaworski, 1994). Acting proactively, Pennsylvania is already in the process of implementing a statewide watershed stormwater management program guided by the state's Stormwater Management Act. The Pennsylvania experience indicates that the GIS and modeling integration is an effective technical tool in implementing watershed-wide stormwater management. In Pennsylvania, despite 75% state funding, stormwater management plans for only 56 of the 356 designated watersheds were completed in the first 13 years (1984–97) of the program. The major reason of this slow progress is an inadequate state budget of approximately $600,000 a year for funding the stormwater management plans as compared with the typical project cost of $20,000 to $350,000 per watershed. Other potential reason might be the unwillingness or unaffordability of the counties to share the remaining 25% of the project cost. Thus, another lesson learned from the Pennsylvania experience is that sufficient, and preferably complete, state financial assistance is necessary to implement the stormwater management plans throughout the state in a timely manner.

This chapter has demonstrated use of a planning-level watershed GIS to estimate input parameters of a hydrologic model. However, to a modeler, there is little difference in the tedium of hand measurements from paper maps or using a GIS, unless there is a clear evidence of the superiority of GIS results to more traditional methods (DeVantier and Feldman, 1993). The cost-effectiveness and advantages of GIS integration have been documented previously both in a vector (Ross and Tara, 1993; Pearson and Wheaton, 1993) and a raster (Meyer et al., 1993; Stuebe and Johnston, 1990) GIS. For example, an ArcInfo interface with HEC-1 and HEC-2 resulted in substantial cost savings (Phipps, 1995). It was found that if the models were not linked to the GIS, manual calculations would be required, which would take at least five times longer than using the ArcInfo linkage. Experience gained through the above case studies agrees with this finding. It was found that the PSRM and GIS integration offers cost-effective and technically sound solutions to Pennsylvania's watershed-wide stormwater management implementation. The cost savings are realized for five reasons:

1. Frequently, GIS-based watershed physical parameters—such as area, overland flow width, slope, percentage imperviousness, and runoff curve number—are sufficiently accurate due to the small size of single-reach subbasins and do not require further adjustment during calibration. This capability reduces the model calibration time and reduces the cost of modeling.

2. Traditionally, physical parameters of a model were estimated by creating manual overlays of paper maps and manual measurements of length, width, and area using map wheels and planimeters. Manual estimation of physical parameters, especially for large watersheds like Turtle Creek and Lake Erie, is both tedious and subjective. For example, even in a small watershed like the 8.4-mi^2 (21.8-km^2) Bull Run watershed subbasin, hydrologic soil groups and land use maps should be manually superimposed to delineate and measure approximately 525 runoff curve number polygons, as shown in Figure 10-14(c), followed by assigning curve numbers from the SCS runoff curve number table (Table 7-1), followed by taking aerial averages of polygon curve numbers over the subbasins. This manual procedure may take several days as compared with the proposed GIS technique, which may take only a few hours. The time difference will be even more pronounced for larger watersheds.

3. Perceptions of most modelers changes with time, fatigue, and mood. Thus, multiple manual measurements of the same subbasin by the same modeler may be different each time. Perceptions of the different modelers to do the same task are also different and depend on their experience, skill, and vision. For example, manual measurements of the area of a subbasin by different modelers may be different. GIS eliminates such human subjectivity in model parameter estimation.

4. GIS users have constant access to the most current data available. In the future, the stormwater control criteria and the release rate map may be easily and cost-effectively updated simply by reclassifying the land use from the latest aerial photograph or satellite imagery to reflect the contemporary development conditions, updating the subbasin data file, and rerunning the model.

5. The GIS database developed to support watershed modeling can be expanded to accommodate other applications and departments. Additional layers can be added and/or the accuracy of the existing GIS can be increased to keep up with the emerging applications, such as, asset inventory, work order management, and water quality monitoring. Regardless of the application, enhancing an existing GIS will be definitely cheaper than developing a new GIS from scratch.

SUMMARY

A study of six Pennsylvania watersheds indicates that a planning-level watershed GIS consisting of subbasins, streams, soils, land use, and DEM layers can be used to estimate physical input parameters of typical lumped-parameter hydrologic models. Typical physical parameters include subbasin area, centroid coordinates, overland flow width (or length), slope, percent imperviousness, and runoff curve number; and stream slope, size, capacity and/or travel time. Both the vector and the raster GIS formats can be utilized to take advantage of the best features of each. For small watersheds, manual digitization and vector polygon overlay is cost-effective. For large watersheds, automated scan digitization, raster polygon overlay, and DEM-based automatic delineation are more appropriate. In the six study watersheds, model calibration did not require adjustment of model physical parameters. This is believed to have been caused by the single-reach watershed subdivision scheme and accuracy of the GIS estimates. PSRM provides three stormwater management tools: a peak-flow presentation table, a release rate table, and a release rate map. The release rate map is a practical tool to implement a watershed-wide stormwater management plan. The GIS and PSRM integration was found to be ideally suited to develop watershed-wide stormwater management plans required by the Storm Water Management Act of Pennsylvania. Pennsylvania experience indicates that sufficient financial assistance is a key factor for the success of statewide stormwater management implementation. It is also true that under certain conditions GIS-based modeling may not be cost-effective, especially if the watershed is very small (less than 1 mi^2 (2.6 km^2) and 20 subbasins), or if the model is intended for one time use only (e.g., designing a detention pond).

SELF-EVALUATION

1. How was GIS used to develop watershed stormwater management plans in Pennsylvania?

2. What is the difference between primary and secondary layers of GIS data?

3. How does GIS linkage reduce the cost of H&H modeling?

4. What is a release rate map and how can you use it to control the development in watersheds?

5. What method of GIS and H&H modeling linkage was used to develop PSRM input data for the case study watersheds described in this chapter? Which other methods are available?

Chapter 11

CASE STUDIES

GIS applications are growing throughout the world. This chapter shows how people around the world are applying GIS in their water, wastewater, and stormwater projects.

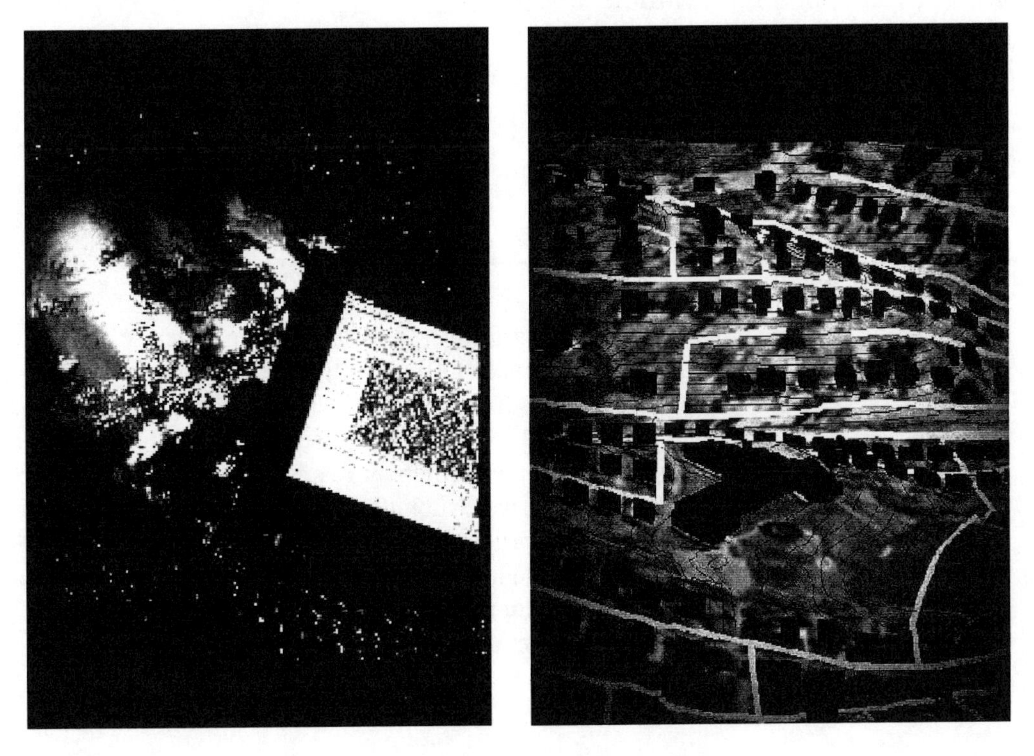

*GIS = **G**et **I**ntelligent **S**olutions*

LEARNING OBJECTIVE

The learning objective of this chapter is to document GIS application projects around the world for water, wastewater, and stormwater systems. Major topics discussed in this chapter include

- ☑ Combined Sewer Overflow (CSO) management
- ☑ Demand estimation for water distribution system modeling
- ☑ Flood inundation mapping
- ☑ Groundwater recharge
- ☑ Hydrologic and hydraulic (H&H) modeling
- ☑ Long-term hydrologic impact assessment
- ☑ Nonpoint source pollution assessment
- ☑ Permit compliance
- ☑ Real-time flood information mapping
- ☑ Sanitary sewer modeling
- ☑ Sewer system design
- ☑ Storm Water Management Model (SWMM)
- ☑ Water quality modeling integration

AN APPLICATIONS SAMPLER

This chapter presents a collection of recent case studies on GIS applications for water, wastewater, and stormwater systems[1]. These case studies were written specially for publication in this book by 29 GIS and water professionals from five countries (Canada, Fiji Islands, Germany, Turkey, and the United States). For the names and organizational affiliations of the case study authors, please see the "Acknowledgments" section. The case studies were submitted in response to my "Call for Case Studies" distributed to various Internet discussion groups. Most of the case studies were submitted by the members of the SWMM-Users Internet discussion group. For more information on SWMM-Users, please see Chapter 12 (GIS Resources).

[1] Two case studies on H&H modeling by D. Waye and on CSO management by S. Byun and B. Marengo were included in Chapter 7 (Modeling Integration) and Chapter 9 (Wastewater System Applications), respectively.

NODE WATER DEMAND ESTIMATION IN SURREY

Application	Estimation of demands for water distribution system modeling
Project Status	Completed in 2000
Reference	Johnston et al., 2001
GIS software	ArcView
Other software	WaterCAD, Microsoft Excel
GIS data	Legal lots, zoning, official community plan (OCP), community boundaries, water mains, population, and model nodes
Hardware	Pentium 450 PC
Study Area	Surrey, British Columbia, Canada
Organization	Kerr Wood Leidal Associates Ltd., North Vancouver, British Columbia

This project involved the recalibration of an existing water model. Although the model was capable of estimating bulk water flows and fire flows, it was not able to predict some phenomena observed in the field, such as areas of low pressure and the flow split between various areas within each pressure zone.

Using a GIS, the accuracy of the existing model was improved by developing a new demand flow database that models domestic, industrial-commercial-institutional, and irrigation demands, for each of the approximately 50,000 lots within the system. This database was developed by combining zoning, development, and census population layers with a property lot layer. The use of GIS allowed this work to be completed efficiently and in a cost-effective manner. The recalibrated model's predictions of maximum day and peak-hour demands and pressures matched well with observed values from field measurements and the SCADA system.

The model was further refined to allow for predictions of future demands. The city's property lot layer was combined with spatial population projections and the Official Community Plan, allowing for analysis of the system to 2021.

The new WaterCAD model correctly predicts current conditions and forecasts future flows using the city's most up-to-date planning information. To assist in capital planning, a system optimization study was then conducted using the model to accurately assess system upgrading options.

Figure 11-1 is from ArcView, and shows the various layers that were used in the water model demand allocation study. The figure shows the spatial relationship between the information used to calculate demand allocation (legal lots, zoning/OCP codes), and the water system information (water mains and model nodes). In all, demand for more than 50,000 lots was calculated and spatially connected to the water system model.

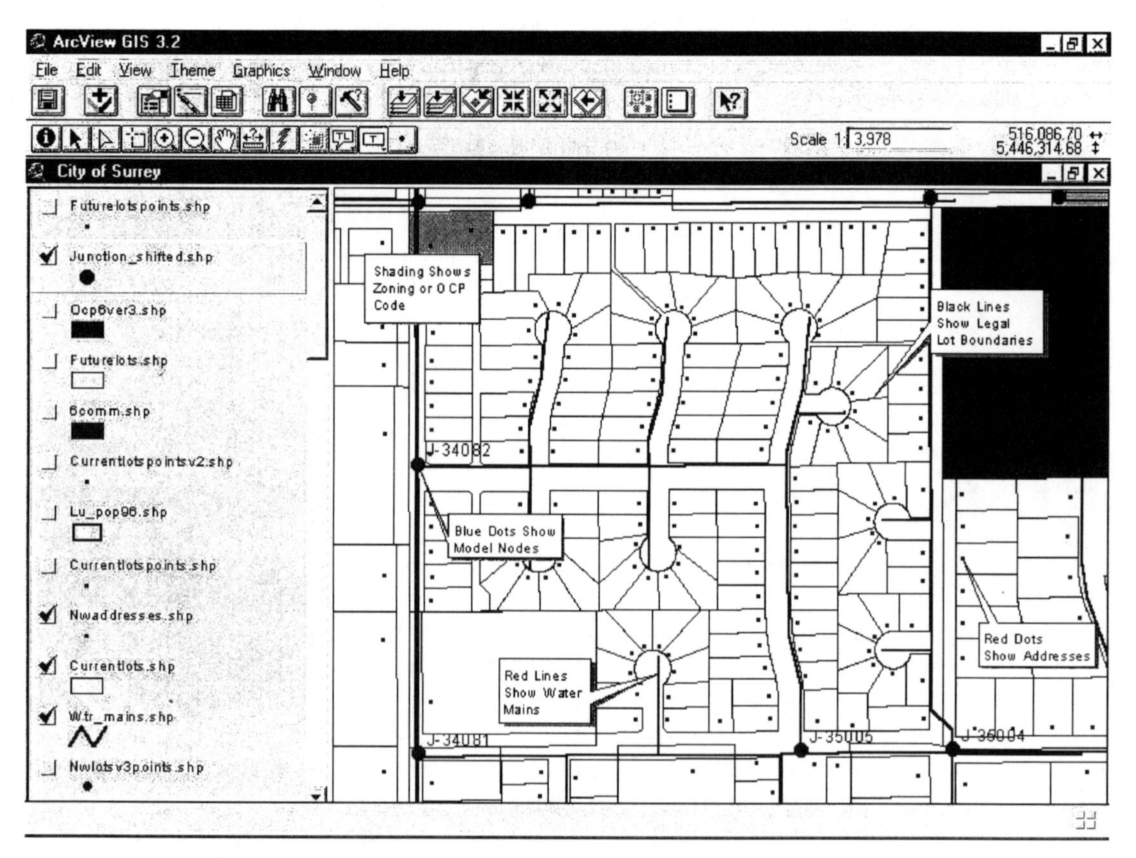

Figure 11-1. ArcView GIS Layers Used for Estimating Water Demand

MAPINFO AND WATERCAD LINKAGE IN RAROTONGA

Application	GIS and hydraulic model linkage for data transfer
Project Status	Completed in 2000
Reference	Dawe and Schölzel, 2001
GIS software	MapInfo Professional 5.5, MapBasic, Vertical Mapper 2.5
Other software	Microsoft Access (database software), WaterCAD (hydraulic modeling software), AutoCAD (drafting software)

GIS data	Digital Terrain Model (DTM), land use, user demand, network physical characteristics, aerial photographs
Hardware	PC, server, printer
Study Area	Rarotonga, Cook Islands
Organization	South Pacific Applied Geoscience Commission (SOPAC), Fiji Islands

Public water supplies in many South Pacific island countries are too often an exercise in disaster management. Inadequate personnel, inadequate training, and mountainous workloads are just some of the day-to-day problems Pacific water utilities have to face. Tools such as GIS, and hydraulic models of the distribution systems they are responsible for, can make management of these systems much easier. Data needed for hydraulic modeling are often found in a variety of different forms—network information and layout as AutoCAD drawings or blueprints, meter readings in spreadsheets, DTM and aerial photographs in a GIS, and pressure and flow data as hard copy.

Disorganization of data was the biggest obstacle that had to be overcome in setting up the WaterCAD hydraulic model for Rarotonga. The first step was then to locate relevant data followed by cleaning them up and getting them stored in some kind of digital format. The MapInfo GIS software made this easier by being able to link easily to spreadsheet tables and AutoCAD .DXF interchange files. Before, the relevant data were in a variety of different forms and formats. Now, they can be retrieved from a centralized location within the MapInfo GIS.

With the data in digital form, the model was then created importing various pieces of information, such as elevation and demand data into the model. Once the model was up and running, information such as pressure and flow data could then be easily exported back to the MapInfo GIS for better visualization of results down to the exact house where low pressures would be experienced. This kind of readily available information (e.g., inventory of network assets, analysis of system operation, identification of problem areas, and effectiveness of proposed upgrades) because of the open linkage between the GIS and the hydraulic modeling software is invaluable. It gives true ownership of the distribution system to the people who have to manage it, especially in places like the Pacific. Figure 11-2 shows the calibrated hydraulic model of the Turangi Zone, Rarotonga, Cook Islands.

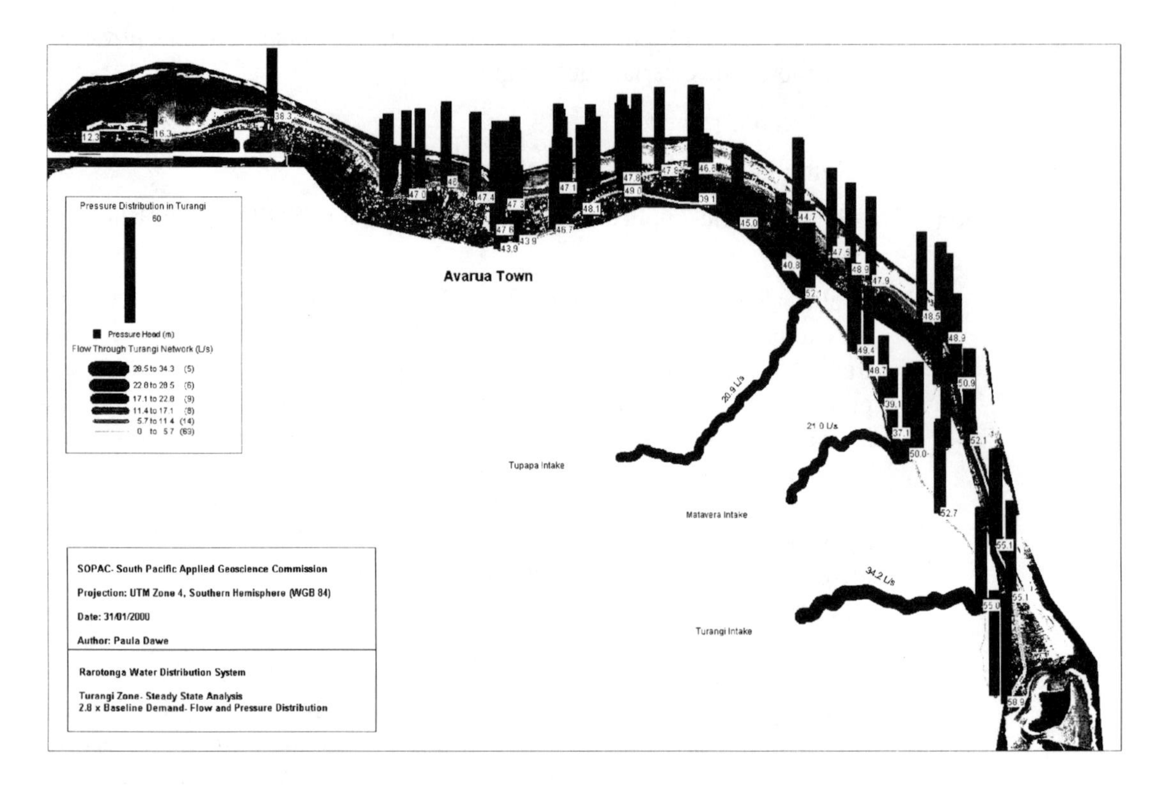

Figure 11-2. Calibrated Hydraulic Model of the Turangi Zone, Rarotonga, Cook Islands

PITTSBURGH'S NPDES PERMIT COMPLIANCE

Application	Improving H&H modeling for CSO permit compliance
Project Status	Completed in 1999
Reference	Smullen and Angell, 2001
GIS software	ArcInfo and ArcView
Other software	SWMM and Storage Treatment Overflow Runoff Model (STORM)
GIS data	Land use, sewersheds, TIGER/Line, and planimetric vector basemaps from low altitude aerial photography
Hardware	Dell Precision Workstation 610 PC, Pentium III 550/512K XEON-TANN with 32 gigabytes of hard disk space and Diamond Fire GL1 video card
Study Area	Nine Mile Run watershed, Pittsburgh, Pennsylvania, USA
Organization	Camp Dresser & McKee, Inc., Edison, New Jersey, USA

The EPA National CSO Policy requires that CSO dischargers perform a system inventory and hydraulic characterization as part of their National Pollution Discharge Elimination System (NPDES) permit compliance programs. Detailed H&H models of the sewer system and tributary drainage areas are typically developed to help understand the existing system, to characterize CSO frequency and volume, and to evaluate the effectiveness of alternative interceptor and wastewater treatment plant operating scenarios. The models also are used in later phases of the compliance program to document the Nine Minimum Controls (NMC) and develop a Long Term Control Plan (LTCP).

A GIS is used to store, manage, and display spatial information for the hydrologic modeling of a sewer service area. The U.S. Bureau of Census TIGER/Line files described in Chapter 5 are a typical source often used to develop basemaps. Political boundaries, hydrological features, and population and housing statistics are a few key items incorporated into the GIS database for the study area. Land use data, often developed from spectral analysis of satellite imagery, and slopes developed from digital elevation model (DEM) data are used to further characterize the sewer service area. These are some of the typical geographic data available to begin H&H modeling of a combined sewer system.

The ability to perform overlays of such data is essential in H&H modeling of sewer systems and in particular is useful as a comparison method to estimate impervious cover. Providing initial estimates of the percent imperviousness of sewersheds is an example where the GIS easily facilitates an otherwise burdensome task. Often, these estimates are performed using standard land use–impervious cover relationships. Another approach to estimating impervious cover with the GIS uses planimetric vector basemaps developed from low altitude aerial photography. This project used the latter approach to improve the hydrologic modeling capabilities needed for a successful NPDES permit compliance program.

Figure 11-3 shows the level of detail provided by the planimetric vector basemaps. The figure shows the buildings, 5-ft contour lines, roads, and parking lots, to name a few.

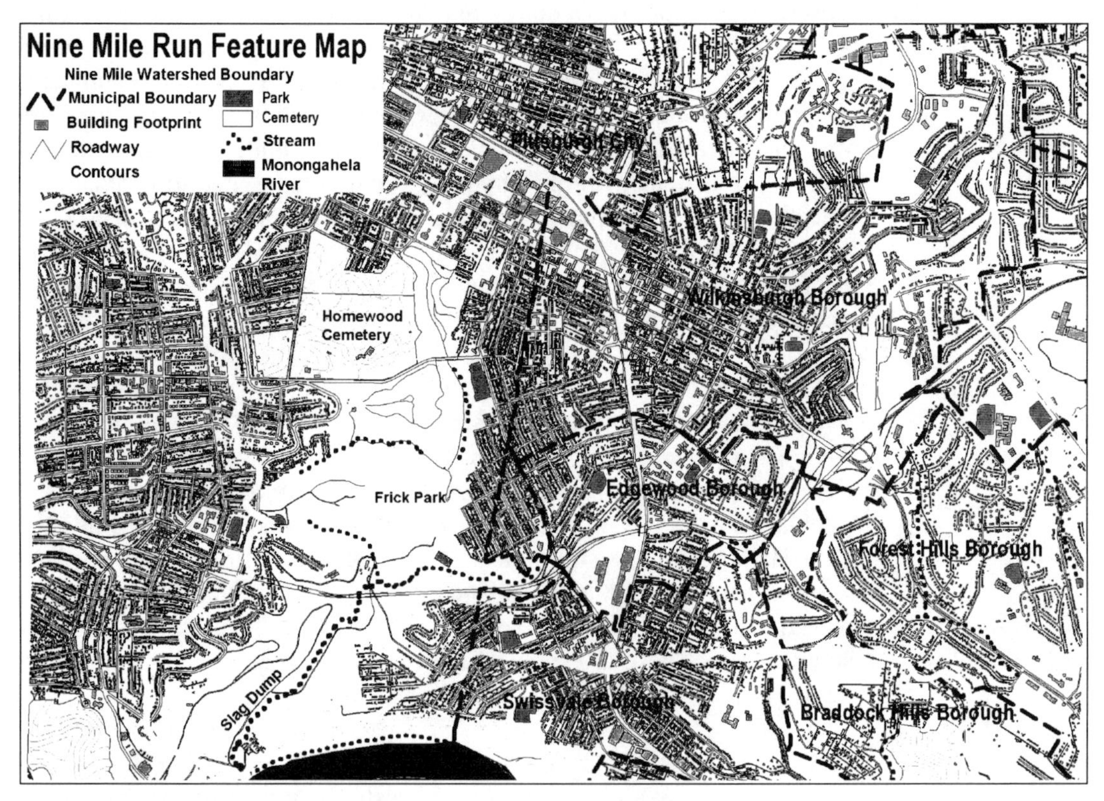

Figure 11-3. Planimetric Vector Basemap for the Nine Mile Run Watershed of Pittsburgh

Blacksburg's Use of GIS in Sewer System design

Application	GIS in sewer systems design
Project Status	Completed in 1996
Reference	Agbenowosi, 2001
GIS software	ArcInfo
Other software	Gravity sewer design program (GSDPM3) and Cost Estimation Module
GIS data	DEM, manholes, building locations, streets
Study Area	Blacksburg, Virginia, USA
Organization	Virginia Polytechnic Institute and State University, Blacksburg, Virginia

In the planning and design of sewer networks, decisions are spatially dependent because of the right-of-way considerations and the desire to have flow by gravity. This research involved the application of combined optimization–GIS technology in the sewer network design process. The spatial analysis capabilities of GIS were utilized for the detailed design of sewer systems, including pump station locations and the force main path to minimize pipe costs and the cost of pumping. The program uses the selected manhole locations to generate the candidate potential sewer networks and delineates the area into sub-sewersheds.

In generating the network layout, a gravity-seeking algorithm developed in C language is used to navigate through the initial network and determine the optimal grouping of manholes into subsewersheds. The algorithm starts from the selected outfall of the system and then travels up each adjoining link (pipe). Each time, the active link is inspected for direction of ground slope along that link. The links on the positive slopes are kept, whereas those on the negative slopes are marked for possible deletion. Before the deletion, the links on negative slopes are tested for the possibility of laying the pipe against the ground slope. The excavation depths/costs are used to determine whether or not a marked link is permanently deleted. This procedure leads to the creation of a single network layout or a series of network layouts in sub-sewersheds.

The path and destination of each force main in the system is determined by the Dijkstra's shortest-path algorithm from the set of potential paths. The force main path-selection procedure seeks to minimize cost of pumping by finding the path that has the least total dynamic head. A modified length is used to represent the length of each link and the force main segment. The modified length is the physical length of the link (representing the friction loss) plus an equivalent length (representing the static head). ArcInfo GIS software is coupled with gravity sewer pipe sizing software (GSDPM3) along with a number of modules developed in C language to accomplish the design. The resulting tool provides an efficient means of designing and estimating the cost of a sewer system for an area. This includes the cost of pipes, excavation, and force main. The terrain-modeling capabilities of the GIS are exploited to determine the sewer network layout, the location of pump stations, and the least-cost path for the force main. What-if scenarios can be easily simulated by altering the constraints related to manhole locations and right-of-way restrictions. Further annexation possibilities or system expansion needs can be analyzed long before their need arises. The graphics capability of the GIS enhances the comprehension of the design procedure and visually displays the impacts of the existing structures or features that serve as obstacles along the possible paths.

Figure 11-4, illustrates the interaction between the user and the GIS-based design system. Initial manhole locations selected by the user within ArcInfo are combined with topography and surface feature data. The manhole elevations are extracted, and the distances between adjacent manholes (pipe lengths) are also determined. A set of external programs is then used to determine the appropriate layout and site pump stations and to determine the force main path. The results are presented to the user both in graphic and tabular formats.

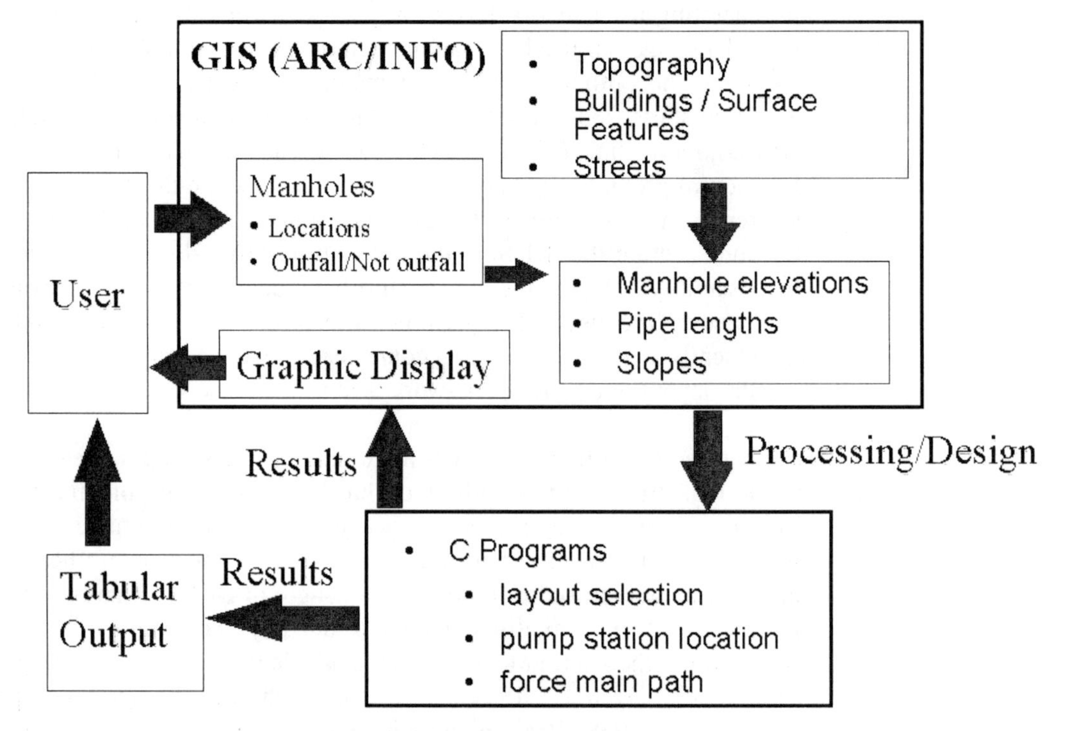

Figure 11-4. GIS-Based Optimal Sewer System Design

DAVENPORT'S SEWER SYSTEM MASTER PLAN

Application	H&H modeling and planning
Project Status	Completed in 2000
Reference	Roeth and Roth, 2001
GIS software	ArcView
Other software	Visual SWMM, Excel, MicroStation
GIS data	Land use, sewers, manholes, topography, city limits, city streets, streams, water bodies, bedrock, and USGS digital raster graphics (DRG)
Hardware	Dell 400 MHz Pentium II PC, Ethernet network, Canon 1150 and HP 5000 printers, and HP 650C plotter

Study Area Cities of Davenport and Bettendorf, Iowa, USA

Organization Stanley Consultants, Inc., Muscatine, Iowa

This project developed a sanitary sewer collection system master plan for the cities of Davenport and Bettendorf, Iowa, that have a combined population of approximately 130,000. The master plan looked at the next 50 years of projected growth in the study area and identified capacity inadequacies of the existing collection system. A capital improvements plan was developed, outlining projects required during the next 50 years to serve the growing communities.

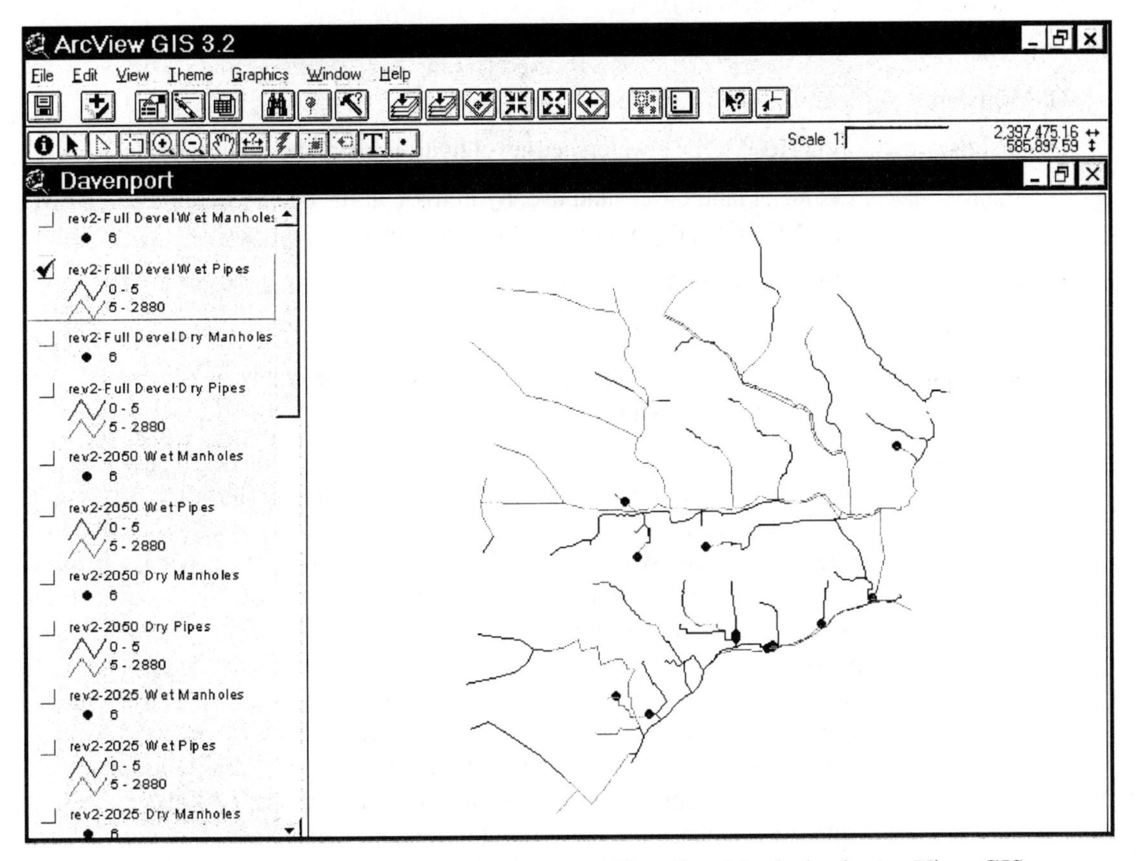

Figure 11-5. Map of Surcharging Pipes and Flooding Manholes in ArcView GIS

Figure 11-5 shows how ArcView GIS software was used to highlight surcharging pipes and flooding manholes identified by the SWMM model.

The communities provided background geospatial data in several different formats. A GIS was used to view and manipulate these data sources, to divide the study area into more than 120 drainage subbasins, and to allocate the proposed land uses among the drainage subbasins. A spreadsheet was used to calculate flow patterns from land use and flow monitoring data. The results of the spreadsheet computations were used as input to the SWMM models. Modeling results were brought back to the GIS for review and analysis. Final report figures were exported to CAD for final production.

VIRGINIA'S NONPOINT SOURCE POLLUTION ASSESSMENT

Application	A statewide nonpoint source (NPS) pollution assessment
Project Status	Completed in 1997
Reference	McBride et al., 2001
GIS software	ArcInfo and ArcView
Other software	EUTROMOD, a watershed-level hydrologic/water quality model
GIS data	Landsat land cover/land use, hydrologic units, county boundaries, DEM, STATSGO soils, weather station locations
Hardware	SUN Microsystems and PC
Study Area	Entire state of Virginia, USA
Organization	Patrick Center for Environmental Research, Academy of Natural Sciences, Philadelphia, Pennsylvania, USA

This study developed a new methodology for performing Virginia's statewide, watershed-level nonpoint source pollution assessment. The amended Clean Water Act of 1987, Section 319, provides specific mandates for NPS pollution control, which require states to perform an assessment of water quality problems and to develop a management program to address diffuse pollution sources. States must determine high-priority areas and focus their NPS control efforts in those watersheds. There were several problems with Virginia's previous assessment methodology: (1) the watershed rankings for NPS pollution were inconsistent from assessment to assessment, (2) the rankings were not truly based on estimated pollutant loads, but rather on indices and weighting factors, and (3) statewide spatial data were not used.

The revised methodology utilizes GIS technology, a database management system, and computer modeling. The size of the project area (106,000 km^2), the complexity of the natural system, and the large volume of data necessitated the use of a GIS. With a GIS, it was possible to compile the statewide digital data (land cover, soils, topography, annual precipitation) and non-spatial data (crop management information, livestock inventories, erosion and sediment control projects). From this spatial database, input parameters were derived for a simple nutrient and sediment loading model, EUTROMOD. EUTROMOD was originally developed to provide guidance and information for managing watersheds and eutrophication in lakes and reservoirs. For this study, EUTROMOD was modified by removing the lake portion of the model and by integrating tables and macros that allowed for modeling a large number of watersheds.

The modeling resulted in an assessment based on estimated loadings for each of the predefined 493 hydrologic units in the state. The model output includes runoff volume, sediment loading, and dissolved and sediment-attached nitrogen and phosphorus. The estimated loads were used to rank the hydrologic units for potential NPS pollution into high-, medium- and low-priority groups. In summary, the revised methodology is an attempt to better identify high-priority watersheds using the best available data and a statewide modeling approach. Figure 11-6 shows 493 hydrologic units of the Commonwealth of Virginia ranked by total nitrogen load.

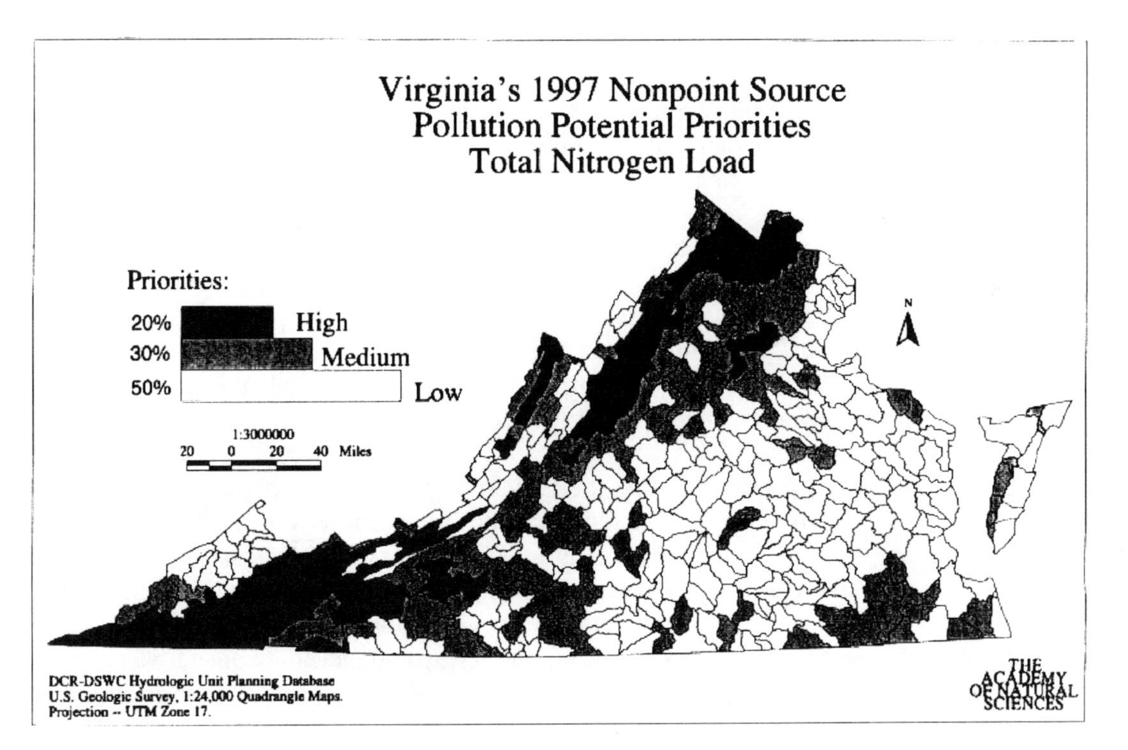

Figure 11-6. Hydrologic Units of the Commonwealth of Virginia Ranked by Total Nitrogen Load

WATER QUALITY MODELING INTEGRATION IN PENNSYLVANIA

Application	GIS and water quality modeling integration
Project Status	Completed in 1996
Reference	Byun et al., 2001
GIS software	ArcInfo
Other software	1. AGNPS, Agricultural Non-Point Source Pollution Model, 2. ArcInfo GRID, an ArcInfo extension used for data characterization on a cell-by-cell basis, and 3. Microsoft Excel, used for data conversion.
GIS data	Topography, soils, Landsat land use / land cover, watershed boundary, hydrology
Hardware	SUN Microsystems, PC, and digitizer
Study Area	Upper White Clay Creek watershed (715 hectares), Chester County, Pennsylvania, USA
Organization	Academy of Natural Sciences, University of Pennsylvania, Philadelphia, USA

Because the White Clay Creek watershed is a predominantly agricultural area, nonpoint source pollution poses a threat to the water quality. NPS pollution often occurs in agricultural areas where storm runoff can wash sediments and nutrients to the nearest water body. This study used GIS technology to characterize the upper White Clay Creek watershed and to develop input parameters for AGNPS modeling. AGNPS is a distributed-parameter, event-based model that analyzes NPS pollution on a cell-by-cell basis within a watershed. For any specified storm event, runoff and pollutants (sediment, nutrients) are routed through the grid cells in a stepwise manner to the watershed outlet.

The project team first used ArcInfo to compile a spatial database. They imported some data sources (land use / land cover and hydrology) and digitized others (topography, soils, watershed boundary). Vector data were converted to raster data using a 100-m cell size. Grids were derived to represent several of the 22 required AGNPS input parameters such as slope, soil erodibility factor, and fertilization level. Parameter values were transferred to the AGNPS model by way of Excel. The model was then used to simulate the effects of various storms, which produced output in terms of total runoff volume, peak runoff rate, upland erosion, channel erosion, sediment yield, and estimates of nitrogen, phosphorus, and chemical oxygen demand (COD) in concentration and mass units. Model's output was then compared to sampled data from three storm events, as observed by the Stroud Water Research Center. Because sampling data were limited, model could not be fully verified; however, the model did provide results consistent with recorded peak flows and observed sediment, nitrogen, and phosphorus concentrations.

The calibrated model was manipulated to determine how certain circumstances would affect the model output. Three alternative scenarios were simulated: a 100% forested watershed, a nutrient management plan, and a suburbanized watershed. Using GIS, it was easy to recharacterize the watershed in these three ways and generate the input parameters for each. This facilitated a comparison of the "what-if" scenarios and demonstrated how the integration of GIS and AGNPS could act as a management tool. In summary, GIS was vital for organizing spatial data, generating the model's input parameters, and allowing quick and easy changes to evaluate alternative watershed conditions.

Figure 11-7.shows topography, roads and streams of the upper White Clay Creek watershed. Figure 11-8. shows land use/land cover distribution for the upper White Clay Creek watershed.

HYDROLOGIC IMPACT ASSESSMENT IN OHIO

Application	GIS-based long-term hydrologic impact assessment
Project Status	Completed in 2000
Reference	Muthukrishnan et al., 2001
GIS software	ArcView with spatial analyst and L-THIA/NPS extension
GIS data	Land use map, hydrologic soil groups, watershed boundaries, river network, and long-term daily precipitation data
Study Area	Chagrin River Watershed, Geauga County, Ohio, USA
Organization	Chagrin River Watershed Partners, Inc., Ohio, USA

The Long-Term Hydrologic Impact Assessment and Non-Point Source Pollution (L-THIA/NPS) model is a simple and easy to use tool that watershed managers and urban planners can use to assess the long-term effects of land use change. L-THIA/NPS uses land use, hydrologic soil type, and long-term precipitation data, all of which can be easily obtained from the local government agencies.

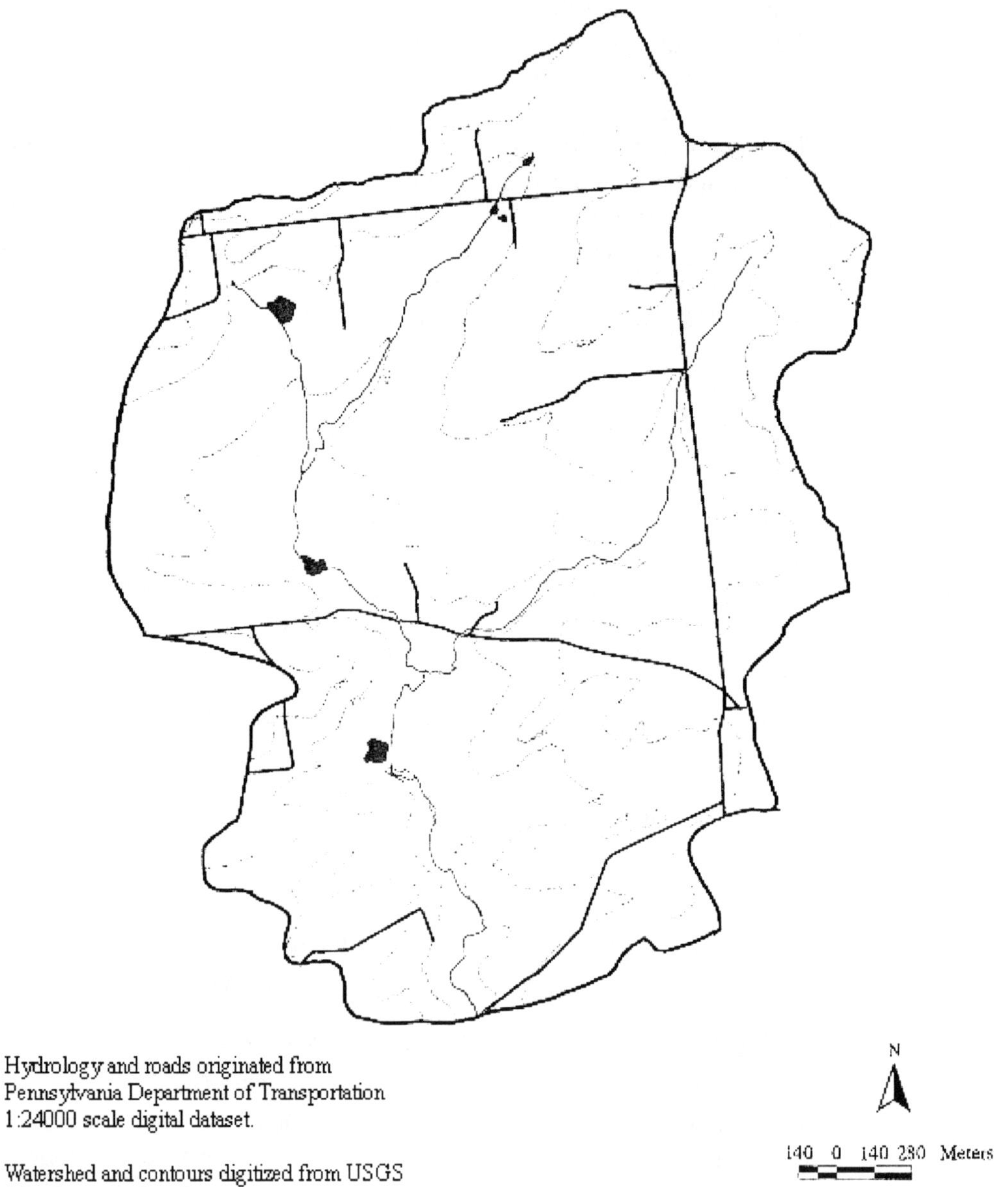
White Clay Creek Watershed
Hydrology and roads originated from
Pennsylvania Department of Transportation
1:24000 scale digital dataset.
Watershed and contours digitized from USGS
7.5 minute quadrangle maps.
N
140 0 140 280 Meters

Figure 11-7. Topography, Roads and Streams of the Upper White Clay Creek Watershed

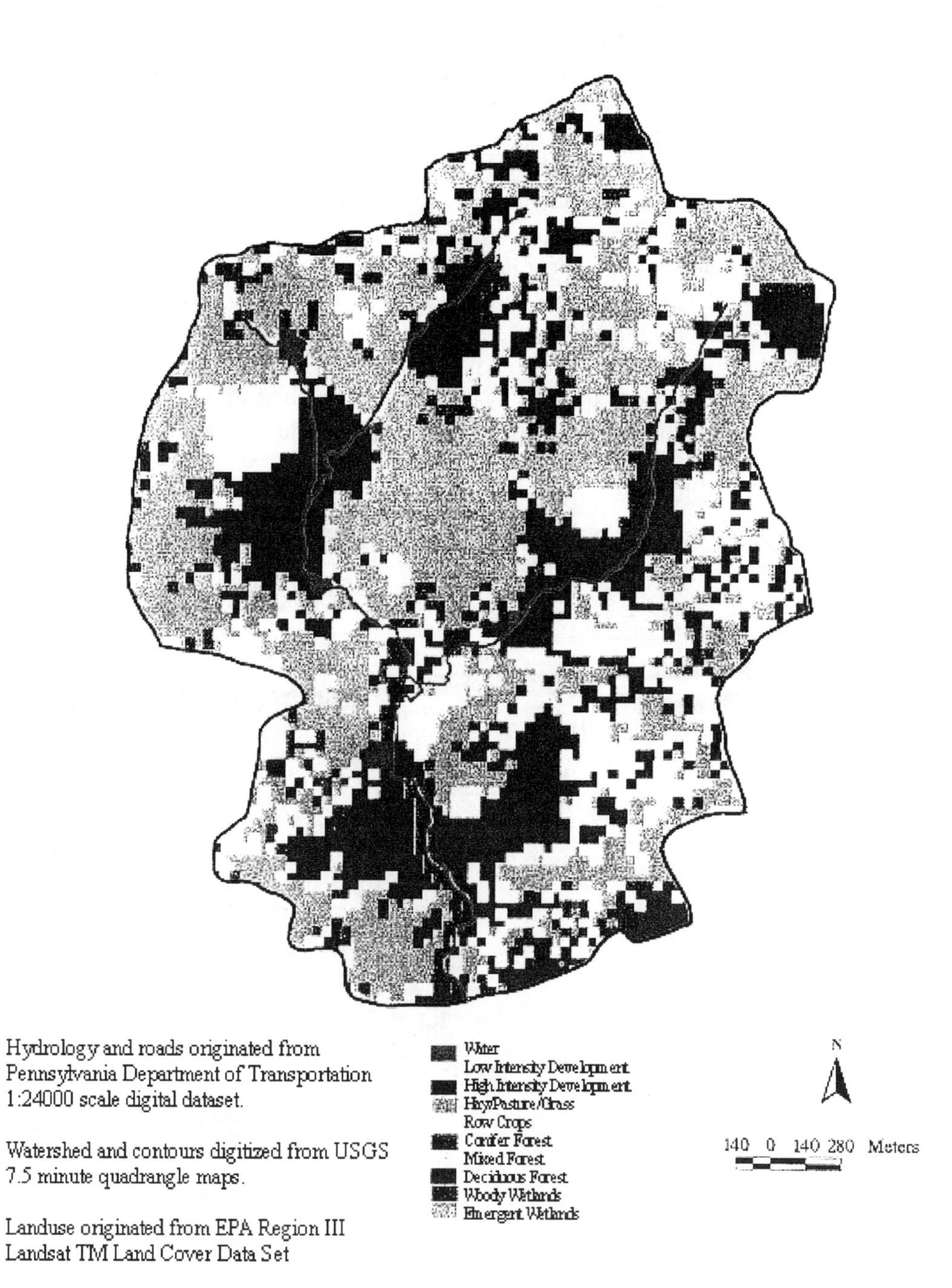

Figure 11-8. Land Use/Land Cover Distribution For the Upper White Clay Creek Watershed.

A Web–based L-THIA/NPS model is also available at the *www.ecn.purdue.edu/runoff* Web site for public use, which does not require ArcView GIS software and can be run using any standard Web browser. An L-THIA/NPS extension for ArcView can be downloaded for free from the same Web site.

The rapid development in the Chagrin River watershed along with the need to prevent environmental degradation prompted the watershed managers and planners in Geauga County to look for an effective and easy to use model to assess the impacts of future land use changes. The L-THIA/NPS model, developed at Purdue University, was chosen to assess the long-term hydrologic impacts in Silver Creek and Griswold Creek sub-watersheds of the Chagrin River watershed. Figure 11-9 shows final runoff depth map generated by the model and the L-THIA/NPS extension menu.

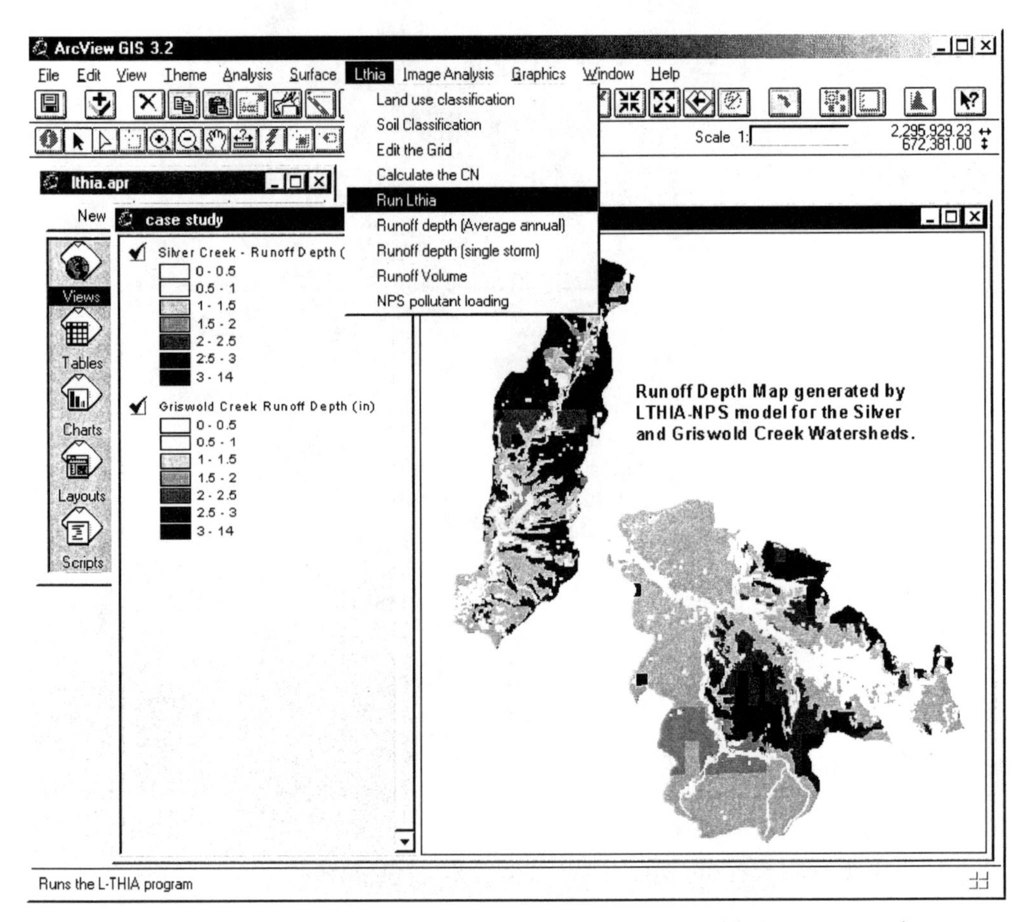

Figure 11-9. Runoff Depth Map Generated By The L-Thia/NPS Extension

Three different land use change scenarios—full buildout, partially modified buildout, and modified buildout, were developed from the current zoning regulations and the results were compared with the existing land use. The results indicate that the future land uses in the proposed buildout scenarios will cause an increase in average annual runoff of approximately 5% to 20% compared with present-day conditions. These increases are due primarily to the conversion of forest into residential land. To minimize the effects of development, development should be concentrated in areas with greatest existing development (i.e., areas with the highest existing curve number). This implies that a minimal amount of development of existing forests is preferred. If conservation design is used in the development of low-density and high-density residential areas, annual runoff will be decreased by approximately 3% to 7% compared with similar development without conservation design. Overall, the analysis suggests that carefully applied management practices can eliminate much of the projected increase in average annual runoff that would otherwise occur as these areas undergo land use change.

Real-time Flood Mapping in Fort Collins

Application	Real-time flood information mapping
Project Status	Completed in 2000
Reference	Townsley and Ford, 2001
GIS software	ArcView, Spatial Analyst and 3D Analyst extensions, and Arc Explorer
Other software	Microsoft Visual Basic; Digital Fortran; Visual Basic, Perl, and Java scripts and Windows Scripting Host; Microsoft Access; EPA-SWMM; and HEC's PRECIP, HEC-2, HEC-RAS, HEC-DSS programs.
GIS data	Digital terrain model; rain zones, stream channels; watershed boundaries; digital aerial photographs
Hardware	Personal computers and servers
Study Area	Fort Collins, Colorado, USA
Organization	David Ford Consulting Engineers, Inc., Sacramento, California, USA

The City of Fort Collins suffered a devastating flood in July 1997. To reduce the city's vulnerability to future floods, the city installed an extensive ALERT data collection system to monitor rainfall and stage in real time. But the flashy nature of floods in Fort Collins means that a warning based upon observations alone may be too late. City emergency managers and floodplain managers realized this and sought a system that would permit forecasts and mapping of areas that would be inundated in the future.

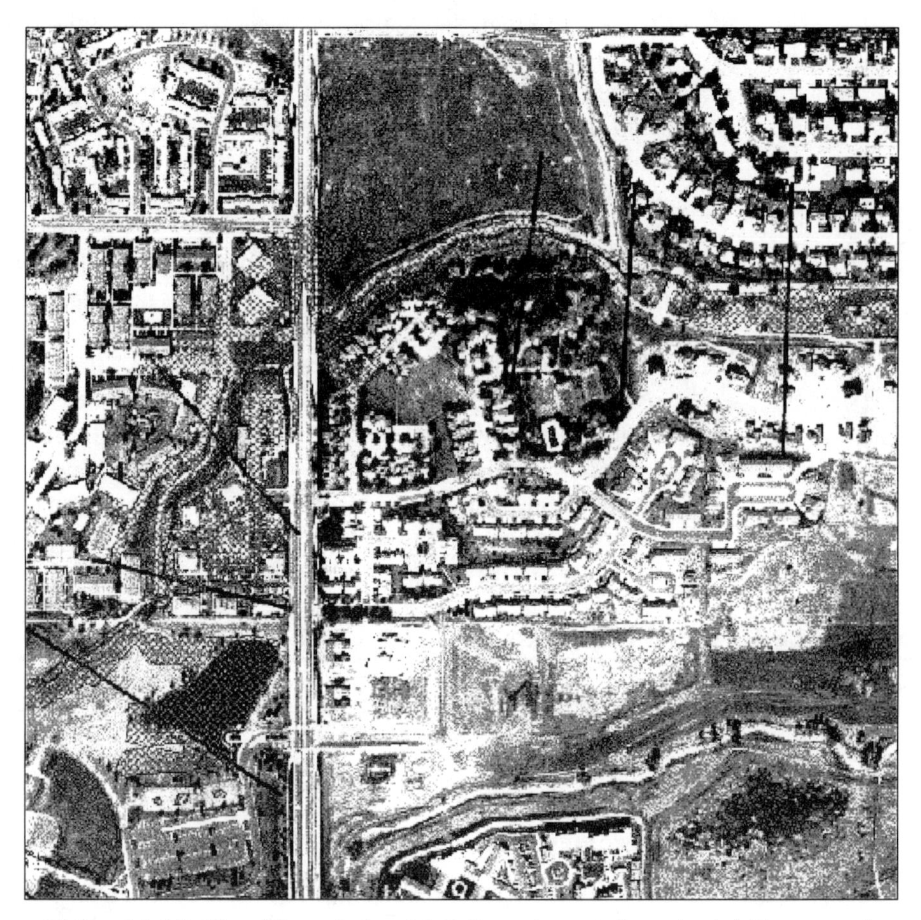

Figure 11-10. Flood Inundation Map Superimposed on Aerial Photograph

A real-time flood inundation mapping system (RTFIM) was developed that integrates (1) a watershed model (SWMM), (2) channel models (HEC-2 and HEC-RAS) developed for drainage studies in Fort Collins, (3) real-time data collected by the city's ALERT system, (4) topographic data in GIS format, and (5) aerial photographs and other GIS layers that show vulnerable property and people. CFS, the Catchment Forcasting System, developed by David Ford Consulting Engineers, integrates these components into a system capable of forecasting future flow and stage and generating and displaying maps of potentially inundated areas. To produce the maps, RTFIM was developed as a series of scripts that gathers ALERT data and then executes the watershed and channel models in the proper sequence for the complex, interconnected watersheds of Fort Collins. Other scripts and small component programs parse the output from these programs and store results in a relational database. Then

ArcView GIS scripts solve the spatial analysis problem—intersecting planes that represent the water surface with a digital elevation model to create the inundated area maps.

Figure 11-10 shows an example of the application. Here, the inundated area is superimposed on aerial photographs and other GIS layers that show vulnerable property and people. This map can be used by city personnel in targeting responses to eminent flooding, or can be posted to the Web to increase general public awareness. However, the primary benefit is that response lead time is increased which minimizes the likely flood damage.

Rainfall–Runoff Modeling in Amman

Application	Rainfall–runoff modeling, with the purpose to estimate flood volume and groundwater recharge
Project Status	Completed in 2000
Reference	Jacobi, 2001
GIS software	ArcView
Other software	Surfer, Delphi, and Microsoft Access
GIS data	Hydrological soils; vegetation; delineations of surface and groundwater basins; and locations of climatic stations, rainfall stations and reservoir sites
Hardware	Windows 98 PC with 128 MB RAM, 400 MHz processor, and 2 GB free disk space; optional network server with Oracle database
Study Area	Entire Hashemite Kingdom of Jordan
Organization	Ministry of Water and Irrigation, Amman, Jordan

Direct runoff and groundwater recharge are hardly sufficient to fulfil the water demands of Jordan. Their scarcity makes the need for an accurate quantification even more important. However, as in many other countries of the region, rainfall records are the only reliable long-term data source.

The presented rainfall–runoff model is raster-based and uses algorithms derived from the SCS runoff curve number (CN) method described in Chapter 7 (Modeling Integration). When averaging CN-values over larger catchments, the influence of smaller areas of high runoff characteristics (e.g., areas sealed through urbanization) would be lost. By applying a raster concept, a high spatial variation is achieved, and the effects of change of land use can be easily simulated.

The model requires daily rainfall records as input. Information on soil type and vegetation is taken from existing maps in vector format and is then transformed into a nationwide raster layer using ArcView's Spatial Analyst extension. The hydrological soil types are strongly related to the originating geology: marl, sand/gravel, chert-limestone, and so on. The vegetation classes were limited to brush, forest, orchards, and cultivated; the largest part of Jordan is barren. Thus, the CN values could not use the existing SCS tables but had to be estimated by the modeler's experience and later verified through calibration runs. The grid layers are then converted into Surfer grids for faster and more stable processing using Golden Software's Surfer software.

A rainfall preprocessing module written in Access reads rainfall data from a central database and rearranges them to suit the model for off-line operation. Daily data for totalizer stations are simulated with appropriate algorithms, and near the borders so-called virtual stations are inserted to exert some semi-manual control over the otherwise automatic spatial extrapolation of rainfall height.

The main processing is done in a module programmed with Delphi and calling Surfer commands through OLE. For each day, daily rainfall is read in and regionalized using the Kriging method. In several grid-math steps, the base grid of CN values is transformed to calculate the actual CN value (considering the previous rainfall) and the resulting direct runoff. Instead of differentiating growing/non-growing seasons, monthly potential evapotranspiration is considered in the soil–water balance. For deep infiltration, a similar nonlinear approach is used as for surface runoff, of course with different CN values. The results are stored as monthly grid files in mm units.

In the post-processing, monthly volumes of runoff and recharge are summarized for various set of subbasins. Under ArcView, the subbasins are delineated and saved in ArcInfo generate format, from where they are transformed into sets of surfer blanking files. An Access application is then called to perform the blanking and volume calculation by Surfer commands through OLE calls. The results are than stored in data tables and can be used for further processing like calculation of water balances. The principal advantage of the model is that new water balances can be quickly calculated once data from the recent water year are available. In addition, the available rainfall records permit calculations of the past 30 years, and thus significant statistical calculations are made including trend analyses. Figure 11-11 shows the model results under ArcView.

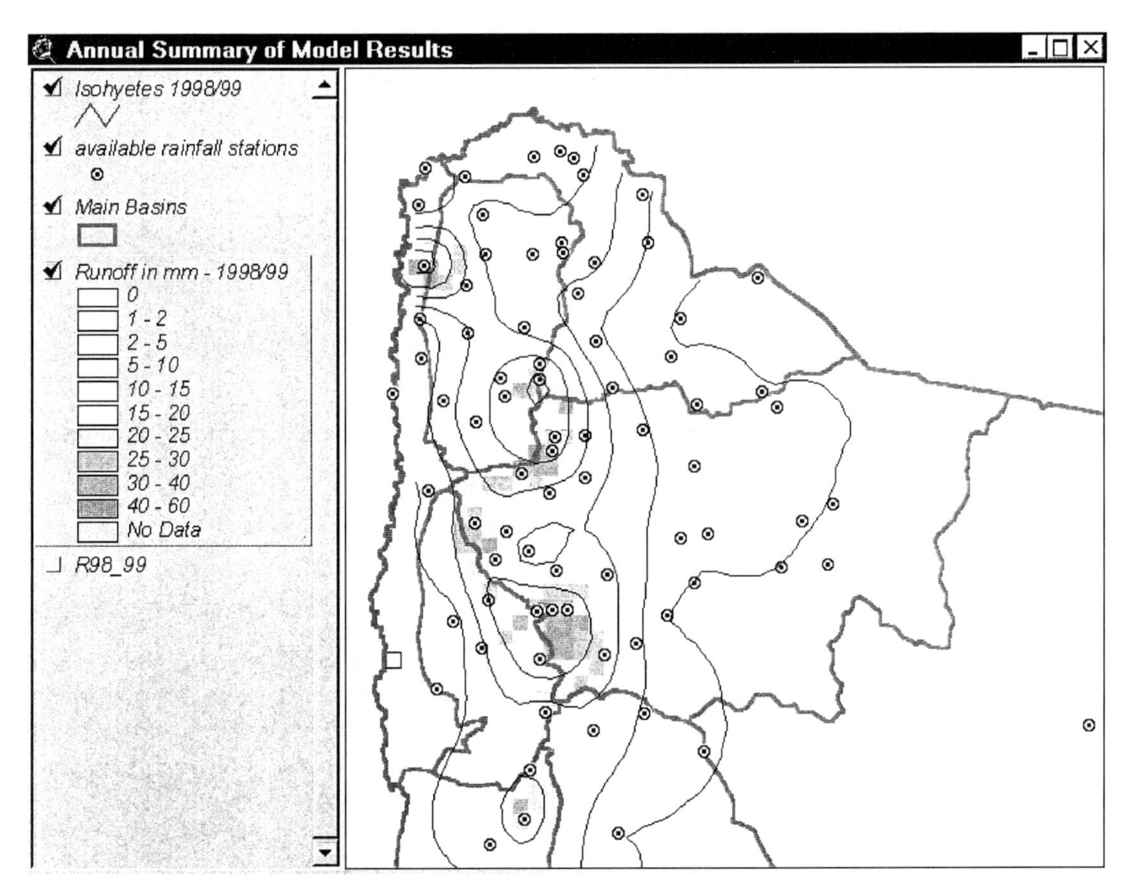

Figure 11-11. ArcView Screenshot Showing Model Results for Amman, Jordan

Flood Inundation Maps in Ankara

Application	Determination of inundation maps by GIS
Project Status	Completed in 1996
Reference	Aktas, 2001
GIS software	ArcInfo
GIS data	Land use, soils, contour lines
Hardware	Windows NT workstation
Study Area	Lake Mogan, Ankara, Turkey
Organization	Middle East Technical University, Ankara

GIS was used to determine inundation maps under certain rainfall conditions. Inundation mapping used contour maps, land use, and soil type for a study area near Lake Mogan. Additional information is available in Aktas's master's thesis (Aktas, 1996).

SUMMARY

This chapter presented 11 case studies on GIS applications for water, wastewater, and stormwater systems in Canada, Cook Islands, Jordan, Turkey, and the United States. The case studies indicate that GIS is being used in all aspects of water, wastewater, and stormwater management from planning and H&H modeling to mapping and permit compliance.

SELF-EVALUATION

1. On the basis of the case studies presented in this chapter, which GIS application seems to be the most prevalent?

2. On the basis of the case studies presented in this chapter, which GIS software seems to be the most common?

3. Using the case study presentation format of this chapter, write a case study summary of a GIS project completed by your organization.

Chapter 12

GIS RESOURCES

This chapter presents a list of representative GIS information resources, such as Web sites, software, data, books, periodicals, and conferences.

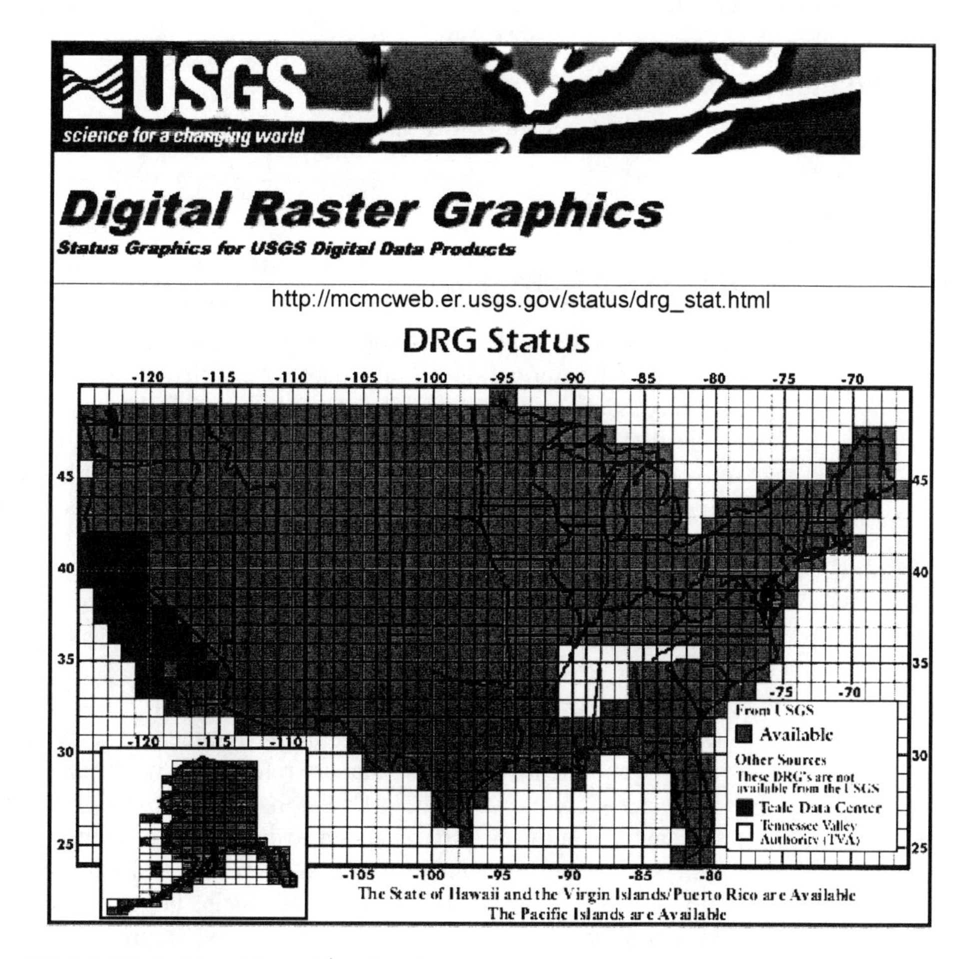

USGS Web Site Showing the Status of Digital Raster Graphics Data

LEARNING OBJECTIVE

The learning objective of this chapter is to compile a list of GIS information resources, such as the Internet, software, data, journals, books, conferences, and magazines. Major topics discussed in this chapter include

- GIS Web sites
- Search engines
- Geographic data clearinghouses and geopartals
- Public-domain data sources
- Commercial data sources
- GPS receiver manufacturers
- GIS software vendors
- Professional organizations
- Internet discussion groups and lists
- GIS magazines, newsletters, and journals
- GIS books
- GIS conference proceedings

IMPORTANCE OF GIS RESOURCES

The field of GIS applications is growing at a rapid pace. The number of resources for information on GIS hardware, software, data, and services therefore is enormous. For example, Geospatial Solutions' 2002 Buyers Guide lists 160 GIS-software-related companies that provide 54 types of software products from CAD to image processing to work management (Geospatial Solutions, 2001). The same guide also lists 78 hardware companies providing 24 types of hardware products (e.g., computers, GPS, plotters, digitizers), 110 data vendors offering 26 types of data products (e.g., digital elevation models or DEMs, digital orthophoto quadrangles or DOQs, census, satellite imagery), and 184 service firms providing 23 types of GIS-related services (e.g., data acquisition, computer programming, scanning, photogrammetry). It is, therefore, obvious that there is an overwhelming number of GIS information resources and identifying those that are relevant to water, wastewater, and stormwater systems would be a time-consuming process. That is where this chapter comes in—it is intended to provide a condensed list of GIS resources that are relevant to water, wastewater, and stormwater system professionals. However, because of their large number, all the information resources cannot be listed here—only representative examples of each type of resources will be presented.

GIS Web Sites

The Internet can overwhelm you with information, but just getting tons of new information might not be a good idea. Too much information can bog you down, and though the Internet brings a world of geographic information to your desktop, it is still up to you to select the right information. Useful GIS websites include

- Directions Magazine — *www.directionsmag.com*
- ESRI's GIS information — *www.gis.com*
- GeoCommunity — *www.geocomm.com*
- GeoSpatial Solutions — *www.geospatial-online.com*
- GIS Applications — *www.GISApplications.com*
- GIS Hydro (University of Texas) — *www.ce.utexas.edu/prof/maidment/*
- GIS Water Resources Consortium — *www.crwr.utexas.edu/giswr/*
- GIS WWW resources list — *www.geo.ed.ac.uk/home/giswww.html*
- GISHydro (University of Maryland) — *www.gishydro.umd.edu/welcome.htm*
- GISLinx (categorized GIS links) — *www.gislinx.com*
- Spatial Hydrology — *www.spatialhydrology.com*
- Spatial Odyssey — *wwwsgi.ursus.maine.edu/gisweb/*
- Virtual GIS Library — *campus.esri.com/campus/library/*

Search Engines

- GeoCommunity — *search.geocomm.com*
- GIS INFOMINE — *infomine.ucr.edu/search/mapssearch.phtml*

Geographic Data Clearinghouses and Geoportals

One of the best ways to discover GIS data is through geographic data clearinghouses on the Web. Clearinghouses promote a more open market in which GIS data with superior characteristics can be selected over inferior data. State, regional, nonprofit, and professional organization GIS data clearinghouse Web sites are the most valuable source of free local spatial data. Popular clearinghouse sites include

	Directions Magazine Data Center	*www.directionsmag.com/datacenter*
	Geography Network	*www.geographynetwork.com*
	GIS Data Depot	*www.gisdatadepot.com*
	List of clearinghouses	*www.geographynetwork.com/data/clearinghouses.cfm*
	National Geospatial Data Clearinghouse	*www.fgdc.gov/data/data.html*

Public-Domain Data Sources

	U.S. Census Bureau: TIGER/Line data for census blocks and geographic features	*www.census.gov/main/www/cen2000.html* *www.census.gov/geo/www/tiger/index.html*
	U.S. Department of Housing and Urban Development: E-Maps	*www.hud.gov/emaps*
	U.S. Environmental Protection Agency: Reach files and LULC	*www.epa.gov/enviro/index_java.html* *www.epa.gov/ost/basins/* *www.epa.gov/nsdi/*
	U.S. Geological Survey: DEM, DRG, DLG, DOQ, LULC, and so on	*edc.usgs.gov* *earthexplorer.usgs.gov* *www-nmd.usgs.gov* *water.usgs.gov/GIS/*
	U.S. Natural Resources Conservation Service: SSURGO and STATSGO data	*www.ftw.nrcs.usda.gov/ssur_data.html* *www.ftw.nrcs.usda.gov/stat_data.html*

Commercial Data Sources

	American Digital Cartography: ETAK and TIGER data	*www.adci.com*
	ChartTiff: collarless/seamless USGS data	*www.charttiff.com*
	DeLorme: seamless 2D and 3D topo maps for USA	*www.delorme.com*
	ESRI: ArcData Online	*www.esri.com/data/online*

	ETAK: accurate and true shape geocoded streets	*www.etak.com*
	GeoLytics: census blocks	*www.censuscd.com*
	Geowarehouse: GIS/GPS data and products	*www.geowarehouse.com*
	Horizons Technology: Sure Maps: seamless topographic maps	*www.suremaps.com*
	LANDINFO International	*www.landinfo.com*
	Map Mart: USGS DEMs, DRGS, and DOQs	*www.mapmart.com*
	Maptech: cropped, or seamless USGS topographic maps	*www.maptech.com*
	Micropath Corporation: full/collarless DRGs and DEMs	*www.micropath.com*
	National Geographic: TOPO!	*www.nationalgeographic.com/topo/*
	TopoDepot: vector topographic quadrangles for the U.S.A.	*www.topodepot.com*
	TopoZone: a la carte (custom) topographic maps	*www.topozone.com*
	USGS Earth Science Information Center	*ask.usgs.gov/products.html*

GPS RECEIVER MANUFACTURERS

	Ashtech Precision Products	*www.ashtech.com*
	Leica Geosystems	*www.leica.com*
	Magellan Corporation	*www.magellangps.com*
	Sokkia Corporation	*www.sokkia.com*
	Topcon	*www.topconps.com*
	Trimble Navigation	*www.trimble.com*

GIS Software Vendors

Software	Vendor	Website
ArcInfo, ArcGIS (ArcView, PC ArcInfo, ArcCAD, Map Objects, ArcIMS, etc.)	ESRI, Redlands, California	*www.esri.com*
AutoCAD Map	Autodesk, San Rafael, California	*www.autodesk.com/gis*
ENVI	Research Systems, Boulder, Colorado	*www.rsinc.com*
ER Mapper	Earth Resources Mapping, Australia	*www.ermapper.com*
Geo/SQL	Geo/SQL Technologies, Calgary, Canada	*www.geosql.com*
GeoGraphics	Bentley Systems, Exton, Pennsylvania	*www.bentley.com*
GeoMedia Pro, MGE	Intergraph, Huntsville, Alabama	*www.intergraph.com*
GRASS	Baylor University, Waco, Texas	*www.baylor.edu/~grass/index2.html*
Idrisi, CartaLinx	Clark Labs, Worcester, Massachusetts	*www.clarklabs.org*
IMAGINE	ERDAS, Atlanta, Georgia	*www.erdas.com*
MapInfo Professional	MapInfo Corp. Troy, New York	*www.mapinfo.com*
MrSID	LizardTech, Seattle, Washington	*www.lizardtech.com*
GE Smallworld	Smallworld Systems, Cambridge, U.K.	*www.smallworld-us.com*

Professional Organizations

ⓘ	American Society for Photogrammetry and Remote Sensing (ASPRS)	*www.asprs.org*
ⓘ	Association of American Geographers	*www.aag.org*
ⓘ	Geospatial Information & Technology Association (GITA), formerly AM/FM International	*www.gita.org*
ⓘ	International Society for Photogrammetry and Remote Sensing (ISPRS)	*www.isprs.org*
ⓘ	Urban and Regional Information Systems Association (URISA)	*www.urisa.org*

Internet Discussion Groups and Lists

Do not limit Internet use to downloading GIS data. The benefits of the Internet do not stop at providing access to Web sites. Use the Internet to ask questions and get answers. An important benefit of the Internet is special interest discussion groups, such as Usenet and Listservs. Usenet is a network of more than 5,000 special interest groups or discussion forums in which electronic messages and articles are posted for anyone who cares to review or reply to them. Listservs or E-mail lists are available for the distribution of notices, information, and other documents to large audiences. Listservs are similar to Usenet groups but administered through E-mail and usually moderated.

The most beneficial use of the discussion forums and lists is to post questions and get answers. It is not uncommon to receive a large number of replies and expert opinions within a matter of hours. List members can point you to sources of GIS data and software, help you with your application questions or problems, or even share with you their own custom application scripts and source code. An E-mail question is instantly distributed to hundreds or even thousands of Listserv subscribers around the world. Internet discussion lists are also an excellent means of providing GIS community service and addressing hot GIS topics. For example, two Internet discussion lists, ArcGIS and ARCVB (*lists.directionsmag.com/discussion/*), have been formed to help ESRI users make the transition to the new ArcGIS software. The ArcGIS discussion list centers on using the ArcGIS software. The ArcVB list focuses on customizing and programming the new software and helping Avenue and AML programmers to adjust to the new VBA and component object model environment.

A subscription can be started simply by sending the "subscribe" command to the Listserv. This is generally done by sending an E-mail message, usually with "subscribe Listserv name" in the body and/or the subject of the E-mail. For example, to subscribe to the ArcView discussion group called ArcView-L, send an E-mail to "arcview-l-request@esri.com" with "subscribe arcview-l" in the body of the E-mail message. It is important to note that the E-mail address for processing Listserv commands is different from the Listserv address itself; thus sending a Listserv command, such as "unsubscribe" to the Listserv does nothing but irritate hundreds or even thousands of Listserv subscribers. A list of popular GIS discussion groups is given in Table 10-1.

Table 10-1. Internet GIS Discussion Groups

List Name	*Special Interest*	*Web site*
AGIS-L	Atlas GIS Software	*lists.directionsmag.com/discussion/*
ArcView-L	ArcView GIS, ESRI	*support.esri.com/listserve/*
BASINS	EPA's BASINS software	*www.epa.gov/OST/BASINS/listserv.htm*
ERDAS-L	ERDAS Software	*www2.erdas.com/supportsite/lists/newsgroups.html*
ERMAPPER-L	ER Mapper Software	*www.ermapper.com/technicl/ermapperl/listserv_body.htm*
ESRI-L	ESRI Software	*support.esri.com/listserve/*
GeoMedia-L	GeoMedia, Intergraph	*www.intergraph.com/gis/joinlists.asp*
GIS-L	General GIS	*www.hdm.com/urisa3.htm*
GISList	General GIS	*spatialnews.geocomm.com/community/lists/*
GRASSLIST	GRASS GIS Software	*www3.baylor.edu/grass/support.html*
Idrisi-L	Idrisi software	*www.clarklabs.org/Forum.asp*
GeoSpatial Users Community	Intergraph	*www.intergraph.com/gis/community/*
MapInfo-L	MapInfo Software	*spatialnews.geocomm.com/community/lists/*
MapObjects	MapObjects, ESRI	*lists.directionsmag.com/discussion/*
Metadata-L	Metadata	*spatialnews.geocomm.com/community/lists/*
NPSINFO	Nonpoint source pollution	*www.epa.gov/owow/nps/changes.html*
PCI Geomatics	PCI Geomatics	*www.pcigeomatics.com/support/disc.html*
SWMM-Users	EPA's SWMM Software	*www.chi.on.ca/swmmusers.html*
SW-GIS	Smallworld Software	*groups.yahoo.com/group/sw-gis/*

Directions Magazine (*www.directionsmag.com*) maintains a list of the major Internet-based mapping discussion lists. You can use this facility to keep up with discussions, and you can use its search engine to find topics of interest from particular lists or from all lists.

For those who do not like to be bogged down with frequent E-mails, some groups also distribute a "digest" version delivering all postings bundled into a single message once a day. If available, the digest versions can be generally subscribed by adding "digest" to the list name. Thus, for subscribing to GeoMedia-L digest, you will use the "subscribe geomedia-l-digest" command.

Some Listservs like SWMM-Users (*www.chi.on.ca/swmmusers.html*) maintain searchable archives of past questions and answers. Some Listservs, like ArcView-L, encourage members to post a summary of the responses they have received. A summary posting is usually preceded by "SUM" in the subject line of the E-mail. Summary postings on archive sites are the quickest means of obtaining condensed information.

The following example, based on a question posted on SWMM-Users in August 2001, demonstrates the power of Listservs.

Question Posted on August 15, 2001

Dear Colleagues:

The regional government I work for is in the position of purchasing new dynamic sewer modeling software. We are looking for a model that can conveniently allow us to import new pipes info from a GIS database and export simulated data back to GIS. It should have the following features:

- Able to solve dynamic wave equation,
- Minimal constraint on time step size and pipe length,
- Minimal numerical instability,
- Sophisticated control logic,
- User-friendly,
- Graphical interface with animation capability, and
- Any other features you feel indispensable to make your modeling task easier.

Although I can contact the vendors, I hope to hear independent evaluations from those of you already using them privately. If there are some software that make your life difficult, I definitely want to hear from you to avoid repeating your mistakes.

For your information, we have both combined and separated sanitary system in a total sewerage area of 350 km^2, and serves a population of 2.5 million. The model I have currently is on the order of 1,000 to 2,000 pipes with sophisticated control structures.

Thank you very much in advance for your precious time.

Excerpts from Responses

Response No. 1 Received the Same Day: My advice to you is to do it yourself. Do a general software function and feature enquiry to the vendors of the comprehensive list. Select a short list of those you see potential. Request from the vender(s) for a full capacity copy of the software (it should be free). Do some tests for the area you are familiar with, maybe some real project. I believe that this is the best way to select software for your particular needs, not someone else's needs. Good luck.

Response No. 2 Received the Same Day: I have used XP-SWMM and PC-SWMM, and both have different strengths. I don't believe that either one has a good GIS interface. Both use essentially the same type, which allows graphics like shapefiles to be mapped along with the model network. This can be done better just by using ArcView. Neither summarizes GIS layers (e.g., land use, soils) to produce RUNOFF input. Both allow the conduits and nodes to be entered via a GIS database, but if changes are made in the model input file, they are not updated in the database. I hear SWMM is being included in the next release of BASINS by the U.S. EPA, which may improve the GIS linkages.

Response No. 3 Received Next Day: Haestad Methods' SewerGEMS and StormGEMS products are based on ESRI's new ArcGIS geodatabase technology and reportedly have "completely integrated" GIS and hydraulic modeling capability. Beware that the GIS and modeling integration capability really depends on how you define "integration." For more information on this subject, please read my article "GIS and Modeling Integration" on page 46 of this month's (July 2001) *CE News* magazine.

GIS Periodicals

The World Wide Web is a dazzling information resource, but often GIS users overlook the value of the nearest public or university library (Thoen, 2001). A list of GIS magazines and books is provided below. For magazines, the U.S. annual subscription cost is included, if available. Some magazines, like *Professional Surveyor*, are free to qualified subscribers in the United States, especially those who specify, recommend, or influence the purchase of GIS-related products or services. Vendor-specific magazines, like *ArcUser*, are generally distributed for free and typically contain a mix of useful articles and promotional outreach. Some magazines also post selected content from each issue on their Web sites. Representative GIS magazines are listed below.

GIS Magazines

Magazine	*ArcUser*
Publisher	Environmental Systems Research Institute (ESRI) 380 New York Street, Redlands, CA 92373-8100, USA
Phone/Fax	909-793-2853 / 909-793-5953
E-mail	arcuser_editor@esri.com
WWW	www.esri.com/news/arcuser/
Comments	Quarterly magazine for the ESRI user community

Magazine	*Geospatial Solutions*
Publisher	Advanstar Communications PO Box 6139, Duluth, MN 55806-6139, USA
Phone/Fax	218-723-9477 / 218-723-9417
WWW	www.geospatial-online.com
Comments	Former Geo Info Systems, applications of GIS and related spatial information technologies, $64/year

Magazine	*GeoWorld*
Publisher	Adams Business Media 2101 S. Arlington Heights Rd, Suite 150 Arlington Heights, IL 60005, USA
Phone/Fax	847-427-9512 / 847-427-2079
WWW	www.geoplace.com
Comments	Former *GIS World* magazine, $72/year

Magazine	*Imaging Notes*
Publisher	Space Imaging 12076 Grant St., Thornton, CO 80241, USA
E-mail	Tmcclendonpugh@spaceimaging.com
WWW	www.imagingnotes.com
Comments	Commercial remote sensing information

Magazines that Publish GIS Application Articles

Magazine	*CE News*
Publisher	Civil Engineering News, Inc. 1255 Roberts Blvd., Kennesaw, GA 30144-3694, USA
Phone	770-499-1857, Ext. 11
E-mail	subscriptions@cenews.com
WWW	www.cenews.com

Magazine	*Civil Engineering Magazine*
Publisher	American Society of Civil Engineers 1801 Alexander Bell Drive, Reston, VA 20191-4400, USA
Phone/Fax	703-295-6213 / 703-295-6278
E-mail	cemag@asce.org
WWW	www.pubs.asce.org

Magazine	*Professional Surveyor*
Publisher	Professional Surveyors Publishing Company, Inc. 1713-J Rosemont Avenue, Frederick, MD 21702-4170, USA
Phone/Fax	301-682-6101 / 301-682-6105
E-mail	profsurv@profsurv.com
WWW	www.profsurv.com

Magazine	*Water Environment and Technology*
Publisher	Water Environment Federation (WEF) 601 Wythe St., Alexandria, VA 22314-1994, USA
Phone/Fax	703-684-2400 / 703-684-2492
E-mail	pubs@wef.org
WWW	www.wef.org

GIS Newsletters

Newsletter	*ArcNews*
Publisher	Environmental Systems Research Institute (ESRI) 380 New York Street, Redlands, CA 92373-8100, USA
Phone/Fax	909-793-2853 / 909-793-5953
E-mail	requests@esri.com
WWW	www.esri.com/news/arcnews/arcnews.html
Comments	Quarterly newsletter for the ESRI user community as well as others interested in mapping and GIS. It contains material of interest to planners, foresters, scientists, cartographers, academicians, geographers, engineers, business professionals and others who use spatial information. Free subscription.

Newsletter	*WaterWrites*
Publisher	Environmental Systems Research Institute (ESRI) 380 New York Street, Redlands, CA 92373-8100, USA
Phone/Fax	909-793-2853 / 909-793-5953
E-mail	info@esri.com
WWW	www.esri.com/industries/water/
Comments	Free quarterly newsletter for ESRI business partners and water and wastewater user community.

GIS AND RELATED TECHNOLOGY JOURNALS

These journals publish peer-reviewed technical papers on GIS and related technologies, such as remote sensing, photogrammetry, and cartography.

- *Annals of the Association of American Geographers*
- *Business Geographics* (business)
- *Cartographica*
- *Cartography and GIS*
- *Computer* (algorithms and visualization)
- *Computers and Geoscience*
- *Computers, Environment, and Urban Systems*
- *Geocarto*
- *GrassClippings*
- *IEEE Geosciences*
- *IEEE Trans. on Comp. Graphics and Applications* (visualization)
- *International Journal of GIS*
- *International Journal of Remote Sensing*
- *Landscape Ecology*
- *Photogrammetric Engineering and Remote Sensing*
- *Remote Sensing Review*

JOURNALS THAT PUBLISH GIS ARTICLES

These periodicals are not strictly GIS journals, but they do publish GIS application articles on a routine basis. Some of them also publish special annual issues on information technology (IT) and infrastructure management that contain GIS application case studies and technical papers. For example, *Journal AWWA*'s November 2001 (vol. 93, no. 11) IT-focused issue published three articles on GIS/GPS applications. One of these articles showed how to use ESRI's MapObjects software for infrastructure capital asset management.

Journal	*Journal of American Water Resources Association*
Publisher	American Water Resources Association (AWRA) 4 West Federal Street, PO Box 1626 Middleburg, VA, 20118-1626, USA
Phone/Fax	540-687-8390 / 540-687-8395
E-mail	info@awra.org
WWW	www.awra.org

Magazine	*Journal AWWA*
Publisher	American Water Works Association (AWWA) 6666 W. Quincy Ave., Denver, CO 80235, USA
Phone/Fax	303-794-7711 / 303-794-7310
E-mail	Journal@awwa.org
WWW	www.awwa.org

Journal	*Journal of Water Resources Planning and Management*
Publisher	American Society of Civil Engineers 1801 Alexander Bell Drive, Reston, VA 20191-4400, USA
Phone/Fax	703-295-6300 / 703-295-6211
E-mail	marketing@asce.org
WWW	www.pubs.asce.org
Comments	On-line subscription also available

GIS Books

Should you be interested in pursuing an in-depth study of topics related to GIS applications, here are some recommendations. Some of these books are from a list compiled by U.S. Bureau of the Census (Nyman, 2000).

General

- *Dictionary of GIS Terminology*, ESRI Press, 2000, 160 pages, $19.95. Compilation of definitions of hundreds of terms from GIS and related technologies from the editors of ESRI Press.
- *Fundamentals of Geographic Information Systems*, Michael DeMers, John Wiley & Sons, 1996, 504 pages, $76.00. Introductory insights for students.
- *Fundamentals of Geographic Information Systems: A Compendium*, ACSM, 1990, $60.00.

- *Fundamentals of Spatial Information Systems*, Robert Laurini and Derek Thompson, Academic Press, London, 1992, ISBN 0-12-438380-7, $49.95. Approximately 700 pages with nice illustrations,

- *Geographic Information Systems: An Introduction*, Jeffrey Star and John Estes, Prentice Hall, 1990, $51.00. Introductory textbook for students and professionals.

- *Geographic Information Systems: An Introduction*, Tor Bernhardsen, John Wiley & Sons, Inc., second edition, 1999, ISBN 0-471-32192-3, 372 pages. Comprehensive presentation of various GIS technologies.

- *Geographical Information Systems*, edited by Paul A. Longley, Michael F. Goodchild, David J. Maguire, and David W. Rhind, John Wiley & Sons, second edition, 1991, $295.00. Two-volume boxed set containing 72 papers. Volume 1: *Principles and Applications*, and Volume 2: *Management Issues and Applications*. One of the most ambitious, extensive, and authoritative GIS book.

- *Geographical Information Systems and Science* by Paul A. Longley, Michael F. Goodchild, David J. Maguire, and David W. Rhind, John Wiley & Sons, ISBN 0-471-89275-0, 2001. Investigates the role of the Internet in GIS development, delivery, and use. Provides case studies from around the globe about, for example, local government tax assessment, network logistics, environmental modeling, and utility asset management.

- *Geographic Information Systems and Science* by Paul Longley, Michael Goodchild, and David Rhind, Taylor & Francis, 1990, $39.00. Selection of papers and articles on various aspects of GIS.

- *Introductory Readings in Geographic Information Systems*, edited by Donna Pequet and Duane Marble, Taylor & Francis, 1990, $39.00. Selection of papers and articles on various aspects of GIS.

- *Time in Geographic Information Systems*, by Gail Langran, Taylor & Francis, 1992.

GIS Applications

- *Environmental Modeling with GIS*, edited by Michael Goodchild, Bradley Parks, and Louis Steyaert, Oxford University Press, 1993, $65.00. Collection of invited interdisciplinary papers on the integration of GIS and environmental modeling, focusing on global change research, land and water resource management, and environmental risk assessment.

- *Environmental Planning for Communities—A Guide to the Environmental Visioning Process Utilizing a Geographic Information System (GIS)*, EPA/625/R-98/003, Technology Transfer and Support Division, Office of Research and Development, U.S. Environmental Protection Agency, Cincinnati, 2000.

- *Geographic and Land Information Systems for Practicing Surveyors: A Compendium*, ACSM, 1991, $45.00

- *Geographic Information Systems for Resource Management*, edited by William Ripple, ACSM, 1986, $60.00. Papers on land suitability; water, soil, and vegetation resource management; and urban and global GIS applications.

- *GIS for Water Resources and Watershed Management*, edited by John G. Lyon, Ann Arbor Press, Chelsea, Mich., ISBN 1-57504-090-5, 224 pages, $69.95.

- *GIS Modules and Distributed Models of the Watershed*, Task Committee Report, Paul A. DeBarry and Rafael G. Quimpo, committee co-chairs, ASCE, 1999. Comprehensive information about GIS and DEM applications in hydrologic modeling.

- *Hydrologic and Hydraulic Modeling Support with Geographic Information Systems*, edited by David Maidment and Dean Djokic, ESRI Press, 2000, 228 pages, $24.95. Compilation of invited papers in water resources at the 1999 ESRI International User Conference.

- *Hydrologic Applications of GIS*, edited by A.M. Gurnell and R.D. Montgomery, John Wiley & Sons, 2000, 176 pages, $85.00. Collection of papers focused on the application of GIS to the solution of hydrologic problems.

- *Introduction to Urban GIS*, by William Huxhold, Oxford University Press, 1991, $32.50.

- *Landscape Ecology and Geographic Information Systems*, edited by Roy Haines-Young, David Green, and Stephen Cousins. Taylor and Francis, 1993, $79.00. Review of new GIS and remote sensing tools in the growing field of landscape ecology.

- *Mapping the Next Millennium: The Discovery of New Geographies*, Stephen Hall, Random House, 1992, $30.00. Contains reports from the scientific frontiers where virtually every aspect of the physical cosmos is being mapped, including the floor of the ocean and the hole in the ozone layer.

- *Principles of Geographical Information Systems for Land Resources Assessment*, P.A. Burrough, Oxford University Press, 1986, $44.00. Advanced-level GIS textbook.

- *Resource Management Information Systems: Process and Practice*, Keith McCloy, Taylor and Francis, 1995, 415 pages, $50.00. Monograph on the technology required for effective management of spatially distributed resources, be they environmental, agricultural, urban, or water.

- *Spatial Analysis, GIS, and Remote Sensing Applications in the Health Sciences*, edited by Donald P. Albert, Wilbert M. Gesler, and Barbara Levergood, Ann Arbor Press, Chelsea, Mich., ISBN 1-57504-101-4, 300 pages, $69.95.

- *Three Dimensional Applications in Geographic Information Systems*, edited by Jonathan Raper, Taylor & Francis, 1989, $66.00. Survey of approaches and problems in modeling real geophysical data.

- *Wetland and Environmental Engineering Applications of GIS*, edited by John Lyon, Lewis Publishers, Boca Raton, Fla., 1996.

- *Wetland Landscape Characterization—GIS, Remote Sensing and Image Analysis*, by John Grimson Lyon, Ann Arbor Press, Chelsea, Mich., ISBN 1-57504-121-9, 224 pages, $69.95.

- *Zeroing In*, Andy Mitchell, Environmental Systems Research Institute, 1997. Twelve chapters on different GIS applications, including one on water and sewers.

MANAGEMENT

- *Geographic Information Systems: A Management Perspective*, Stan Aronoff, WDL Publications, 1989, $57.00. An excellent introduction to GIS principles and applications for users and managers.

- *How to Choose A GIS Consultant.* Available free from EI Technologies, LLC, 2620 S. Parker Road, Suite 150, Aurora, CO 80014, USA.

- *Managing Geographic Information Systems*, Nancy Obermeyer and Jeffrey Pinto, Guilford Publications, 1994, 226 pages, $36.00. Practical guide to key organizational issues for implementing GIS technology.

- *The GIS Book*, George B. Korte, Onword Press, third edition, 1999. A manager's guide to purchasing, implementing, and running a GIS.

- *Workbook*, UGC Corporation, 1990. Management-level primer on GIS with accompanying video.

GIS Software

- *Getting to Know ArcGIS Desktop: The Basics of* ArcView, ArcEditor, and ArcInfo, ESRI, 2001, ISBN 1-879102-89-7, $49.95. Ideal classroom text, lab manual, or self-study workbook. Includes a CD-ROM with a trial version of ArcView Version 8.1 plus data for completing the hands-on exercises.

- *Getting to Know ArcView GIS, Version 3.0 Edition*, ESRI, 1997, 660 pages, $49.95. Suitable for teaching an ArcView GIS software course or as a self-study workbook. Includes a CD-ROM with a demonstration copy of ArcView Version 3.0, sample data, tutorials, exercises, and multimedia presentations.

- *Inside ArcFM Water*, ESRI, 2000. Information on ESRI's ArcFM Water object model for water distribution systems.

- *Modeling Our World*, Michael Zeiler, ESRI, 1999. A primer for understanding the various models used to represent geographic information in ArcInfo 8 software, including object modeling and geodatabases.

- *Understanding GIS—The ARC/INFO Method, Version 7.1*, ESRI, 1996, 602 pages, $54.00. Basics of ArcInfo GIS.

Database and Data Structures

- *Accuracy of Spatial Databases*, edited by Michael Goodchild, Taylor & Francis, 1989, $77.00. Detailed information about error and accuracy, particularly of modelling uncertainty and reliability, testing accuracy, and the practical implications for use of spatial data.

- *Applications of Spatial Data Structures*, Hanan Samet, Addison-Wesley, 1989, $45.25. Applications of hierarchical data structures in computer graphics, image processing, and GIS.

- *Database Design for Mere Mortals*, Michael J. Hernandez, Addison-Wesley Developers Press, second edition, 1997, ISBN 0-201-69471-9, 440 pages, $27.95. A hands-on guide to relational database design.

- *Design and Analysis of Spatial Data Structures*, Hanan Samet, Addison-Wesley, 1990, $43.25. Information on hierarchical data structures.

- *Elements of Spatial Data Quality*, edited by Stephen Guptill and John Morrison, Elsevier Science, 1995, 200 pages, $149.50. Comprehensive discussion on data quality.

- *GIS Database Concepts*, URISA, 1999. Part of URISA Quick Study series, 27 pages, $19.00.

- *Handbook of Relational Database Design*, Candace Fleming and Barbara von Halle, Addison-Wesley, 1989, $46.00. This book provides a practical approach to designing relational databases. It contains two complementary design methodologies: logical data modeling and relational database design. The methodologies are independent of product-specific implementations and have been applied to numerous relational product environments.

- *Introduction to Database Systems*, volumes 1 and 2, C.J. Date, Addison-Wesley, 1990. $46.25.

- *Reactive Data Structures for Geographic Information Systems*, Peter van Oosterom, Oxford University Press, ISBN 0-19-823320-5, 1994. This 200-page book is part of the Spatial Information Systems series of Oxford University Press. In addition to an overview of GIS technology, the book contains many recent research results. It is illustrated with 80 figures and includes 296 references, which are easily accessible from an index with 872 entries.

- *Statistics for Spatial Data*, by Noel Cressie, John Wiley & Sons, 1991, $90.00. One of the most comprehensive and readable text to date on the analysis of spatial data through statistical models. It unifies a previously disparate subject under a common approach and notation.

MAPS AND CARTOGRAPHY

- *Analytical and Computer Cartography*, Keith Clarke, Prentice Hall, Englewood Cliffs, N.J., 1990, $52.00.

- *Cartographic Design: Theoretical and Practical Perspectives*, edited by Clifford H. Wood and C. Peter Keller, John Wiley & Sons, 1996, 306 pages, $125.00.

- *Geographic Information Systems and Cartographic Modelling*, Dana Tomlin, Prentice Hall, 1990, $51.00. A comprehensive approach to analysis and modeling using raster systems. An excellent introduction to GIS-based analysis emphasizing environmental decisions.

- *Map Appreciation*, Mark Monmonier, Prentice Hall, 1988, $41.00. This book teaches how to work with maps and promotes graphic literacy.

- *Map Generalization: Making Rules for Knowledge Representation*, edited by Barbara Buttenfield and Robert McMaster, John Wiley & Sons, 1991, $95.00. This book is one of the first to focus on the development of a rule base for digital mapping. It provides a framework to help solve mapping problems, improve efficiency, preserve consistency, and incorporate sound principles into digital mapping.

- *Map Projections—A Working Manual*, John Snyder, USGS Professional Paper 1395, USGS Distribution Center, Box 25286, MS 306, Denver Federal Center, Denver, CO 80225, $32.00.

- *Spatial Accuracy Assessment—Land Information Uncertainty in Natural Resources*, edited by Kim Lowell and Annick Jaton, Ann Arbor Press, Chelsea, Mich., ISBN 1-57504-119-7, 450 pages, $89.95.

- *Quantifying Spatial Uncertainty in Natural—Theory and Applications for GIS and Remote Sensing*, edited by H. Todd Mowrer and Russell G. Congalton, Ann Arbor Press, Chelsea, Mich., ISBN 1-57504-131-6, 350 pages, $74.95.

GPS

- *The Global Positioning System and GIS*, Michael Kennedy, Ann Arbor Press, Chelsea, Mich., ISBN 1-57504-017-4, 268 pages, $59.95. A classroom or self-teaching educational tool intended as an introduction for mapping professionals, managers, and students.

- *GPS for Land Surveyors*, by Jan Van Sickle, Ann Arbor Press, Chelsea, Mich., second edition, ISBN 1-57504-075-1, 270 pages, $69.95.

Remote Sensing

- *Mathematical Principles of Remote Sensing—Making Inferences from Noisy Data*, by Andrew S. Milman, Ann Arbor Press, Chelsea, Mich., ISBN 1-57504-135-9, 400 pages, $74.95.

- *Remote Sensing and Image Interpretation*, T.M. Lillesand and R.W. Kiefer, John Wiley & Sons, fourth edition, 1999, 724 pages. Remote sensing and satellite imagery textbook.

- *Remote Sensing Change Detection—Environmental Monitoring Methods and Applications*, edited by Ross S. Lunetta and Christopher D. Elvidge, Ann Arbor Press, Chelsea, Mich., ISBN 1-57504-037-9, 350 pages, $69.95.

Conference Proceedings

- *32nd Annual Conference and Symposium on GIS and Water Resources*, AWRA, Ft. Lauderdale, Florida, September 22–26, 1996.

- *Building Databases for Global Science*, edited by Helen Mounsey, Taylor & Francis, 1988, $93.00. Papers from the first meeting of the International Geographical Union's Global Database Planning Project.

- ESRI User Conference Archives at www.esri.com/library/userconf/archive.html.

- *Geographic Information Systems in Public Utilities, Specialty Conference Proceedings*, Water Environment Federation, Orlando, December 1–4, 1991.

- *GIS Hydro '98—Introduction to GIS Hydrology*, Pre-Conference Seminar, 18th Annual User Conference, ESRI, San Diego, July 1998.

- *Proceedings of the National Conference on Environmental Problem-Solving with Geographic Information Systems*, EPA/625/R-95/004, Office of Research and Development, U.S. Environmental Protection Agency, Cincinnati, September 1995.

- *Proceedings: 5th International Symposium on Spatial Data Handling*, IGU Commission on GIS, Charleston, S.C., August 1992, $50.00. Two volume set contains more than 70 selected papers representing the state of the art in geographical information processing.

Summary

This chapter presented a handy compilation of representative GIS information resources that should be helpful for your water, wastewater, and stormwater applications, such as the Internet, data, software, periodicals, books, and conferences. This chapter is intended to serve as your reference guide.

Self-Evaluation

1. What GIS resources are most important for the needs of for your organization?

2. Compile a list of GIS resources (Web sites, Listservs, and magazines) specific to the needs of for your organization.

3. Identify a GIS issue in your organization, formulate a question, subscribe to a relevant Internet GIS discussion group or Listserv, post the question, and prepare a summary of responses received.

REFERENCES

Abbott, M.B., J.C. Bathurst, J.A. Cunge, P.E. O'Connell, and J. Rassmussen. (1986). "An Introduction to The European Hydrological System--Systeme Hydrologique Europeen, SHE, 1: History and Philosophy of A Physically-Based, Distributed Modeling System." *Journal of Hydrology*, 87(1/2), 45-59.

Adams, D. and B. Johnson (1994). "Building a Cadastral Map." *American City & County*, April 1994, Page 14.

Agbenowosi, N. (2001). "GIS and sewer System Design." Case Study Submission in Response to a Call for Case Studies distributed by the Author (U. M. Shamsi) for this book.

Aktas, M. (1996). *GIS in Water Resources*. Masters Thesis, Dr. Nuri Merzi Thesis Supervisor, Middle East Technical University, Ankara, Turkey.

Aktas, M. (2001). "Determination of Inundation Maps by GIS." Case Study Submission in Response to a Call for Case Studies distributed by the Author (U. M. Shamsi) for this book.

Alston, R and D. Donelan (1993). "Weighing the Benefits of GIS." *American City & County*, October 1993, Vol. 108, No. 11, Page 14.

Aron, G. (1987). *Penn State Runoff Model for IBM-PC*. User's Manual, Department of Civil Engineering, Pennsylvania State University, University Park, Pennsylvania.

Aron, G., B.A. Dempsey, T.A. Seybert, T.A. Smith III, T.J. Ostrowski, S.A. Closson. (1996). *Penn State Runoff Quality Model PSRM-QUAL*, User's Manual, Version 95.0a, Department of Civil and Environmental Engineering, Pennsylvania State University, University Park, Pennsylvania.

Aron, G., D.F. Lakatos (1990). *Penn State Runoff Model for IBM-PC*. User's Manual, Department of Civil Engineering, Pennsylvania State University, University Park, Pennsylvania.

Aron, G., D.J. Wall, E.L. White, C.N. Dunn, D.M. Kotz (1986). *Field Manual of Pennsylvania Department of Transportation Storm Intensity-Duration-Frequency Charts*. Department of Civil Engineering, Pennsylvania State University, University Park, Pennsylvania.

Aron, G., J.V. Radziul, D.F. Lakatos, and D. Blair (1979). "Penn State Urban Runoff Model to Pinpoint Flood Peak Source Locations." *Water Resources Bulletin*, AWRA, 15(5), 1250-1264.

ASCE (1999). *GIS Modules and Distributed Models of the Watershed.* Task Committee Report, Paul A. DeBarry and Rafael G. Quimpo Committee Co-chairs, American Society of Civil Engineers, Reston, Virginia.

Ball, M. (2001). "GEOWorld Examines Reader Survey Results." *GEOWorld*, October 2001, Vol. 14, No. 10, Page 8.

Barnes, S. (2000). "One Last Thing – GIS Market Revenues Top $1.5 Billion." *Geospatial Solutions*, June 2000, Page 58.

Bell, J. (1999). "AutoCAD Map 3." *Professional Surveyor*, Vol. 19, No. 1, January/February 1999, 52-53.

Bell, J. (1999a). "ArcView GIS." *Professional Surveyor*, Vol. 19, No. 3, March 1999, 48-50.

Bell, J. (2001). "XmapGeographic With 3-D TopoQuads from DeLorme." *Professional Surveyor*, Vol. 21, No. 3, March 2001, 28-32.

Bentley Systems (2001). "Arizona Water Opens the Floodgates on Next-Level Utility Management with MicroStation GeoGraphics." Bentley Systems website at www.bentley.com/products/geproducts/discovery/casestudies/arizona/arizona.htm.

Bernhardsen, T. (1999). *Geographic Information Systems–An Introduction.* Second Edition, John Wiley and Sons, Inc., New York, NY, 372 Pages.

Berry, J.K. (1994). "What Does Your Computer Really Think of Your Map." *GIS World*, 7(11), Page 30.

Bhaskar, N.R., P.J. Wesley, and R.S. Devulapalli (1992). "Hydrologic Parameter Estimation Using Geographic Information System." *Journal of Water Resources Planning and Management*, ASCE, 118(5), 492-512.

Binford, M. W. (2000). Clark Lab's Idrisi32, *Geo Info Systems*, April 2000, 40-42.

Byun, S.A. and B. Marengo (2001). "GIS and CSO Program Support." Case Study Submission in Response to a Call for Case Studies distributed by the Author (U. M. Shamsi) for this book.

Byun, S.A., J. LeBlanc, M. McBride, T. Okeowo, and W.C. Hession (2001). "GIS and Water Quality Modeling Integrations." Case Study Submission in Response to a Call for Case Studies distributed by the Author (U. M. Shamsi) for this book.

Cannistra, J.R. (1999). "Converting Utility Data for a GIS." *Journal AWWA*, American Water Works Association, February 1999, 55-64.

Carlson, B. (2001). "MapInfo Professional 6.5." *GEOWorld*, October 2001, Vol. 14, No. 10, Page 59.

Chester Engineers (1990). *Turtle Creek Watershed Act 167 Stormwater Management Plan.* Chester Engineers, Pittsburgh, Pennsylvania.

Chester Engineers (1993). *Bull Run Watershed Act 167 Stormwater Management Plan.* Chester Engineers, Pittsburgh, Pennsylvania.

Cheves, M. (2000). "GIS and Surveying." *Professional Surveyor*, Vol. 20, No. 9, October 2000, Page 24.

Cleveland, R. and V. L. Clair (2001). "Mapping System Provides Efficient Sewer Management." *ArcNews Online*, ESRI, Spring 2001.

Corbley, K.P. (2000). "Plugging In." *Imaging Notes*, Space Imaging, Thornton, Colorado, March/April 2000, 14-15.

Corbley, K.P. (2001). "Keep County Information Compatible." *GEOWorld*, October 2001, Vol. 14, No. 10, 38-41.

Cowden, R.W. (1991). "Stormwater/wastewater Management: How An application Can Drive a GIS." *Proceedings of the Water Resources Planning and Management and Urban Water Resources 18th Annual Conference and Symposium*, ASCE, New York, NY, 913-917.

Curtis, T.G. (1994). "SWMMDUET: Enabling EPA SWMM with the ARC/INFO Paradigm." *Arc News*, ESRI, Vol. 16, No. 2, Spring 1994, Page 20.

Dangermond, J. (1994). "ArcView Status and Direction." *Arc News*, ESRI, Vol. 16, No. 2, Spring 1994, 1-4.

Daratech Inc. (2000). *Geographic Information Systems – Markets and Opportunities.* Cambridge, MA.

Dawe, P. and H. Schölzel (2001). "GIS and Hydraulic Model Linkages for Data Transfer." Case Study Submission in Response to a Call for Case Studies distributed by the Author (U. M. Shamsi) for this book.

DeMartino, S. and E. Hrnicek (2001). "Object-Oriented GIS 101." *CE News*, November 2001, Vol. 13, No. 10, 58-63.

Department of Environmental Protection (1985). *Storm Water Management Guidelines and Model Ordinances.* Division of Waterways and Storm Water Management, Harrisburg, Pennsylvania.

DeVantier, B.A., A.D. Feldman (1993). "Review of GIS Applications in Hydrologic Modeling." *Journal of Water Resources Planning and Management*, ASCE, 119(2), 246-261.

DeVries, J.J., and T.V. Hromadka (1993). "Computer Models for Surface Water." Chapter 21 in *Handbook of Hydrology*, Edited by D.R. Maidment, McGraw Hill, New York, NY, 21.1-21.39.

Djokic, D., M.A. Meavers, and C.K. Deshakulakarni (1993). *ARC/HEC2: An ARC/INFO - HEC-2 Interface*. User's Manual, Center for Research in Water Resources, The University of Texas at Austin, Austin, TX.

Dodson, R.D. (1993). "Advances in Hydrologic Computation." Chapter 23 in *Handbook of Hydrology*, Edited by D.R. Maidment, McGraw Hill, New York, N.Y. 23.1-23.24.

Dougherty, L. (2000). "Census 2000, Census Sense, Part 2, Geographic Products for One and All." *Geospatial Solutions*, September 2000, 36-39.

Dueker, K.J., (1987). "Geographic Information Systems and Computer-Aided Mapping." *APA Journal*, Summer 1987, 383-390.

Edelman, S., T. Dudley, and C. Crouch (2001). "Much ADO About Data." *CE News*, January 2001, 52-55.

Engelhardt, J. (2001). "Oracle Brings Spatial Functions to Mainstream Business With 9i." *Geospatial Solutions*, September 2001, Page 16.

Engman, E.T. (1993). "Remote Sensing." Chapter 24 in *Handbook of Hydrology*, Edited by D.R. Maidment, McGraw Hill, New York, NY, 24.1-24.23.

EPA (2000). *Environmental Planning for Communities – A Guide to the Environmental Visioning Process Utilizing a Geographic Information System (GIS)*. EPA/625/R-98/003, Technology Transfer and Support Division, Office of Research and Development, U.S. Environmental Protection Agency, Cincinnati, Ohio, September 2000.

ERDAS (2001). "There's More to a Pixel Than Meets The Eye." *ERDAS News*, ERDAS Inc., January 2001, Vol. 2, No. 1, 18-19.

ESRI (1990). *Understanding GIS – The ARC/INFO Method*. Environmental System Research Institute, Redlands, California.

ESRI (1992). *ArcView User's Guide*. Second Edition, Environmental System Research Institute, Redlands, California.

ESRI (1995). *Introduction to Avenue*. ESRI Educational Services, Environmental System Research Institute, Redlands, California.

ESRI (1996). *ArcView Spatial Analyst.* User's Manual, Version 1, Environmental System Research Institute, Redlands, California.

ESRI (1996a). *City of Glendale Database Design.* Environmental System Research Institute, Redlands, California, July 1996.

ESRI (1997). *Getting to Know ArcView GIS.* Second Edition, Environmental System Research Institute, Redlands, California.

ESRI (1998). "California Water Districts Link GIS to Document Management System." *ARC News*, ESRI, Spring 1998, Page 19.

ESRI (1999). "EPA's EnviroMapper Provides Access to Environmental Information." *Environmental Solutions*, Vol. 1, No. 1, Page 4.

ESRI (2000). "Sharing Geographic Knowledge: Providing Access to Geographic Information." *Arc News*, ESRI, Spring 2000, Page 6.

ESRI (2000a). *Building a Geodatabase.* Environmental System Research Institute, Redlands, California, 492 Pages.

ESRI (2000b). *Inside ArcFM Water.* Environmental System Research Institute, Redlands, California, 119 Pages.

ESRI (2001). "ArcGIS 8.1 Nears Release." *Arc News*, ESRI, Vol. 22, No. 4, Winter 2000/2001, 1-2.

ESRI (2001a). "ArcEditor Manages Geodatabase Schema and Includes Advanced Editing of Geodatabases and Coverages." *Arc News*, ESRI, Fall 2001, Vol. 23, No. 3, Page 6.

ESRI (2001b). *Introduction to ArcGIS II.* Course Lectures, Version 2.3, Environmental System Research Institute, Redlands, California, Page 1-8.

Estes-Smargiassi, S. (1998). "Massachusetts Water Resources Authority Uses GIS to Meet Objectives Cost-Effectively." *Water Writes*, ESRI.

Federal Emergency Management Agency (1979). *Flood Insurance Study.* Municipality of Monroeville, Pennsylvania, Flood Map Distribution Center, Baltimore, M.D.

Federal Emergency Management Agency (1984). *Flood Insurance Study.* Township of East Buffalo, Pennsylvania, Flood Map Distribution Center, Baltimore, M.D.

Fitzpatrick, T. (2002). "Making the Switch from AML to VBA." *ArcUser*, ESRI, April-June, 2002, Vol. 5, No. 2, 36-39.

Fortin, J.P., J.P. Villeneuve and B. Seguin (1986). "Development of A Modular Hydrological Forecasting Model Based on Remotely Sensed Data for Interactive Utilization on A Microcomputer." *Hydrologic Applications of Space Technology*, Edited by A.I. Johnson, 307-319.

Garland, E., N. Marth and V.O. Shanholtz (1990). "Integrating Land Use / Land Cover Into a GIS." *1990 International Summer Meeting*, American Society of Agricultural Engineers, Columbus, Ohio, June 24-27, 1990.

Geospatial Solutions (2000). "Geography Network." *Geospatial Solutions*, December 2000, Vol. 10, No. 12, gn1-gn16.

Geospatial Solutions (2001). "2002 Buyers Guide." *Geospatial Solutions*, December 2001, Vol. 11, No. 12, 42-57.

GEOWorld (2001). "GIS in 3-D—Visualization Shines in Diverse Applications." *GEOWorld*, October 2001, Vol. 14, No. 10, 30-36.

Gilbrook, M.J. (1999). "GIS Paves the Way." *Civil Engineering*, ASCE, Vol. 69, No. 11, November, 1999, 34-39.

Goldstein, H. (1997). "Mapping Convergence: GIS Joins the Enterprise." *Civil Engineering*, ASCE, June, 1997, 36-39.

Goodchild, M. (1998). "Uncertainty: The Achilles Heels of GIS." *Geo Info Systems*, November 1998, Page 50.

Goodchild, M. (2002). "MapFusion for GIS Interoperability." *Geospatial Solutions*, Vol. 12, No. 4, April 2002, 48-51.

Gorokhovich, Y., R. Khanbilvardi, L. Janus, V. Goldsmith, and D. Stern (2000). "Spatially Distributed Modeling of Stream Flow During storm Events." *Journal of the American Water Resources Association*, Vol. 36, No. 3, June 2000, 523-539.

Graybill, G. R. (1998). "Database Design Issues for the Integration of ARC/INFO with Hydraulic Analysis Applications." *Proceedings of the 1998 International User Conference*, ESRI, San Diego, California, July 27-31, 1998.

Griffin, C.B. (1995). "Data Quality Issues Affecting GIS Use for Environmental Problem-Solving." *Proceedings of the National Conference on Environmental Problem-Solving with Geographic Information Systems*, EPA/625/R-95/004, Office of Research and Development, U.S. Environmental Protection Agency, Cincinnati, Ohio, September 1995, 15-30.

Hansen, R. (1998). "Importing U.S. Census TIGER Files into ArcView GIS." *ArcUser*, ESRI, July-September, 1998, Page 46.

Hay, L.E. and L.K. Knapp (1996). "Integrating a Geographic Information System, A Scientific Visualization System, and a Precipitation Model." *Water Resources Bulletin*, American Water Resources Association, Vol. 32, No. 2, April 1996, 357-369.

Heaney, J.P., Sample, D., and Wright, L. (1999). *Geographical Information Systems, Decisions Support Systems, and Urban Stormwater Management*. Cooperative Agreement Report No. CZ826256-01-0, U.S. Environmental Protection Agency, Edison, New Jersey.

Heineman, M. (2000). SWMMTools, ArcScripts Web Site (arcscripts.esri.com), ESRI.

Hernandez, M.J. (1997). *Database Design for Mere Mortals*. Addison-Wesley Developers Press, Second Edition, ISBN 0-201-69471-9, 440 Pages.

Hsu, Y.M. (1999). 3D.Add_XYZ Script, ArcScripts Web Site (arcscripts.esri.com), ESRI.

Huber, W.C., R.E. Dickinson (1988). *Storm Water Management Model*. User's Manual, Version 4, Environmental Research Laboratory, Office of Research and Development, U.S. Environmental Protection Agency, Athens, Georgia.

Hurlbut, K. (1999). "Unlocking the Power of Image Data." *ArcUser*, ESRI, Vol. 2, No. 2, April-June, 1998, 18-20.

Huse, S.M. (1995). *GRASSLinks: A New Model for Spatial Information Access in Environmental Planning*. Ph.D. Dissertation, University of California at Berkley.

Hutchinson, S., and L. Daniel (1995). *Inside ArcView*. Onward Press, 329 pp.

Hydrologic Engineering Center (1974). *An Assessment of Remote Sensing Applications in Hydrologic Applications*, Res. Note No. 4, U.S. Army Corps of Engineers, Davis, California.

Hydrosphere Data Products (1996). *GeoSelect Pennsylvania*. CD ROM and User's Guide, NCDC Hourly Precipitation GIS, Data Vol. 6.0, Boulder, Colorado.

Irrinki, S. (2000). "The Digital Utility." *Water Environment & Technology*, Vol. 12, No. 12, December 2000, Water Environment Federation, 29-33.

Jacobi, K. (2001). "Rainfall-Runoff Modeling With the Purpose to Estimate Vol.s of Flood Flow and Groundwater Recharge." Case Study Submission in Response to a Call for Case Studies distributed by the Author (U. M. Shamsi) for this book.

Jaworksi, L. (1994). "Wrestling with Clean Water Act Reauthorization Continues to Meander through Congress." *Water Environment and Technology*, WEF, 6(10), 58-63.

Jenkins, D. (2002), "Making the Leap to ArcView 8.1." *Geospatial Solutions*, January 2002, 46-48.

Johnson, R.C., J.C. Imhoff, and H.H. Davis (1980). *Hydrological Simulation Program–Fortran.* User's Manual, Environmental Research Laboratory, Office of Research and Development, U.S. Environmental Protection Agency, Athens, Georgia.

Johnston, C., A. Boyland, and J. Vine S.A. (2001). "Estimation of Demands for Water Distribution System Modeling." Case Study Submission in Response to a Call for Case Studies distributed by the Author (U. M. Shamsi) for this book.

Joint Planning Commission (1988). *Little Lehigh Creek Watershed Act 167 Stormwater Management Plan.* Lehigh-Northampton Counties, Pennsylvania.

Jostenski, D.B. (1988). *Pennsylvania's SWM Program.* Department of Environmental Resources' Update, Computational Methods in Stormwater Management, 1988 Stormwater Short Course and Symposium, June 13-15, 1988, Pennsylvania State University, University Park, Pennsylvania.

Kavaliunas, J. and L. Dougherty (2000). "Census 2000, Census Sense, Part 1, What Geospatial Pros Need to Know." *Geospatial Solutions,* July 2000, 40-43.

Kibler, D.F., and G. Aron (1980). "Urban Runoff Management Strategies." *Journal of the Technical Councils,* ASCE, Vol. 106, No. TC1, 1-12.

Kindleberger, C. (1992). "Tomorrow's GIS." *American City & County,* April 1992, 38-48.

Korte, G.B. (1994). *GIS Book.* Third Edition, Onword Press, Santa Fe, NM. 213 pp.

Kouwen, N. (1988). "WATFLOOD: A Micro-Computer Based Flood Forecasting System Based on Real-Time Weather Radar." *Canadian Water Resources Journal,* 13(1), 62-77.

Kouwen, N., E.D. Soulis, A. Pietroniro, and R.A. Harrington (1993). "Group Response Units for Distributed Hydrologic Modeling." *Journal of Water Resources Planning and Management,* ASCE, 119(3), 289-305.

Kuppe, J.B. (1999). "Imagery Technology Helps County Make GIS Data Accessible." *ArcUser,* ESRI, October-December, 1999, 10-11.

Lanfear, K. J. (2000) "The Future of GIS and Water Resources." *Water Resources Impact,* American Water Resources Association, Vol. 2, Number 5, September 2000, 9-11.

Lee, C.V. (1998). "Selecting a Geographic Information System." *Proceedings of the 1998 National Conference on Environmental Engineering,* Edited by T. E. Wilson, ASCE, June 7-10, 1998, Chicago, Illinois, 669-674.

Leick, A. (1992). "There Is More to Learn - GPS." *ACSM Bulletin,* May/June, Page 59.

Lewis, R. (1993). "Searching for Hidden GIS Treasures." *American City and County*, 108(1), Page 20.

Limp, W.F. (2001). "Millennium Moves – Raster GIS and Image Processing Products Expand Functionality in 2001." *GEOWorld*, April 2001, Vol. 14, No. 4, 39-42.

Limp, W.F. (2001a). "From the Back Room to the GIS Room." *GEOWorld*, August 2001, Vol. 14, No. 8, 30-35.

Longley, P.A., M.F. Goodchild, D.J. Maguire, and D.W. Rhind (1999). *Geographic Information Systems*, Second Edition, John Wiley & Sons.

Lowe, J. W. (2000). "Building a Spatial Web Site? Know Your Options." *Geo Info Systems*, April 2000, 44-49.

Lowe, J. W. (2000a). "Utilities Databases Internet." *Geo Info Systems*, March 2000, 46-48.

Lowe, J. W. (2002). "GIS Meets the Mapster." *Geospatial Solutions*, Vol. 12, No. 2, February 2002, 46-48.

Maguire, D.J. (1999). "ARC/INFO Version 8: Object-Component GIS." *Arc News*, ESRI, Vol. 20, No. 4, Winter 1998/1999, 1-2.

Maidment, D.R. (2000). *ArcGIS Hydro Data Model.* GIS Hydro 2000 Pre-Conference Seminar, 2000 International User Conference, ESRI, San Diego, California, June 2000. http://www.crwr.utexas.edu/giswr/

Marsalek, J., T.M. Dick, P.E. Winser, and W.G. Clarke (1975). "Comparative Evaluation of Three Urban Runoff Models." *Water Resources Bulletin*, 11(2), 306-328.

Maslanik, J.M., U.M. Shamsi (1991) "Effect of Storm Distribution on Watershed Stormwater Management." *ASCE National Conference on Water Resources Planning and Management*, New Orleans, Louisiana, May 1991.

McBride, M., W.C. Hession and M. Bennett (2001). "A Statewide Non-point Source (NPS) Pollution Assessment." Case Study Submission in Response to a Call for Case Studies distributed by the Author (U. M. Shamsi) for this book.

McKee, L. (1998). "Open GIS Specification Conformance – What Does It Mean?" *Geo Info Systems*, November 1998, 36-40.

Meyer, S.P., T.H. Salem, and J.W. Labadie (1993). "Geographic Information Systems in Urban Storm-Water Management." *Journal of Water Resources Planning and Management*, ASCE, 119(2), 206-228.

Michael, J. (1994). "Creating Digital Orthophotos Requires Careful Consideration of Project Design Elements." *Earth Observation Magazine*, 3(2), 34-37.

Miotto, A. (2000). "A Better Image." *Civil Engineering*, ASCE, January 2000, 42-45.

Moglen, G.E. (2000). "Effect of Orientation of Spatially Distributed Curve Numbers in Runoff Calculations." *Journal of the American Water Resources Association*, AWRA, 36(6), 1391-1400.

Morgan, T.R., D.G. Polcari (1991). "Get Set! Go for Mapping, Modeling, and Facility Management." *Proceeding of Computers in the Water Industry Conference*, AWWA, Houston, Texas.

Murphy, C. (2000). "Digital Cartographic Data – The Bottom Line." *CE News*, July 2000, 52-56.

Murphy, C. (2000a). "GIS Trends." *CE News*, December 2000, 63-65.

Muthukrishnan, S., M. Doyle, S. Pandey, N. Jokay, and J. Harbor (2001). "GIS Based Long-Term Hydrologic Impact Assessment." Case Study Submission in Response to a Call for Case Studies distributed by the Author (U. M. Shamsi) for this book.

Nyman, L. (2000). The Geographic Information Systems FAQ Website, www.census.gov/geo/www/faq-index.html, U.S. Bureau of Census.

Ono, T., L. Cote, S. Dodge, and E. Pier (1998). "The Wastewater Management System (WIMS)." *ArcUser,* Calendar Insert, ESRI, July 1998.

PaMAGIC (2001). *Local Government Handbook for GIS Implementation Within the Commonwealth of Pennsylvania, Part II: GIS Data*. Interim Draft Document, March 2001, Pennsylvania Mapping and Geographic Information Consortium, Lewisburg, Pennsylvania, www.pamagic.org, 51 Pages.

Pearson, M., S. Wheaton (1993). "GIS and Storm-Water Management." *Civil Engineering*, ASCE, 63(9), 72-73.

Peng, Z. (1998). "Internet GIS." *CE News*, December 1998, 42-47.

Phipps, S. P. (1995). "Using Raster and Vector GIS Data for Comprehensive Storm Water Management." *Proceedings of the 1995 International User Conference*, ESRI, San Diego, California, May 22-26, 1995.

Poertner, H.G. (1988). *Stormwater Management in the United States.* Stormwater Consultants, Bolingbrook, IL.

Robinson, R. (1993). "The Integrated City." *Civil Engineering*, ASCE, September 1993, 2A-16A.

Roeth B. and B. Roth (2001). "Sanitary Sewer Modeling." Case Study Submission in Response to a Call for Case Studies distributed by the Author (U. M. Shamsi) for this book.

Ross, M.A., P.D. Tara (1993). "Integrated Hydrologic Modeling with Geographic Information Systems." *Journal of Water Resources Planning and Management*, ASCE, 119(2), 129-140.

Rubin, D.K., M.B. Powers, H. Carr, and D.B. Rosenbaum (1993). "A Whole Lot of Planning Going On." ENR Cover Story, *ENR*, September 20, 38-44.

Schock, M.R., and J.A. Clement (1995). "You Can't Do That With These Data! Or: Uses and Abuses of Tap Water Monitoring Analyses." *Proceedings of the National Conference on Environmental Problem-Solving with Geographic Information Systems*, EPA/625/R-95/004, Office of Research and Development, U.S. Environmental Protection Agency, Cincinnati, Ohio, September 1995, 31-41.

Schultz, G.A. (1988). "Remote Sensing in Hydrology." *Journal of Hydrology*, Vol. 100, 239-265.

Shamsi, U.M. (1988). *Chester's Penn State Runoff Model (CPSRM)*. User's Manual, Chester Environmental, Pittsburgh, PA.

Shamsi, U.M. (1989). *Rainfall Analysis Program (RAP)*. User's Manual, Chester Engineers, Pittsburgh, Pennsylvania.

Shamsi, U.M., (1990). *West Deer Township - Master Water Plan.* Final Report, submitted by The Chester Engineers to the West Deer Township, Pittsburgh, Pennsylvania.

Shamsi, U.M. (1991). "Three Dimensional Graphics in Distribution System Modeling." *Proceedings of the Specialty Conference on Computers in the Water Industry*, AWWA, Houston, TX, April 1991, 637-647.

Shamsi, U.M. (1992). "GIS, Remote Sensing, and Master Water Plan: A Case Study." *Symposium on Geographic Information Analysis*, ASCE, Dallas, Texas, June 1992.

Shamsi, U.M. (1993a). "GIS Forecasts Sewer Flows." *GIS World*, 6(3), 60-64.

Shamsi, U.M. (1993b). "GIS Based Hydrographs." *Advances in Hydro-Science and –Engineering, Proceedings of the First International Conference on in Hydro-Science and –Engineering,* Edited by Sam S.Y. Wang, Center for Computational Hydroscience and Engineering, Washington, D.C., June 7-11, 1993, Vol. I, 564-570.

Shamsi, U.M. (1996). "Stormwater Management Implementation through Modeling and GIS." *Journal of Water Resources Planning and Management*, ASCE, Vol. 122, No. 2, March/April, 114-127.

Shamsi, U.M. (1998). "ArcView Applications in SWMM Modeling." Chapter 11 in *Advances in Modeling the Management of Stormwater Impacts*, Edited by W. James, Vol. 6. Computational Hydraulics International, Guelph, Ontario, Canada. 219-233.

Shamsi, U.M. (1999). "GIS and Water Resources Modeling: State-of-the-Art." Chapter 5 in *New Applications in Modeling Urban Water Systems*, Edited by W. James, Computational Hydraulics International, Guelph, Ontario, Canada, 93-108.

Shamsi, U.M. (2001). GIS and Modeling Integration. *CE News*, Vol. 13, No. 6, July 2001, p 46-49.

Shamsi, U.M. and B.A. Fletcher (1994). "GIS Based Urban Drainage Modeling." Chapter 19 in *Current Practices in Modeling the Management of Stormwater Impacts*, Edited by W. James, Lewis Publishers, Boca Raton, Florida. 293-307.

Shamsi, U.M. and B.A. Fletcher (1995). "GIS in Stormwater Management." Chapter 20 in *Modern Methods for Modeling the Management of Stormwater Impacts*, Edited by W. James, Computation Hydraulics International, Guelph, Ontario, Canada, 315 pp.

Shamsi U.M. and C. Scally (1998). "An Application of Continuous Simulation to Develop Rainfall Versus CSO Correlations." *Proceedings of WEFTEC Asia–Conference and Exhibition on International Wastewater and Water Quality Technology*, WEF, Singapore, March 7-11, 1998.

Shamsi, U.M., S.P. Benner, and B.A. Fletcher (1996). "A Computer Mapping Program for Sewer Systems." Chapter 7 in *Advances in Modeling the Management of Stormwater Impacts*, Edited by W. James, Computation Hydraulics International, Guelph, Ontario, Canada, 97-114.

Shea C., W. Grayman, D. Darden, R.M. Males, and P. Sushinsky (1993). "Integrated GIS and Hydrologic Modeling for Countywide Drainage Study." *Journal of Water Resources Planning and Management*, ASCE, 119(2), 17-128.

Simonovic, S.P. (2002). "The Decision is Clear with Clark Lab's Idrisi32 Release 2." *Geospatial Solutions*, Vol. 12, No. 5, May 2002, 46-48.

Singh, V.P. (Editor) (1995). *Computer Models of Watershed Hydrology*. Water Resources Publications, 1130 Pages.

Slawecki, T., C. Theismann and P. Moskus (2001). "GIS-A-GI-GO." *Water Environment & Technology*, Vol. 13, No. 6, June 2001, Water Environment Federation, 33-38.

Smith, M.B., A. Vidmar (1994). "Data Set Derivation for GIS-Based Urban Hydrological Modeling." *Photogrammetric Engineering and Remote Sensing*, 60(1), 67-76.

Smullen, J.T. and J. Angell (2001). "GIS to improve Hydrologic Modeling for CSO Permit Compliance." Case Study Submission in Response to a Call for Case Studies distributed by the Author (U. M. Shamsi) for this book.

Soil Conservation Service (1983). *Computer Program for Project Formulation Hydrology.* Technical Release 20, U.S. Department of Agriculture, Springfield, Virginia.

Somers, R. (1999). *Framework Data Survey.* Preliminary Report, A Supplement to Geo Info Systems, September 1999, 35 Pages.

Stuebe, M.M. and D.M. Johnston (1990). "Runoff Vol. Estimation Using GIS Techniques." *Water Resources Bulletin*, AWRA, 26(4), 611-619.

Swalm, C., J.T. Gunter, V. Miller, D.G. Hodges, J.L. Regens, J. Bollinger, and W. George (2000). "GIS Blossoms on the Mighty Mississippi." *Geo Info Systems*, April 2000, 30-34.

Terstriep, M.L., M.T. Lee (1989). "Regional Stormwater Modeling, Q-Illudas and Arc/Info." *Computing in Civil Engineering: Computers in Engineering Practice*, ASCE, New York, N.Y.

Thoen, B. (2001). "Maximizing Your Search For Geospatial Data." *GEOWorld*, Vol. 14, No. 4, April 2001, 34-37.

Thomas, G. (1995). "A Complete GIS-based Stormwater Modeling Solution." *Proceedings of the 1995 International User Conference*, ESRI, San Diego, California, May 22-26, 1995.

Thrall, G. I. (1999). "Low Cost GIS." *Geo Info Systems*, April 1999, 38-40.

Thrall, G. I. (1999a). "Want to Think Spatially? Raster GIS with Mfworks." *Geo Info Systems*, September 1999, 46-50.

Thrall, G. I. (2001). "Business Geography Data Resources - Part 1." *Geospatial Solutions*, Vol. 11, No. 5, May 2001, 42-49.

Thrall, G. I. (2001a). "Business Geography Data Resources – Part 2." *Geospatial Solutions*, Vol. 11, No. 6, June 2001, 44-49.

Thrall, G. I. (2002). "2000 Census – It's a CACI One." *Geospatial Solutions*, Vol. 12, No. 2, February 2002, 49-52.

Thuman, A. and K. Mooney (2000). "TMDLs: How Did We Get Here and Where Are WE Going?" *Environmental Perspectives*, Newsletter, HydroQual, Inc., Fall 2000, Vol. 5, 1-4.

Townsley, S. and D, Ford (2001). "Real-time Flood Information Mapping (RTFIM)." Case Study Submission in Response to a Call for Case Studies distributed by the Author (U. M. Shamsi) for this book.

Turtle Creek Watershed Association, Inc. (1977). *Urban Development and Small Watershed Flooding.* East Pittsburgh, Pennsylvania.

U.S. Army Corps of Engineers (1963). *Turtle Creek Basin, Pennsylvania, Flood Protection.* Design Memorandum No. 4, General Design, Pittsburgh, Pennsylvania.

U.S. Army Corps of Engineers (1977). *STORM - Storage, Treatment, Overflow, Runoff Model.* User's Manual, Hydrologic Engineering Center, Davis, California.

U.S. Army Corps of Engineers (1990). *HEC-1 Flood Hydrograph Package.* User's Manual, Hydrologic Engineering Center, Davis, California.

U.S. Department of Agriculture (1986). *Urban Hydrology for Small Watersheds.* Technical Release 55, Soil Conservation Service, U.S. Department of Agriculture, Washington, D.C., 2.5-2.8.

Walski, T.M., B. Toothill, D. Skoronski, J. Thomas, and S. Lowry (2001). "Using Digital Elevation Models." *Current Methods*, Haestad Press, Vol. 1, No. 1, 91-99.

Walski, T.M. and J.W. Male (2000). "Maintenance and Rehabilitation/Replacement." Chapter 17 in *Water Distribution Systems Handbook*, Edited by L.W. Mays, McGraw-Hill, 17.1-17.28.

Walter, C.R. (1998). "Development of Planning Tools for Storm Water Management in Charlotte County, Florida." *Proceedings of the 1998 International User Conference*, ESRI, San Diego, California, July 27-31, 1998.

Wanielista, M.P. (1978). *Stormwater Management, Quantity and Quality.* Ann Arbor Science, Ann Arbor, MI, 221 pp.

Waye, D. (2001). "Hydrologic and Hydraulic Modeling of Urban Watershed." Case Study Submission in Response to a Call for Case Studies distributed by the Author (U. M. Shamsi) for this book.

WEF (1989). *Combined Sewer Overflow Pollution Abatement.* Manual of Practice FD-17, Water Environment Federation, Alexandria, VA, 43-73.

Wells, E. (1991). "Needs Analysis: Pittsburgh's First Step to a Successful GIS." *Geo Info Systems*, October 1991, 30-38.

White, C. and T. McConnell (1998). "Combining GIS with Document Image Management." *ArcUser*, ESRI, Vol. 1, No. 4, October-December, 1998, 47-51.

Zeiler, M. (1999). *Modeling Our World.* ESRI Press, Redlands, California, 199 Pages.

Zimmer, R.J. (2000). "Parcel Mapping Part I." *Professional Surveyor*, Vol. 20, No. 9, October 2000, 52-55.

Zimmer, R.J. (2001). "Is the Sky Really Falling." *Professional Surveyor*, Vol. 21, No. 5, May 2001, 38-39.

APPENDIX A
ACRONYMS

Acronyms generally confuse readers rather than enlighten. Unfortunately, GIS and related technologies are infested with acronyms. Use this Appendix to escape the agony of acronym-mania.

ACE	Army Corps of Engineers
ACP	Asbestos Cement Pipe
ADRG	Arc Digitized Raster Graphics
AIRS	Aerometric Information Retrieval System
AM/FM	Automated Mapping / Facilities Management
AM/FM/GIS	Automated Mapping / Facilities Management / Geographic Information System
AML	Arc Macro Language
API	Application Programming Interface
ASCE	American Society of Civil Engineers
ASCII	American Standard Code for International Interchange
AWRA	American Water Resources Association
AWWA	American Water Works Association
BASINS	Better Assessment Science Integrating Point and Nonpoint Sources
BIL	Band Interleaved by Line
BIP	Band Interleaved by Pixel
BMP	Best Management Practice
BRS	Biennial Reporting System
BSQ	Band SeQuential
CAD	Computer Aided Drafting
CADD	Computer Aided Drafting and design

CAM	Computer Aided Mapping
CASE	Computer Aided Software Engineering
CFS	Cubic Feet per Second
CD-ROM	Compact Disc-Read Only Memory
CGM	Computer Graphic Metafile
CIP	Cast Iron Pipe
CMP	Corrugated Metal Pipe
COE	Corps of Engineers
COM	Component Object Model
CORBA	Common Object Request Broker Architecture
CPSRM	Chester's Penn State Runoff Model
CSO	Combined Sewer Overflow
CTG	Composite Theme Grid
DAK	Data Automation Kit
DBMS	Data Base Management System
DDE	Dynamic Data Exchange
DEM	Digital Elevation Model
DEP	Department of Environmental Protection
DFIRM	Digital Flood Insurance Rate Map
DIB	Device Independent Bitmap
DIME	Dual Independent Map Encoding
DIP	Ductile Iron Pipe
DLG	Digital Line Graph
DRG	Digital Raster Graphic
DTED	Digital Terrain Elevation Model
DOQ	Digital Orthophoto Quadrangle
DOQQ	Digital Orthophoto Quarter Quadrangle
DR3M	Distributed Routing Rainfall Runoff Model
DTM	Digital Terrain Model
DXF	Drawing Exchange Format
EDC	EROS Data Center

EPA	U.S. Environmental Protection Agency
EOS	Earth Observation System
EROS	Earth Resources Observation Systems
ESDLS	EPA Spatial Data Library System
ESRI	Environmental Systems Research Institute
FAP	File of Arcs by Polygon
FAT	Feature Attribute Table
FEMA	Federal Emergency Management Agency
FGDC	Federal Geographic Data Committee
FIRM	Flood Insurance Rate Map
FIS	Flood Insurance Study
FMSIS	Flood Map Status Information Service
FTP	File Transfer Protocol
GDT	Geographic Data Technologies
GIF	Graphics Interchange Format
GIRAS	Geographic Information Retrieval and Analysis System
GIS	Geographic Information System or Systems
GPD	Gallons Per Day
GPS	Global Positioning System
GRASS	Geographic Resources Analysis Support System
GUI	Graphical User Interface
H&H	Hydrology and Hydraulics, Hydrologic and Hydraulic
HEC	Hydrologic Engineering Center
HoLIS	Honolulu Land Information System
HSG	Hydrologic Soil Group
HSPF	Hydrologic Simulation Program Fortran
HTTP	Hyper Text Transfer Protocol
HUC	Hydrologic Unit Code
HUD	Housing and Urban Development
IAC	Interapplication Communication
ICR	Information Collection Rule

ISE	Integral Square Error
IT	Information Technology
JFIF	JPEG File Interchange Format
JPEG	Joint Photographic Experts Group
LANDSAT	Land Satellite
LIDAR	Light Imaging Detection and Ranging
LOMC	Letter of Map Change
LULC	Land Use/Land Cover
MA	Metropolitan Area
MGD	Million Gallon per Day
MSC	Map Service Center
MrSID	Multi Resolution Seamless Image Database
NAD-27	North American Datum 1927
NAD-83	North American Datum 1983
NCDC	National Climatic Data Center
NEXRAD	Next Generation Weather Radar
NGDC	National Geospatial Data Clearinghouse
NHD	National Hydrography Dataset
NLCD	National Land Cover Database
NOAA	National Oceanic and Atmospheric Administration
NPDES	National Pollution Discharge Elimination System
NPS	Non Point Source
NRCS	Natural Resources Conservation Service
NSDI	National Spatial Data Infrastructure
NWI	National Wetlands Inventory
NWS	National Weather Service
ODBC	Open Data Base Connectivity
OIRM	Office of Information Resources Management
OLE	Object Linking and Embedding
P2P	Peer to Peer
PASDA	Pennsylvania Spatial Data Access

PC	Personal Computer
PCS	Permit Compliance System
PFPT	Peak Flow Presentation Table
POTW	Publicly Owned Treatment Works
PRR	Practical Release Rate
PSRM	Penn State Runoff Model
PVC	Polyvinyl Chloride
RCP	Reinforced Concrete Pipe
RCRIS	Resource Conservation and recovery System
RDBMS	Relational Data Base Management System
RF	Reach File
RGB	Red Green Blue (scientific hues or primary colors)
RPC	Remote Procedure Call
RR	Release Rate
RTB	Retention Treatment Basin
SCS	Soil Conservation Service
SDE	Spatial Database Engine
SDTS	Spatial Data Transfer Standard
SHE	System Hydrologique Europeen
SML	Simple Macro Language
SQL	Structured Query Language
SSO	Sanitary Sewer Overflow
STATSGO	State Soil Geographic
STORM	Storage Treatment Overflow Runoff Model
STP	Sewage Treatment Plant
SSURGO	Soil Survey Geographic
SWMM	Storm Water Management Model
TCP	Terra Cotta Pipe
TCP/IP	Transmission Control Protocol/Internet Protocol
TIGER	Topologically Integrated Geographic Encoding Referencing System
TIFF	Tag Image File Format

TIN	Triangular Irregular Network
TM	Thematic Mapper
TMDL	Total Maximum Daily Load
TMS	TIGER Map Service
TR	Technical Release
TRI	Toxic Release Inventory
TRR	Theoretical Release Rate
TVP	Topological Vector Profile
UML	Unified Modeling Language
URL	Uniform (or Universal) Resource Locator
USGS	United States Geological Survey
UTM	Universal Transverse Mercator
USACE	United States Army Corps of Engineers
USACERL	United States Army Construction Engineering Research Laboratory
USBC	United States Bureau of Census
VB	Visual Basic
VBA	Visual Basic for Applications
VCP	Vitrified Clay Pipe
VPF	Vector Product Format
WEF	Water Environment Federation
WIMS	Wastewater Information Management System
WISE	Watershed Information System
WMS	Watershed Modeling System
WWW	World Wide Web

APPENDIX B
GLOSSARY

Automated Mapping / Facilities Management. CADD technology applied to manage utility system data.

AM/FM/GIS. Integrated GIS and AM/FM systems with both the mapping and the facility management functions.

Attribute. Feature characteristics (data) stored in tabular format and linked to the feature by a user assigned identifier.

Basemap. The underlying common geographic reference for all the layers of a map.

Calibration. The determination, checking, or rectifying the model outputs to observed conditions.

Computer aided drafting (CAD). Software that assists the operator in engineering and architectural design and drafting operations.

Computer aided mapping (CAM). computer aided drafting and design technology applied to produce digital maps.

Coverage. The features and attributes of thematically associated data, such as streams, subbasins, and land use forming the basic unit of storage. Coverage was an ESRI terminology for a layer of ArcInfo version 7.x or older.

Database. A collection of information related by a common fact or purpose. Databases store information about features and their relationships with each other.

Database management system (DBMS). A computer program for organizing the information in a database.

Data conversion. Conversion of hard copy maps into digital files using digitization or scanning.

Data dictionary. A list that defines the specific categories of descriptive information required for each map feature. The data dictionary provides the physical description of each layer.

Datum. A set of parameters defining a coordinate system and a set of control points whose geometric properties are known either through measurement or calculation.

Design storm. A rainfall of specified amount, intensity, duration, pattern over time, and to which a frequency is assigned, used as a design basis.

Digitizing. A process of converting an analog image or map into a digital format by a computer.

Drainage. To provide channels, such as open ditches or closed conduits, so that excess water can be removed by surface flow or internal flow.

Drainage area. The area served by a sewer system receiving stormwater and surface water.

Dynamic data exchange (DDE). A special client–server architecture supported by Microsoft that enables two applications (running on the same computer) to interact by exchanging data.

Dynamic link library (DLL). An executable module that contains functions other applications can use to perform tasks. DLLs link to an application at run-time compared with static libraries, which link at compile time.

Features. The graphical representation of real world objects in a GIS.

Geocode. The process of identifying the coordinates of a location from its street address.

Geographic information system (GIS). An information system that combines tabular information with graphic data for efficient collection, storage, retrieval, analysis, and display of spatial data.

Georeference. To establish the relationship between the paper map coordinates and the real-world coordinates.

Global positioning system (GPS). A satellite navigation system used to determine terrestrial position, velocity, and time.

Graphical user interface. A computer program that acts as an interpreter between the users and a computer.

Hydrograph. A graph showing variation in flow rate during a period of time.

Hyetograph. A graph showing variation in depth of rainfall during a period of time.

Impervious. Not allowing, or allowing only with great difficulty, the movement of water into the ground.

Land use. The use of the land surface—such as residential, commercial, commercial, open—that determines the amount and character of the runoff.

Layer. A logical set of thematic data stored in a GIS. Refers to the various "overlays" of data, each of which normally deals with one thematic topic.

Network. A link-node representation of a sewer system for modeling purpose. Nodes represent point and polygon features, such as subcatchments and manholes. Links connect two nodes of the network and represent features that have length and direction attributes, such as sewers, channel, or diversion.

Network analysis. Analytical techniques concerned with the relationships among locations in a network.

Overbank flow. That portion of stream flow that exceeds the carrying capacity of the normal channel and overflows the adjoining floodplain.

Overland flow. The flow of water over the ground before it enters some defined channel or inlet.

Overlay. An overlay operation combines two or more themes to create a new theme that contains both the spatial features and the attributes of the input themes.

Pervious. Possessing a texture that permits water to move through perceptibly under normal head differences.

Pixel. A raster or grid-cell format is a regular grid of uniform size cells called pixels, each of which has an associated data value.

Postprocessor. A computer program that analyzes the output files of a program.

Preprocessor. A computer program that prepares the input files of a program.

Projection. A map projection is a mathematical model that transforms (or projects) locations on the curved surface of the earth on a flat sheet or two-dimensional surface according to certain rules.

Raster format. In raster format, objects are represented as an image consisting of a matrix of equally sized cells or pixels organized in rows and columns.

Relational database. A database that allows accessing information from different tables without joining them together physically.

Relational database management system (RDBMS). A database management system that conducts searches by using data in specified fields of one table to find additional data in another table.

Release rate. The watershed factor determined by comparing the maximum rate of runoff from a subbasin to the contributing rate of runoff to the watershed peak rate at specific points of interest.

Spatial analysis. Analytical techniques associated with the study of the locations of geographical entities together with their spatial dimensions.

Storm frequency. A measure of how often a storm of given magnitude should, on an average, be equaled or exceeded.

Subbasin. A portion of the drainage area that has similar hydrological characteristics and drains to a common point.

Theme. Map components logically organized into sets of layers.

Topology. A mathematical procedure for explicitly defining spatial relationships between features. A branch of geometrical mathematics that is concerned with order, contiguity, and relative position, rather than with actual linear dimensions.

Vector format. In a vector format, positions are stored as (x, y) coordinate pairs. Vector data consist of a series of nodes that define line segments, which are in turn joined to form more complex features, such as networks and polygons.

APPENDIX C
CONVERSION FACTORS

Conversion Factors From U.S. Customary to SI Units

To Covert	*To*	*Multiply by*
acres	square meters (m^2)	4,047
acres	hectares (ha)	0.4047
cubic feet	liters	28.32
cubic feet per second (cfs)	cubic meters per second	0.02832
cubic yards	cubic meters (m^3)	0.7646
feet	meters	0.3048
U.S. gallons (water)	liters	3.7843
U.S. gallons (water)	cubic centimeters (cm^3)	3,785
U.S. gallons (water)	milliliters	3,785
U.S. gallons (water)	grams	3,785
U.S. gallons (water)	cubic meters (m^3)	0.0038
gallons per minute (GPM)	liters per second	0.0631
gallons per minute (GPM)	liters per minute	63.08
inches (in.)	centimeters (mm)	2.54
inches (in.)	millimeters (mm)	25.4
miles (mi)	kilometer (km)	1.6093
millions of gallons per day (MGD)	cubic meters per day	3,785
millions of gallons per day (MGD)	cubic meters per second	0.044
pounds (lb)	kilograms (kg)	0.4536
square feet	square meters (m^2)	0.0929
square miles (mi^2)	square kilometers (km^2)	2.5921
yards	meters	0.9144
square miles (mi^2)	square kilometers (km^2)	2.5899

APPENDIX D
NOTATIONS

Symbol	*Description*
M_i	Modeled hydrograph value at time i
N	Number of hydrograph values
O_i	Observed hydrograph value at time i
Q_i	Pre-development peak discharge of subbasin i
Q_{ij}	Pre-development discharge contribution of subbasin i to the peak discharge of a downstream reach j
Q_{pi}	Allowable post-development peak discharge of subbasin i
R_{ij}	Release rate of subbasin i at a downstream reach j
R_j	The release rate of subbasin j from the release rate map

Index

Note: Page numbers in italics refer to figures or tables.